American Political Cultures

AMERICAN
POLITICAL CULTURES

RICHARD J. ELLIS

NEW YORK   OXFORD
OXFORD UNIVERSITY PRESS
1993

Oxford University Press

Oxford   New York   Toronto
Delhi   Bombay   Calcutta   Madras   Karachi
Kuala Lumpur   Singapore   Hong Kong   Tokyo
Nairobi   Dar es Salaam   Cape Town
Melbourne   Auckland   Madrid

and associated companies in
Berlin   Ibadan

Copyright © 1993 by Oxford University Press, Inc.

Published by Oxford University Press, Inc.,
200 Madison Avenue, New York, New York 10016

Oxford is a registered trademark of Oxford University Press

Library of Congress Cataloging-in-Publication Data
Ellis, Richard (Richard J.)
American political cultures / Richard Ellis.
p. cm. Includes index.
ISBN 0-19-507900-0
1. United States—Politics and government.
2. Political culture—United States—History.
I. Title.
E183.E44 1993   973—dc20

3 5 7 9 8 6 4 2

Printed in the United States of America
on acid-free paper

For my mother and father

Preface

David Riesman once described a historian as "a person whose job it is to destroy the other fellow's generalization."[1] Much the same could be said of anthropologists, who often seem to enjoy nothing more than invoking their anthropologists' veto: not in my tribe. Little wonder, then, that many historians have affirmed their affinity with anthropology.

The work of one anthropologist has proven particularly congenial to the anti-theoretical bias of historians: Clifford Geertz and his method of "thick description." Historians have used Geertz selectively, largely ignoring the early Geertz, who, taking a cue from his teacher, Talcott Parsons, forcefully criticized the anthropologists' practice of "spiteful ethnography,"[2] while embracing the later Geertz, who argues that "the shapes of knowledge are always ineluctably local, indivisible from their instruments and their encasements."[3] Geertz's writings have provided historians with a justification for their natural inclination to retreat from theory to interpretation, from the general to the particular, from determinism to indeterminacy, from science to art. Embracing Geertz has, in short, allowed historians to legitimize what many of them have always preferred doing, namely, to avoid explicit typologies and theorizing.[4]

That Geertz has been adopted as the "patron saint"[5] of American cultural historians tells us a great deal about why, as one historian laments, "there is no satisfactory grand synthesis holding together specialized studies of this-and-that and little effort to see if one exists."[6] For "thick description" provides, at best, a general orientation or approach; if we are to synthesize existing knowledge about American cultures into a coherent whole, what is needed is a theory or categorization of cultures.[7] One can agree with Geertz on the need to understand human behavior as "symbolic action" or to interpret culture as "socially established structures of meaning"[8] and yet still be no further along in specifying *types* of structures of meaning and the social relations that support those meanings. If it is not to be an empty slogan, interpretation of meaning must offer a theory or categorization of meanings that tells us what types of meanings (or symbols, narratives, discourses, or conversations) go with what types of contexts or communities.

If historians insist on adopting an anthropologist for a patron saint, they would do better to adopt Mary Douglas. By systematizing and categorizing belief systems, Douglas offers the conceptual tools necessary to override the anthropologists' veto. Beneath the luxuriant diversity of social contexts, customs, and languages, Douglas argues, the basic convictions about life can be usefully grouped into five sets of cultural biases: individualism, egalitarianism, fatalism, hierarchy, and hermitude.[9]

Douglas's theory of culture is promising not only because it offers a way out of the wilderness of detail into which Geertzian thick description leads[10] but also because it offers a way of synthesizing knowledge about American culture without returning to the idea of a national consensus, which historians' painstaking work over the past several decades has discredited. Douglas's categorization rejects the notion of a unified national culture or national character without embracing the idea that every societal group possesses a distinct culture. It stakes out a middle ground between homogeneous consensus and unconstrained diversity.

If focusing on the unconstrained diversity of the American past produces history that "is more chaos than history,"[11] focusing on consensus produces history that cannot be supported by the available evidence. The historical and contemporary record shows that disagreements in the United States over public policy are only "the tip of the iceberg"; beneath the surface exist fundamentally different worldviews, fundamentally different conceptions of human nature, authority, equality, and freedom.[12] Cultural dissensus in the United States is not something new, not a product of the social movements of the 1960s and 1970s, but an enduring and defining attribute of the American past. My aim in this book, then, is both to document the pervasive cultural conflict in American history and to demonstrate that these past conflicts can be meaningfully compared to the conflicts of today. Both past and present, I argue, can be made sense of in terms of Douglas's five categories

My interest in American political culture dates back to my days as an undergraduate at the University of California at Santa Cruz. It was there, in the classes of Robert Hawkinson, that I was first introduced to Louis Hartz's *Liberal Tradition*, as well as to the work of J. David Greenstone, Samuel Huntington, Sacvan Bercovitch, Richard Hofstadter, Michael Kammen, and Clifford Geertz. I am profoundly grateful to Bob for sparking my interest in American political culture, as well as for providing me with a powerful model of what good undergraduate teaching should look like.

In graduate school I had the great good fortune to come within the orbit of Aaron Wildavsky. Over the past five or six years we have shared so many conversations on the subjects explored in this book that it is impossible for me to say where his ideas end and mine commence. It is possible for me to say that without his wise counsel and encouragement this book would not have been written. Aaron not only introduced me to the particular theory that informs this book but, much more important, gave me the opportunity to collaborate on two book-length projects. From these collaborative experiences I gained what cannot be had in the classroom: the chance to observe firsthand how a splendid scholar and the scholarly community work. Words cannot discharge the debt I owe for that privilege.

I also wish to acknowledge my other teachers at the University of California at Berkeley. Mike Rogin and Ken Jowitt both believed, with Weber, that big errors are better than small truths. For my sake I hope they are right. Martin Landau helped clear away a lot of my misconceptions about science, theory, and objectivity. Nelson Polsby and Ray Wolfinger rudely but usefully reminded me that it was occasionally necessary to plant my feet in the real world. None of them, I should add, is in any way responsible for the content of this book; indeed, I suspect each of them would either disagree with the argument or express dismay with the argumentation.

During the writing of this book, a number of people commented in helpful ways on all or parts of the manuscript. My special thanks go to Michael Flusche, George M. Fredrickson, Dean Hammer, Daniel Walker Howe, Michael Kammen, Peter Kolchin, Charles Lockhart, Gary Lee Malecha, Karen Orren, Daniel Rodgers, Brian Weiner, Joe White, and Major Wilson. I also owe thanks to the three anonymous reviewers for Oxford University Press; this book is much better as a result of their help.

A work of synthesis like this is even more than ordinarily reliant on the work of others. Among the historians whose interpretations have most influenced my understanding of the American past are John Ashworth, James Banner, Eric Foner, David Hackett Fischer, Lawrence Friedman, Eugene Genovese, Richard Hofstadter, Daniel Walker Howe, Rhys Isaac, Isaac Kramnick, and Gordon Wood. The kind of synthesis attempted in these pages is a tribute to these and other historians who have painstakingly read through countless letters, newspapers, diaries, and speeches to reach an understanding of American political culture at a particular point in time and space. Some historians, I am sure, will be inclined to view this book not as a tribute to the historian's craft but as an affront to it. The paucity of primary sources may even prompt some to dismiss this book as irrelevant to the concerns of serious, practicing historians. Such a reaction is more than a little ironic, for it reveals a puzzling lack of faith in the worth of their own craft. I believe, with T. H. Marshall, that "it is the business of historians to sift this miscellaneous collection of dubious authorities and to give to others the results of their careful professional assessment." As Marshall adds, "surely they will not rebuke the [social scientist] for putting faith in what historians write."[13] Far from undermining or calling into question the value of the historian's craft, works of synthesis based on secondary sources attest to the cumulative character of historical research. Whether I have relied on the best available sources is a question others must answer.

In its scope this book owes a great deal to other historically minded political scientists, especially Louis Hartz, J. David Greenstone, and Samuel Huntington. Their example persuaded me, for better or worse, that an attempt to generalize about the entire sweep of the American past was both possible and worthwhile. I expect that my interpretation will exhibit the deficiencies that may be found in these authors' works; I only hope that it has some of their virtues.

Completion of this book was facilitated by a summer grant from the National

Endowment for the Humanities as well as an Atkinson grant from Willamette University. My most important source of support came from my wife, Juli, whose labors enabled me to take a year off after graduate school to complete a first draft of this book. For this, and for so much else, I am thankful to her.

Salem, Ore.   R. J. E.
September 1992

Contents

AMERICAN POLITICAL CULTURES

1

Individualism and Community in American Life

Few observers of American political culture have failed to comment on the individualistic character of the American people. Most truisms are not without validity, and this one is no exception. As a starting point the proposition may be satisfactory, but as a conclusion this pat formula disguises far more than it reveals about American political values and social relations. To begin with, the term "individualism" is fraught with ambiguity. Any number of scholars have noted the extraordinary lack of precision that has historically accompanied the word's usage.[1] Steven Lukes, for instance, has identified no less than eleven distinct meanings of the word.[2] The extreme variability of the term's meaning should be enough to give one pause before rushing to affirm the pervasive individualism of Americans.

In this chapter I focus on one particular ambiguity that surrounds the concept of individualism. Following anthropologist Mary Douglas, I distinguish between two forms of individuation. Individual autonomy, Douglas suggests, may be constrained either through group membership or through external prescriptions.[3] Douglas refers to these two dimensions of individual constraint as "group" and "grid," or "boundary" and "structure"; Durkheim labels them "integration" and "regulation."[4] Group restrictions on individual autonomy can be measured by determining whether rules of admission to a group are strong or weak, and whether the life support a group gives to its members is complete or partial. The further one moves along the group dimension, the stronger will be the boundaries identifying and separating members from nonmembers. "The extreme case of a strong group," explains Douglas, "will be one in which the members gain their whole life-support from the group as such."[5]

Knowing the strength of group allegiance gives us only a partial picture of the extent of autonomy enjoyed by individuals, for social environments with similar degrees of group involvement can differ in the degree to which they are open to individual negotiation. At the high end of grid, interaction among individuals is strictly regulated and all members know their station. An example, Douglas tells us, would be "the military regiment with its prescribed behavior and rigid timetabling."[6] As one moves down grid into less structured social environments, roles become progressively more fluid and ill-defined, and individuals are increasingly expected to negotiate their own relationships with others.

Because these two dimensions of individuation (group and grid) are distinct, individual autonomy can be expanded along one dimension while simultaneously being restricted along another. An individual may be relatively free from binding regulations, for instance, yet be tightly integrated into a group. Conversely, an individual may be subject to many prescriptions but be isolated from group life. Douglas's scheme offers a more variegated, and hence more discriminating, conception of the process of individuation than that provided by theories that assume a single dimension of individuation. Theorists who work with a single dimension of individualism discern only two ways of life, modern individualism at one end of the continuum and traditional hierarchy at the other. Douglas's two dimensions, in contrast, yield four types of social life.[7] In addition to the familiar categories of competitive individualism (where group involvement and external prescriptions are minimal) and hierarchical collectivism (in which group allegiance and external prescriptions are high), Douglas alerts us to the existence of egalitarianism or communitarianism (in which external prescriptions are minimal but group commitment is high) and atomized subordination or fatalism (where external prescriptions are high and group involvement is low).

Many interpretations of American political culture, implicitly or explicitly, are grounded in a single dimension of individuation. Following Tocqueville and Louis Hartz, political theorists commonly contrast the aristocratic or hierarchical institutions and mores of the Old World with the democratic or individualistic institutions and customs of the New World. Historians, more sensitive to the hierarchical social and political structures of colonial life, are more likely to trace the gradual but ineluctable process by which cohesive communities of structured inequalities gave way to "typical American individualism, optimism, and enterprise."[8] Either America is born modern or America becomes modern. The dichotomy dictates that the weaker traditional hierarchy is, the stronger modern individualism must be.

Douglas's fourfold typology allows one to see that hierarchy's weakness does not necessarily translate into individualism's unrivaled hegemony. An individual's group involvement may be decreased at the same time that the individual becomes subject to a more restrictive set of social regulations. This combination of individual isolation and social regulation Douglas terms "fatalism" or "atomized subordination." Tocqueville, in a characteristic flash of insight, recognized the same phenomenon in his discussion of the affinity between individualism and despotism. Despotism, Tocqueville explained, "sees the isolation of men as the best guarantee of its own permanence. So it usually does all it can to isolate them."[9] Isolation from social groups can thus result in either an individualistic political culture or a fatalistic and atomized one, depending on the degree to which the individual is free from externally imposed prescriptions.

How prevalent this fatalistic political culture is and has been in America is a matter of some dispute.[10] Egalitarian critics of American society sometimes maintain that fatalism is the predominant political cultural tradition in the United States. A good case can be made that members of today's "underclass" inhabit a universe of atomized subordination, and women have arguably found themselves in such a position in many households. I return to the topic of fatalism in chapter 7, where I discuss the importance of fatalism among southern slaves, but my immediate concern in this chapter is with that way of life which Douglas terms "egalitarian" or "communitarian."

The egalitarian political culture resembles individualism in its aspiration to free the individual from the structured differences of hierarchy. Egalitarianism differs from individualism, however, in its effort to integrate the individual into a caring collectivity. Because most students of American political culture collapse these two distinct dimensions of integration and regulation, there is a tendency to miss the important differences between these two ways of life. Competition is revered by individualists, solidarity is privileged by egalitarians. Differences between people are suspect for egalitarians, while individualists defend inequalities as a reflection of individual worth. Participation is the guiding star for egalitarians; self-regulation is the ideal of individualists. It may be, as Tocqueville, Hartz, and others have argued, that in America hierarchy has never posed a serious challenge to individualism,[11] but this does not mean that individualism has gone unchallenged. On the contrary, the individualist way of life has repeatedly come under fire in the United States from egalitarians who condemn individualism for isolating the individual from the community, dimming the sense of collective purpose, damming the springs of civic virtue, and creating unconscionable inequalities.

Individualism's Contemporary Critics

The intensity and the scope of the communitarian challenge to competitive individualism have varied considerably by historical era. In some periods, as during the early 1950s, this alternative cultural perspective has been relatively muted, a fact that perhaps explains something about how Hartz and Daniel Boorstin reached the conclusions they did about an individualist consensus.[12] As America enters the 1990s, however, the nation's cultural landscape differs strikingly from that of the 1950s. Far from being the unquestioned axiom described by Hartz, competitive individualism is routinely criticized (particularly in the academy) for promoting selfishness, greed, egoism, fragmentation, and injustice. The individualist way of life is dismissed by the undergraduate student, who foresees "the environmental ethic of 'us' . . . replacing the yuppie 'I'"; by the educational reformer, who laments that American public schools "breed competition and isolation among students" and recommends more "cooperative and collaborative [learning] processes"; by the critic of higher education, who scores universities for their "ethic of competitive individualism," which defeats the sense of community needed for fruitful teaching and learning; and by the college administrator, who informs a student that the word "individual" is "a 'Red Flag' phrase today, which many consider racist. . . . Arguments that champion the individual over the group ultimately privilege the 'individuals' belonging to the largest or dominant group."[13]

The contemporary critique of individualism comes not, as it did in nineteenth-century Europe, from counterrevolutionaries like Joseph de Maistre and Louis de Bonald, who defended a hierarchical social order in which all knew their station,

but largely from egalitarian proponents of small-scale community and participation.[14] The aim of these egalitarian critics is to remedy the atomization of contemporary America by integrating individuals into caring, cohesive communities, but
without the hierarchical authority and fixed stations characteristic of past ages.
Individual autonomy is to be reconciled with collective choice.

The communitarian critique of individualism animates the best-seller *Habits
of the Heart*, which takes to task America's dominant "culture of separation." The
authors of this study of contemporary American mores suggest that "though the
processes of separation and individuation were necessary to free us from the
tyrannical structures of the past, they must be balanced by a renewal of commitment and community if they are not to end in self-destruction."[15] An even stronger
plea for combining individual autonomy with communal involvement is articulated in Benjamin Barber's *Strong Democracy*. Modern American individualism,
laments Barber,

> is all too often a world in which men and women do not exist for others; in
> which, although there are no public censors, there can also be no public goods;
> in which monolithic social ends are prudently outlawed by imprudently proscrib-
> ing all social ends; in which altruistic behavior is discouraged in the name of
> bargaining efficiency and utility accounting. In this world, there can be no fra-
> ternal feeling, no general will, no selfless act, no mutuality, no species identity,
> no gift relationship, no disinterested obligation, no social empathy, no love or
> belief or commitment that is not wholly private.

Barber rejects not only the privatization and self-regarding that he attributes to
individualism but also the inequalities associated with "the conservative notion of
community." He repudiates, for instance, Robert Nisbet's assertion that "inequality is the essence of the social bond" because this "radically antidemocratic" claim
"polarizes egalitarianism and communitarianism."[16] What Barber desires to put in
place of atomized individualism is not hierarchical collectivism but egalitarian
communitarianism.

The earnest quest among historians and political theorists for a civic republican alternative to Lockean liberalism is further testimony to the current discontent
among those on the left with competitive individualism. Much of this scholarship
is valid on its own terms as a description of early American political culture, but
its remarkable appeal to intellectuals cannot be fully explained except against the
backdrop of contemporary politics. Much of this literature, as Christopher Lasch
comments, tells "us less about the concept of virtue than about the fear of social
fragmentation, competitive individualism, and self-seeking that underlies attempts
to revive it."[17] In place of liberal individualism, civic republicans advocate a rough
equality of condition, "widespread participation," and "a language of political discourse that can articulate the . . . common good."[18]

Among the most thoroughgoing contemporary expressions of the egalitarian
critique of individualism is that found within the feminist movement.[19] In a political life "dominated by a self-interested, predatory individualism," many feminists
argue, there can be "no substantive sense of civic virtue, no vision of the political
community which might serve as the groundwork of a life in common." Such a

life in common can be achieved only if "an ethic based on individualism, competition, and private profit" is replaced by "an ethic of sharing, cooperation, and
collective involvement."[20]

Opposition to the individualist way of life from a communitarian perspective
is voiced in *The Feminization of America,* by Elinor Lenz and Barbara Myerhoff.
They assert that "by all accounts, there is in America today a desperate and growing loneliness," a loneliness they attribute to the American "cult of self-aggran-
dizing individualism," which "not only sets the individual above the group [but]
disconnects the communal ties that in most societies have bound people together."
In place of this dominant culture of competitive, self-reliant individualism, which
is attributed to "the male principle," Lenz and Myerhoff wish to substitute a
distinctly feminine culture of "cooperative individualism."[21]

The preference for community without hierarchy is evident in many feminists'
plea for "freedom that does not entail isolation and community that does not enforce
uniformity."[22] "Group solidarity without group autocracy" is the way Ann Ferguson
sums up this preferred way of life. Ferguson rejects both "possessive individualism"
and "repressive . . . social hierarchy," and in their place hopes to substitute social
structures that encourage "a sense of community and group solidarity" but at the
same time preserve "an anarchist/individualist value of individual autonomy."[23]

In *Recreating Motherhood* Barbara Rothman insists on the need to replace the
atomization and compartmentalization of modern American society "with a sense
of organic wholeness, roundness, interconnectedness." Since it is "the physical
embodiment of connectedness," Rothman sees motherhood as constituting a frontal
challenge to individualist philosophy. "We have in every pregnant woman the
living proof that individuals do not enter into the world as autonomous, atomistic
isolated beings, but begin socially, begin connected." What Rothman desires is
not a return to the connectedness of the hierarchies of the past, but a connectedness that preserves "a sense of unique self." She acknowledges that "liberal feminists are good and strong critics of patriarchal society," but she also concludes
that "they do not fight the ideology of technology or capitalism." True feminism,
in her view, must set itself against both coercive hierarchy and atomized individualism.[24]

A similar critique of competitive individualism is articulated by Martha Ackelsberg and Kathryn Addelson, who advocate "creating a society based on cooperative, rather than competitive, principles." Competition, in their view, is a divisive
social doctrine that "serves to justify vast differentials of power, authority,
resources, and even human dignity in society." To accept the competitive individualist version of reality, they argue, is to contribute to "the perpetuation of social
hierarchies and to the disempowerment of all people who find themselves dis-
advantaged by those structures." Both hierarchy and fatalism are thus attributed to
the competitive individualist way of life. In place of the divisive and alienating
politics of individualism, Ackelsberg and Addelson wish to substitute "an alternative, egalitarian, nonauthoritarian society" of "cooperatively structured institutions
and practices."[25]

The common thread that unites these various critiques is the alleged destruction of community that results from a life of competitive individualism. Public

purposes must be made to override private desires. None of these critics of individualism sanction hierarchical authority. All agree that large-scale bureaucratic
organizations cannot provide real community. Instead true community must be
local, small-scale, egalitarian, and participatory.

That this egalitarian critique of individualism is widespread in contemporary
American society has been ably documented elsewhere.[26] The more difficult
question is whether today's cultural dissensus is a new phenomenon, or whether
dissensus was characteristic of our past as well. Is the communitarian challenge to
individualism a novel product of the social movements of the 1960s and 1970s, or
does this alternative social vision have roots that extend deep into the nation's
history? I try to answer this question by first examining the political culture of
some of the earliest settlers of the New World, the Puritans.

Puritanism and Community

From Tocqueville's European vantage point, the Puritan settlers in the New World
"presented the unusual phenomenon of a society in which there were no great lords,
no common people, and, one almost may say, no rich or poor." When Tocqueville
remarked that "the whole destiny of America [was] contained in the first Puritan
who landed on those shores," it was largely this "novel phenomenon of a society
homogenous in all its parts" that he had in mind.[27]

Like so many of Tocqueville's sweeping historical generalizations, the proposition, although exaggerated, contains an important element of truth. Most immigrants to Massachusetts did indeed come from the middling strata of English society—prosperous farmers, yeomen, artisans, and merchants. Very few aristocrats
made the trip across the Atlantic. Relatively few, too, came from the bottom of
English society. Once arrived in New England, this middling strata of people, which
in feudal England had the burden of supporting the aristocracy above and the landless poor below, actively discouraged settlement of the top and bottom ranks of
British society. Not that the Puritans did away with social stratification. Far from
it. They carefully preserved the social distinctions between lesser gentry, yeomanry,
and cottagers. If Puritan New England was not the homogeneous social order that
Tocqueville claimed, the removal of England's top (king, peerage, and great gentry) and bottom strata (landless laborers and wandering poor) did make New
England fundamentally different from the complex hierarchies of the Old World.[28]

The Puritans did not, as Tocqueville implies, leave all hierarchical practices
of the Old World behind. In the meetinghouses, for instance, men were seated on
one side of the aisle and women on the other, all arranged in order of age, wealth,
and reputation. Children were required to stand and bow before their parents, and
those at the lower end of the social scale were expected to defer to those of higher
rank.[29] And there is no gainsaying the hierarchical language of a John Winthrop:
"God almighty in his most holy and wise providence has so disposed of the condition of mankind, as in all times some must be rich some poor, some high and
eminent in power and dignity; others mean and in subjection, . . . ordering all these
differences for the preservation and good of the whole." Even Jonathan Edwards

conceded the beauty of a social order in which all have "their appointed office, place and station, according to their several capacities and talents, and everyone keeps his place, and continues in his proper business."[30] Nonetheless, Tocqueville is correct that the Puritans, despite preserving many of the deferential practices of seventeenth-century England, in many ways represented a fundamental departure from and rejection of the hierarchical world they left behind.

Feudal bonds were inherited and lifelong; the Puritans, in contrast, demanded that social relations rest on individual consent that must be periodically renewed.[31] One should not forget, as Perry Miller reminds us, "that the life of the Puritan was completely voluntaristic. . . . The churches of New England were made up of 'saints,' who came into the church because they wanted membership, not because they were born in it, or were forced into it, or joined because of policy and convention."[32] "There can be no necessary tye of mutual accord and fellowship," Puritan divine Thomas Hooker preached, "but by free engagement, free . . . in regard of any human constraint. . . . No man [is] constrained to enter into [a social covenant] unless he will."[33] Even a covenanted marriage was dependent on the free consent of both parents and children. Children who married without their parents' consent could be fined for "self-marriage," but parents who arbitrarily withheld their consent could be sued by the child.[34] In a healthy state, Puritans believed, citizens must "create the society by willing consent and active participation."[35]

By establishing a direct, unmediated relationship between God and the believer, Puritanism directly challenged the medieval notion of the cosmos as a hierarchy of essence and degrees.[36] In the great chain of being, every species and inanimate object had its precise place. In place of the infinite gradations of station characteristic of medieval Europe, the Puritans substituted a dichotomous worldview separating saints and sinners, good and evil. As Michael Zuckerman puts it, the Puritans "distilled dichotomies out of the traditional multiplicities of medieval pluralism."[37] In the Puritan worldview, "All the men in the world are divided into two ranks, Godly or Ungodly, Righteous or Wicked."[38] There were no shades of gray in Puritan cosmology, only a cosmic struggle between darkness and light.

The Puritans' departure from the hierarchical way of finely graded distinctions is evident even in their preference for plain dress.[39] Both the style and the color of their clothing were to be simple and relatively uniform. That the Puritans passed strict sumptuary laws forbidding the wearing of fancy clothing does not distinguish them from most other American colonies or even much of Europe. What distinguished the Puritans was that these statutes applied to everyone, not merely to the common people. This is not to say that all Puritans dressed exactly alike. Leaders and elders did dress differently from ordinary people, but these differences were minor compared to those differences that existed in the southern colonies or Stuart England.[40]

The gross economic inequalities characteristic of the Old World were also systematically undermined by the Puritans. Most important, perhaps, was their rejection of primogeniture. "It is against all equity," John Winthrop complained, "that one [son] should be a gentleman to have all, and the rest as beggars to have nothing." In distributing their land, New England communities exhibited the same concern with avoiding the creation of enormously rich or entirely landless families. Although most towns granted larger shares of land to those of higher social station, and none went as far as the Separatist Roger Williams, who vowed to receive "only unto myself, one single share, equal unto any of the rest," New England's distribution of land in the seventeenth century was, as David Hackett Fischer concludes, "remarkably egalitarian."[41]

If Hartz and Tocqueville are right that America lacked the hierarchy of feudal Europe, they are wrong that America has, in Hartz's words, known only "the reality of atomistic social freedom."[42] For if the hierarchical structure of privilege and inequality was nowhere near as developed in New England as in Old England, the group demands on individual behavior were if anything stronger.

The communal vision of the Puritans was expressed by John Winthrop in a sermon delivered aboard the *Arbella* in 1630: "[W]e must be knit together in this work as one man, we must entertain each other in brotherly affection, we must be willing to abridge ourselves of our superfluities for the supply of others' necessities, . . . we must delight in each other, make others' conditions our own, rejoice together, mourn together, labor and suffer together, always having before our eyes . . . our community as members of the same body."[43] Another Puritan leader told them that they should "look upon themselves, as being bound up in one *Bundle of Love*; and count themselves obliged, in very close and Strong Bonds, to be serviceable to one another."[44] "We are knit together as a body," echoed another, and "we do hold ourselves straightly tied to all care of each other's good, and of the whole by every one, and so mutually."[45]

The strong group orientation of Puritanism is emphasized by Perry Miller. Puritan philosophy, Miller explains, "demanded that in society all men, at least all regenerate men, be marshalled into one united array. The lone horseman, the single trapper, the solitary hunter was not a figure of the Puritan frontier; Puritans moved in groups and towns, [and] settled in whole communities."[46] For the Puritans, it was self-evident that "Holy societies honor our maker more than holy separate persons."[47] The solitary individual was seen as a threat to the cohesion of the community. Single people were forbidden by law to live by themselves. The same fear of centrifugal dispersion resulted in laws prohibiting houses from being built more than a half mile from the central meetinghouse.[48]

Reminding each other of the evils and dangers outside helped affirm the value of cohesion within the group. Those who ventured into the wilderness reported hearing "such terrible roarings" that "must either be Devils or Lions." More terrible still were the Indians, who were believed to be in league with the devil. "Who knows what their rage and Satan's malice may stir them up unto to work us a mischief." The Indians, John Higginson insisted, had been set against the colonists "to make [the colonists] cleave more closely together, and prize each other . . . , and stop their now beginning breaches." The more terrifying the evils on the outside, the greater the justification for heightening "God's Hedge" separating the godly from the ungodly.[49]

Numerous claims were made upon individuals in the name of the group. A statute passed in 1633, for instance, declared that "no man shall give his swine any corn but such, as being viewed by two or three neighbors, shall be thought unfit for any man's meat." The law also required that families be inspected on a regular basis by representatives of the community so as to ensure that group standards were being upheld. Those families that were found wanting had their children and servants removed to other homes.[50]

Individual economic behavior was also tightly monitored in the name of the community. Boston's minister John Cotton formulated a strict code of business ethics for New England. Among his five "false principles" were "that a man might sell as dear as he can, and buy as cheap as he can," and "that, as a man may take advantage of his own skill or ability, so he may take advantage of another's ignorance or necessity." Entrepreneurs who disregarded these precepts were severely disciplined by the community. Witness the fate of Robert Keayne, an extremely wealthy Boston merchant who began life as a butcher's boy in Berkshire. In 1639 several irate customers complained that Keayne had overcharged them for a bridle and a bag of nails. The General Court found Keayne guilty of oppression for having taken "above six-pence in the shilling profit" and fined him 200 pounds. A separate inquiry was undertaken by Boston's Congregational church, which also found Keayne guilty of "selling his wares at excessive rates." Keayne was denounced from the pulpit and threatened with excommunication until he came before the congregation and "did with tears acknowledge and bewail his covetous and corrupt heart." The disgraced merchant tried, to no avail, to retrieve his reputation by giving away large sums of money. Shattered by the incident, he began to drink heavily and was subsequently removed from public office.[51]

The fate of this self-made man should give pause to those who too readily assume an "affinity" between Puritanism and laissez-faire capitalism.[52] To the extent that Puritanism helped tear down the fixed hierarchical statuses and distinctions of the old order, it may indeed have contributed to the growth of competitive capitalism. But in their insistence on subordinating the individual to the group, the Puritans clearly stood opposed to untrammeled, competitive individualism. Their distance from the individualist way of life is evident in their readiness to forbid exports in times of scarcity, to condemn the reckless pursuit of land, to fix just prices and wages, and to curtail individual profits in the name of the whole.[53] As the case of Robert Keayne indicates, what it took to be a good Puritan often conflicted with what was required to be a good capitalist.

Puritan New England, then, was a culture that fits neither the model of Old World hierarchy nor that of modern individualism. Instead the Puritans combined, as Perry Miller puts it, "a piety in which the individual was the end of creation" and "a social philosophy in which he was subordinate to the whole." Michael Walzer draws attention to the same combination of individualism and collectivism when he observes that the Puritan's "voluntary allegiance led him to a collective discipline." Similarly, another scholar is struck by Puritanism's "dual concentration on personal states of mind and communal conduct." Historian Michael Zuckerman, too, stresses the movement's proclivity "simultaneously to individuate and aggregate."[54]

For those scholars working with a single dimension of individuation, the Puritans thus present the puzzling anomaly of a people simultaneously individualist (low grid) and collectivist (high group). Failure to distinguish these two social dimensions has generated interpretations of Puritanism that tend to seize on one side of the picture at the expense of the other or, at best, place the Puritans at a midway point in a linear transition from traditional hierarchy to modern individualism.[55] The interpretation of Puritanism as a halfway house between traditional hierarchy and modern individualism errs not only because it slights the way in which Puritanism fused collectivism with individual consent into a distinctive way of life, but also because it overlooks subsequent movements in American history that have attempted to marry individualism with community.[56]

The succeeding sections in this chapter show how the attempt to combine individual autonomy with community recurs in the political culture of civic republicanism, abolitionism, and progressivism. To lump Puritanism together with classical republicanism, abolitionism, and progressivism is not to deny the existence of crucial differences between these phenomena any more than to lump apples and oranges together as fruit is to deny that important differences exist between apples and oranges.[57] It is rather to focus attention on a property of social organization that these movements have in common, namely, the joining of an intense commitment to community with an equally deep devotion to individual autonomy. In each case the attempt to combine individual consent with collective choice generates (and is in turn sustained by) a distinctive cultural bias that differs systematically from the ethos of competitive individualism.

Classical Republicanism, Commerce, and Civic Virtue

Writing several decades after the Revolution, Mercy Warren wondered whether the intervening years had not shown that Americans were "too selfish and avaricious for a virtuous republic." "The general equality of fortune which had formerly reigned among [Americans]," she lamented, was disappearing amid a prevalent "spirit of public gambling, speculation in paper, in lands, in every thing else." She feared that America, like the ancient republics, would be "corrupted by riches and luxury" and hence cease to be "a simple, virtuous, and free people."[58] The pursuit of private gain would drive out—indeed, was already driving out—public-spirited commitment to the commonweal.

The falling away from republican virtue and simplicity described and prophesied by Warren was an archetypal expression of what scholars have termed "classical republicanism." The key word in the language of classical republicanism was "virtue," by which was meant the willingness of the individual to sacrifice private interests for the good of the community.[59] In the classical republican tradition, Isaac Kramnick explains, "the pursuit of public good is privileged over private interests, and freedom means participation in civic life rather than the protection of individual rights from interference."[60] The precondition for disinterested pursuit of the public interest, according to classical republicans, was an independence from governmental or social superiors.[61]

Classical republicanism thus conforms to neither the hierarchical nor the individualist model. It was, as Forrest McDonald comments, "at once individualistic and communal: individualistic in that no member of the public could be dependent upon any other and still be reckoned a member of the public; communal in that every man gave himself totally to the good of the public as a whole."[62] As such it fits poorly into the conventional tradition–modernity dichotomy: republicanism, J. G. A. Pocock says, sought "a flight from modernity . . . no less than from antiquity."[63] In its deep suspicion of commerce, its elevation of the public good over private interests, and its equation of virtue with sacrifice for the commonweal, classical republicanism is clearly at odds with competitive individualism. But, at the same time, republicanism was far too suspicious of authority to nurture hierarchical institutions, and its celebration of "simplicity of manners" undercut the hierarchical emphasis on distinctions in status, while its insistence on personal autonomy subverted the relations of dependence on which hierarchy depended.

The fullest embodiment of the classical republican tradition opposed to both hierarchy and competitive individualism was Samuel Adams.[64] "A Citizen," Adams believed, "owes everything to the Commonwealth." Like many of his countrymen, Adams feared that when the revolution against the British ended, Americans would so "forget their own generous Feelings for the Publick and for each other, as to set private Interest in Competition with that of the great Community."[65] In their zeal to get ahead, Adams worried, people would lose sight of the good of the whole. But if Adams was appalled by competitive individualism's elevation of private interest over the collective good, he was no less alarmed by the restrictions on individual choice that accompanied hierarchical collectivism. He celebrated that "in these times of Light and Liberty, every man chuses to see and judge for himself."[66] The difficulty for Adams was reconciling a political culture that was simultaneously individualistic and communitarian.

Adams hoped to reconcile community and individual autonomy by transforming America into a "Christian Sparta." Sparta was an attractive political model for many eighteenth-century republicans because it seemed to avoid the invidious distinctions and relations of dependence of hierarchy, while at the same time elevating the collective good over selfish individual interests. By forbidding commerce and the accumulation of wealth, Sparta provided an environment in which individuals could devote themselves to the disinterested pursuit of the public good.[67]

That equality and virtue were inextricably linked was an unquestioned axiom of classical republican thought. Montesquieu only gave voice to a truism when he argued that virtue (i.e., devotion to the collective good) depended greatly on "a frugality of life" and a "mediocrity of . . . abilities and fortunes." "In proportion as luxury gains ground in a republic," Montesquieu reasoned, "the minds of the people are turned towards their particular interests." Conversely, it followed that "the less we are able to satisfy our private passions, the more we abandon ourselves to those of a general nature." Prevented from pursuing private gain, the people will turn to "the only passion left them," namely, virtue.[68]

The dilemma for classical republicanism was how to maintain the cohesive and homogeneous character of the community without destroying individual freedom. William Moore Smith caught this dilemma exactly when reminded members of the Continental Congress that "where property is secure, industry begets wealth; and wealth is often productive of a train of evils naturally destructive to virtue and freedom." If the people exclude wealth, Smith continued, "it must be by regulations intrenching too far upon civil liberty." But if wealth is permitted to flour-

ish unchecked by the community, then "the syren luxury" will gradually undermine the society.[69]

Smith's proposed solution to this "sad dilemma" was "to regulate the use of
wealth, but not to exclude it,"[70] thus following Montesquieu's advice that to preserve the equality necessary in a republic it was "absolutely necessary there should
be some regulation in respect to [all] forms of contracting. For were we once allowed
to dispose of our property to whom and how we pleased, the will of each individual would disturb the order of the fundamental law."[71] Throughout the Revolution, classical republican beliefs about the "baneful influence" of "exorbitant
wealth" on republican virtue as well as the precedence of communal purpose over
private interests were used to justify the enactment of economic regulations (including wage and price controls) as well as moral and sumptuary controls.[72]

Foreign commerce was particularly suspect in the eyes of many classical republicans who feared that such trade "destroys that simplicity of manners . . . and
equality of station, which is the spring and peculiar excellence of a free government." Commercial intercourse with the corrupt Old World would only result in
transmitting the "infectious" disease of "luxury" back to the colonies, thereby
suffocating the delicate germ of virtue. Not a few Americans felt it was necessary
to wall the country off from European luxury, creating what the British radical
Richard Price called "a world within themselves."[73]

Classical republican ideology attained perhaps its most egalitarian expression
in Revolutionary Pennsylvania. The republican skepticism of competition in politics manifested itself in a new constitution that eliminated entirely the executive
branch and the upper chamber of the legislature. As one advocate explained, such
a system of divided power would only be "setting-up distinctions, and creating
separate, and jarring interests in a society" which should possess "but one common interest." Not a few radicals looked back to an idealized image of "the old
Saxon model of government" which "consisted in its incorporating small parcels
of the people into little communities by themselves." Within each of these "small
republics" the people could participate directly in the running of their common
affairs.[74]

The Pennsylvania radicals who framed the 1776 state constitution also displayed
the classical republican fear of unregulated commerce. The new constitution
explicitly authorized local committees to control prices in the name of the public
good. Price controls came under vigorous criticism from Philadelphia merchants,
who argued that "freedom of trade, or unrestrained liberty of the subject to hold
or dispose of his property as he pleases . . . is absolutely necessary to the prosperity of every community." Advocates of price controls defended their actions by
rejecting the idea "that trade will regulate itself" and affirming that "every right
or power claimed or exercised by any man or set of men should be in subordination to the common good." They found the idea that "every man [can] do what he
pleases . . . repugnant to the very principles on which society and civil government are founded." The essence of republicanism, Pennsylvania radicals repeated
over and over again, was devotion to the common good and a rough equality "of
wealth and power." Individuals who possessed "enormous wealth," they reiterated,
always posed a "danger in free states."[75]

Anti-Federalists, too, drew upon classical republicanism when they criticized the proposed federal Constitution on the grounds that it would replace public-spirited citizens with private, self-seeking individuals. Only in small territories and among relatively homogeneous populations, Anti-Federalists agreed, could the virtue requisite to a republic's survival be inculcated among the people. Size was felt to be crucial because, as Montesquieu had argued, "In large republics, the public good is sacrificed to a thousand views; in a small one the interest of the public is easily perceived, better understood, and more within the reach of every citizen."[76] Only in a small community, where one knew and sympathized with one's neighbors, could an individual be expected to prefer the common good over private interests.[77]

Size was also important because smallness was the best guarantee of sameness. In the absence of a similarity of "manners, sentiments and interests," "Brutus" lamented, there would be "a constant clashing of opinions; and the representatives of one part will be continually striving against those of the other."[78] "A diversity of interests," echoed the "Impartial Examiner," meant that "the general good may be lost in a mutual attention to private views."[79] Anti-Federalists were much more likely to agree with "Cato" that the public good would be lost in the clashing of "interests opposite and dissimilar in their nature" than they were to accept Madison's "policy of supplying by opposite and rival interests, the defect of better motives."[80] To Anti-Federalists Madison's doctrines seemed disturbingly reminiscent of Bernard Mandeville's pernicious doctrine that "private vices are public benefits."[81]

Just as political competition was suspect, so too, in the eyes of many Anti-Federalists, was unchecked economic competition. As Cato reminded his fellow citizens of New York, "a commercial society begets luxury, the parent of inequality, the foe to virtue, and the enemy to restraint."[82] The Federalist argument that the new Constitution, by protecting commerce, would make us "rich as individuals at home" was met with Patrick Henry's rejoinder that "you are not to inquire how your trade may be increased . . . but how your liberties can be secured." Henry held up the Swiss model of "republican simplicity" as one worth emulating.[83] "As people become more luxurious," another Anti-Federalist reminded the Massachusetts delegates, "they become more incapacitated of governing themselves."[84] Many believed with the Impartial Examiner that the nation's "excessive fondness for riches" would be its downfall.[85]

Anti-Federalism (and, more generally, classical republicanism) has long puzzled historians because, like Puritanism, it seems to be at once traditional and modern. In defending the "traditional assumption that the state was a cohesive organic entity with a single homogeneous interest" against the Madisonian vision of competing interests, the Anti-Federalists seem to look backward to a politics of the past. Yet in challenging the right of an elite to represent the common people and in their insistence that society "no longer be a hierarchy of ranks or even a division into unequal parts between gentlemen and commoners" they seem to anticipate the democratic politics of the future.[86] If Anti-Federalism (like classical republicanism) is Janus-faced, it is because it combines the group orientation characteristic of hierarchies with the commitment to individual autonomy typical of individualism. This combination generates a communitarian way of life hostile to both competitive individualism and hierarchical collectivism.

Abolitionism, Perfectionism, and Competition

"Be ye therefore perfect even as your Father which is in heaven is perfect." This was the message of perfectibility carried to the country by revivalist preachers like Charles Grandison Finney. By accepting Christ, men and women could become, or at a minimum should strive to become, literally perfect. The reborn would be totally unselfish and altruistic. "All sin," Finney insisted, "consists in selfishness; and all holiness or virtue in disinterested benevolence."[87]

Taken literally, Christ's injunction became a powerful instrument of reform. This optimistic faith in the "infinite worthiness" of man underlay the wave of social reforms that swept America in the 1830s and 1840s.[88] Perfectionism encouraged a belief that it was possible to remake the American people and American society. The spirit of the regenerate, as Finney told his audiences, "is necessarily that of the reformer. To the universal reformation of the world they stand committed."[89] Along with this commitment to "universal reformation" came a refusal to countenance halfway measures that temporized with sin.

The perfectionist ethic radicalized each reform movement it touched. The temperance movement moved from preaching moderation to advocating total abstinence. The American Peace Society, which began as a forum for moderate clergy, was taken over by advocates of nonresistance who repudiated the use of all force or coercion. The most radical change of all came in the antislavery movement, where the American Colonization Society, established in 1816 to secure the gradual and voluntary emancipation of slaves and to send them to Africa, was repudiated in the early 1830s by proponents of immediate, unconditional abolition of slavery. To accept anything short of immediate abolition, they declared, was to compromise with evil.[90]

That the perfectionist's uncompromising call for the "complete and final overthrow" of "great and sore evils" was potentially subversive of such institutions as slavery, the military, even government itself, was obvious even to contemporaries.[91] Less frequently recognized is the fundamental challenge that perfectionism posed to a system of competitive individualism. Competition as a social process assumes a world of imperfect beings, inescapably mired in selfish desires and partial knowledge.[92] For the individualist, it is axiomatic that, in the words of Adam Smith, "nobody but a beggar chuses to depend chiefly upon the benevolence of his fellow-citizens." Human nature, individualists tell us, is what it is (self-seeking) and nothing much can be done to change it. In a world of imperfect beings filled with "self-love," the best that we can do is "truck, barter, and exchange."[93]

A perfect people have no need for competition. Ulterior incentives are unnecessary in a world in which people are capable of acting out of an altruistic desire to serve others. Cooperation would drive out competition, and harmony could replace the jarring conflict of interests. If "human nature is open to total renovation in the twinkling of an eye,"[94] then it becomes a moral imperative to change and not simply channel existing, imperfect human nature. Having renounced the "demon" of selfishness,[95] society might indeed depend on the benevolence of the butcher, the brewer, or the baker.

Armed with the doctrine of perfectionism, abolitionists not only launched an uncompromising war against the "sin of slavery" but also offered a critique of competitive individualism. Very few, to be sure, repudiated private ownership of property; many showed unbounded enthusiasm for the latest technological advances (as did most European socialists, including Marx); and only a small minority followed European socialists (and southern slaveholders) in condemning "wage slavery." But it does not follow from this that abolitionists were, as many historians would have it, simply "one expression of the bourgeois competitive individualism . . . of antebellum Northern society."[96] The abolitionists' social vision combined freedom from institutional restraints with collective fellowship. If in their defense of individual autonomy they made common ground with competitive individualists, in their desire for community, unity, and harmony, they departed fundamentally from the competitive individualist ethos.[97]

The distinctive pattern of values and social relations that set the abolitionists (particularly those associated with William Lloyd Garrison) apart from the dominant competitive individualist culture of the North is described by Lawrence Friedman in his meticulous study *Gregarious Saints*. "At the heart of the immediatist crusade," Friedman finds, were "small groups."[98] Faced with a hostile public and possessed of an intense fear of defilement from having "to meet and mingle" with the morally tainted, immediatists withdraw into "the sanctuary of small informal intimacy clusters" like the Garrison-led Boston Clique, the circle of central New York Liberty men who gathered about Gerrit Smith's Peterboro estate, and Lewis Tappan's intimacy cluster of New York City–Oberlin evangelical associates.[99]

These immediatist groups were characterized, Friedman discovers, by an ineradicable tension between "harmonious collectivity" and "unfettered individuality."[100] Another historian echoes this view that abolitionism, like perfectionism, simultaneously "contained an anarchic appeal and a collectivist call, a command to shun evil and consult only conscience, and a mandate to join with the like-minded and look outward for perfect fellowship."[101] "A world in ourselves and in each other" is the way one abolitionist characterized their life of weak prescriptions and strong group boundaries.[102] The combination of strong group ("harmonious collectivity," "collectivist call," "a world in each other") and low grid ("unfettered individuality," "anarchic appeal," "a world in ourselves") defined the abolitionists' distinctive way of life.

Lewis Tappan's question "Is it right to be rich?" suggests something of the abolitionists' unease with the competitive individualist's untrammeled pursuit of wealth. Among Garrisonians, this ambivalence often slipped over into outright hostility. Lydia Maria Child, for instance, lamented that "in Wall-street, and elsewhere, Mammon, as usual, cooly [*sic*] calculates his chance of extracting a penny from war, pestilence, and famine." Her opinion of bidding and bargaining was equally jaundiced: "Commerce, with her loaded drays, and jaded skeletons of horses, is busy as ever 'fulfilling the World's contract with the Devil.'" Much the same attitude toward the freewheeling bustle of individualism was expressed by Garrison: "Mammon reigns in filthy splendor, and humanity finds none to sympathize with it." Wendell Phillips indicted competitive individualism for its disregard of principle: "It is hard," Phillips said, "to plant the self-sacrifice of a rigid Anti-Slavery, or any other principle, in the heart of . . . a prosperous, industrious, money-loving country, intensely devoted to the love of material gain."[103]

For the abolitionist, explains historian Ronald Walters, slavery was only "the most lurid case of unrestrained greed, . . . a sin that permeated the whole nation." Abolitionists believed that "Northerners as well as Southerners had an excessive love of money and a tendency to warp human relationships into business transactions." Walters continues, "The plantation ideal of defenders of slavery had much in common with the Christian utopia of Garrisonian nonresistants. Each was a counterimage, a form of protest against an emerging society in which relationships between humans seemed dictated more by the market-place than by bonds of morality and feeling."[104] The difference was that abolitionists desired to replace the ethically blind hand of bidding and bargaining with a way of life dedicated to diminishing, rather than legitimating, differences between people.

The marketplace was suspect in the eyes of abolitionists because it subverted their vision of a caring community of equals. Abolitionists sometimes idealized a simpler, freer, and more cohesive past, where people lived together "in perfect equality, and had no need of laws, or labor."[105] They yearned, in the words of Theodore Weld, for a "freer, . . . more harmonious form of human existence."[106] The ideal toward which they strove was not a world of perfect competition but a world in which "heads and hearts unite in working for the welfare of the human race."[107] One detects the same group-oriented social vision in Garrison's hope that the interests of capital and labor could be made to "blend harmoniously together with those of the common weal." Another, admittedly more radical manifestation of this communal vision was Elizur Wright's desire to see "capitalists and laborers . . . integrally associated, made partners, each sharing in the just profits in fixed proportions, and thus each interested in the other's welfare as well as his own."[108] Even when abolitionists praised capitalism, "the capitalism they envisioned," Walters points out, "was not that of J. P. Morgan; it was socially responsible, almost utopian. Lewis Tappan, for instance, saw the corporation as a number of men coming together to pool labor and capital to work in unity for a useful goal."[109]

Because abolitionists elevated cooperation over competition, rejected existing institutions as corrupt (and corrupting), and harbored an almost limitless faith in man's potential for goodness, some were inevitably attracted to utopian communities. Among the most well known of these communitarian experiments was Adin Ballou's Hopedale Community (or Fraternal Community Number One), founded in 1842. Ballou was converted to abolitionism and nonresistance in 1837 by Garrisonian recruiting agents. His views expressed the orthodox Garrisonian line: he inveighed against "selfish, unscrupulous, and heard-hearted Individualism"; he renounced all authority not based on persuasion; he was appalled at society's "revolting extremes of wealth and poverty"; and he was involved in every reform of his age—peace, women's rights, and temperance as well as the abolition of slavery. Where he differed was in his determination that true reform could best be achieved in a formal community that had withdrawn from the unregenerate world.[110]

Less successful, and far more radical, was the attempt by the Garrisonian John A. Collins to found a self-sufficient commune on three hundred and fifty acres in upstate New York. Collins was a close associate of Garrison's and had played an important role in organizing and running various Massachusetts abolition societies. In 1840 Garrison sent the thirty-year-old Collins to England to solicit funds for the financially strapped American Anti-Slavery Society. Collins failed in his assigned mission, but while there he became appalled by the living and working conditions of the English laboring classes. The English system of laissez-faire, Collins explained to Garrison, had created "a vast and complicated system" of slavery which "gives to the poor the ostensible appearance of freedom the more successfully to grind him to powder." Collins left England a convert of Robert Owen's utopian socialism, convinced that the private ownership of property made "man practically an enemy to his species," and that the future of America lay in self-sustaining collectives.[111]

Other followers of Garrison to launch utopian communities included Bronson Alcott in New England, the Brooke brothers in Ohio, and Garrison's own brother-in-law, George Benson, who had been a committed abolitionist since the movement's inception in 1831. Another utopian venture was organized by the perfectionist and abolitionist Marcus Spring (who was married to the daughter of Arnold Buffman, one of the original members of the Garrisonian movement). Among the members of Spring's "loving community" at Raritan Bay, New Jersey, was no less a luminary than Theodore Weld.[112]

These "communitarian experiments," historian John Thomas writes, "in effect were anti-institutional institutions." A less paradoxical way of conveying the same thing is to say that they were efforts at creating a low-grid, high-group way of life. Each of these communities, Thomas finds, "veered erratically between the poles of anarchism and collectivism as they hunted feverishly for a way of eliminating friction without employing coercion."[113] Looked at in this way, the organizational dilemma facing these utopian ventures was essentially similar to that which faced all abolitionists: that of reconciling, in Friedman's words, their "simultaneous cravings for cordial collectivity and pious individuality—for gregarity and sainthood."[114]

Although the great majority of abolitionists doubted the practicality of these self-sufficient communes, many remained intrigued by the models they offered. Garrison, for instance, was a frequent visitor to Benson's community, and Hopedale influenced many abolitionists, serving as it did "as a way-station for all of New England's reformers on their way to or from the endless succession of conventions and conferences for the improvement of American society."[115] Whatever their judgment about the workability of these communal utopias, few Garrisonians doubted the need for a more harmonious form of existence in which benevolence replaced selfishness and cooperation was privileged over competition. In their everyday life, Garrisonians self-consciously tried to build informal, intimate communities of believers and spurned contact with institutions (e.g., political parties and churches) that were morally tainted. Most did not join formal communes because they believed that the "principles of universal Christian brotherhood . . . must be matured in the hearts and lives of individuals, before they can be embodied in any community," but this did not mean that they doubted that "when the new organization commences, it will doubtless be in small communities."[116] To withdraw prematurely into self-sufficient communes, moreover, jeopardized their ultimate goal, which was nothing less than "communityzing the whole."[117]

Progressivism and the Public Interest

The progressive creed, if we are to believe Hartz's account, was reducible to something like "down with trusts, up with competitive individualism." The progressives' "trust-busting and boss-busting," Hartz explains, was simply "a version of the national [Horatio] Alger theme." For "if the trust is what is bad, you can stay at home with Locke simply by smashing it."[118] Some progressives were, to be sure, interested primarily in restoring competitive individualism, but progressive reform was not all of a piece.[119] Important elements within what is commonly referred to as progressivism hoped not to restore competition but to transcend it. Community and fraternity, they hoped, would replace selfish individualism.

This communitarian critique of competitive individualism drew inspiration from both religious and secular sources. The Social Gospel that flourished in the late nineteenth and early twentieth centuries taught people to judge public life by the Christian ideals of brotherhood and love. By this exacting yardstick, the selfish pursuit of material interests characteristic of the marketplace as well as politics was found terribly wanting. Appalled by the "horrible blunder and stupidity of our whole industrial system," Protestant ministers like Charles M. Sheldon asked people to ask themselves the simple question "What would Jesus do?" Christ, Sheldon believed, could not sanction the present system that "does not work according to any well-established plan of a brotherhood of men, but is driven by forces that revolve around some pagan rule of life called 'supply and demand' or a 'market.'"[120] "Heedlessness of others' interests" violated the "Golden Rule," which is "the Rule of Love."[121] It was with the message of the Social Gospel in mind that the editors of the progressive *Independent* suggested that "religion might . . . condemn the entire structure of modern business as radically wrong and hostile to the true life of fellowship."[122]

One Progressive reformer strongly influenced by the Social Gospel was Samuel ("Golden Rule") Jones, a three-term mayor of Toledo, Ohio. Jones denied that bosses and trusts were the source of the nation's difficulties. "Our trouble is not with the bosses, with the aristocrats, with the corporations or the Standard Oil Company," he explained, "but with a system that denies brotherhood and makes a weaker brother the legitimate prey of every strong man." It was not concentration but competition itself that was the root evil. Capitalism in America was "individualism gone mad." "A strictly Christian life, according to the life and teachings of Jesus," Jones believed, was "incompatible with what is known as 'success' in business." Profit and competition were completely counter to the spirit of love and brotherhood upon which a Christian community must be based.[123]

Among Christian progressive reformers, observes historian David Danbom, "it became commonplace . . . to look forward to the replacement of the competitive system with a cooperative one."[124] "In the new society toward which we are now tending," declared a Presbyterian minister from Wisconsin, "men will cease to compete against each other, and . . . will co-operate with one another to make common cause, burying their mutual differences before a common foe."[125] "Our world is becoming more social, in the sense of being less individualistic," another reformer confidently reported. Although some will try "to put the clock back," there seemed nonetheless "a reasonable hope of moving forward, out of the chaos of mere individualism with unrestrained competition and continual strife" and into a new and brighter world of cooperation.[126]

The message that society was, or should be, "evolving from a competitive jungle into a cooperative commonwealth" was spread not only by clergymen but by a new generation of reform-minded social scientists.[127] Social scientists developed a critique of competitive individualism that closely paralleled (and reinforced) the communitarian ethos preached by the reform-minded clergy. Sociologists like Charles Horton Cooley worried that the specialization and division of labor that accompanied modernization made it increasingly difficult for the individual to identify with the whole. The result, Cooley believed, was a loss of that "community spirit" that had held past societies together. Cooley rejected both Spencer's individualist vision of coordination without community and the Hegelian view of coordination through a strong national state. He warned not only against "irresponsible individualism" and the "disorganization and disintegration of laissez-faire" but also against the bureaucratic coercion, conformism, and militarism of the national state. If the marketplace left individuals without a sense of belonging, the large formal institution was no better for it "does not enlist and discipline the soul of the individual but takes hold of him by the outside, his personality being left to torpor." In Cooley's view, the "community ideal" was to be sought not in an organic state but in the face-to-face interaction of families, neighborhoods, and occupations. It was not simply nostalgia for the past that inspired Cooley; he welcomed the new freedom that modern society offered, and he rejoiced that "association" in the modern world was "throwing off the gross and oppressive bonds of time and place, and substituting congenial relations of sympathy and choice."[128]

A similar preference for "cooperation without statism," for "the cohesiveness of the face-to-face community without the curtailment of individual choice," was central to the thinking of political theorist and Boston civic reformer Mary Parker Follett.[129] Like Cooley, Follett believed that the problem with American society was that "the sustaining and nourishing power of the community bond . . . is almost unknown now." The freedom to pursue one's private preferences for Follett was not freedom at all: "We are lost, exiled, imprisoned until we feel the joy of union." In place of the individualistic clash of private preferences, Follett looked to face-to-face communication to transform those preferences so that "the 'claims' of others [become] my desires." Follett's term for this alternative way of life was "cooperative collectivism."[130]

The interpenetration of social science and Social Gospel in producing a progressive communitarian critique of competitive individualism was evident, among other places, in the settlement house movement. Inspired in large part by the Social Gospel's message to apply Christian precepts to public life, the settlement houses served as a convenient site for reformist social scientists to gather empirical data about the conditions of the urban lower classes and provided laboratories in which to experiment with cooperative living. Nowhere was this confluence of influences clearer than at Jane Addams's Hull-House, which, as historian Robert Crumden observes, "stood halfway between the Protestant churches and the University of Chicago."[131]

Addams hoped that Hull-House would provide "little islands of affection in the vast sea of impersonal forces."[132] She lamented that society had become infused with the "cold calculating spirit of selfishness and self seeking" and spoke glowingly of the virtues of "companionship and solidarity," "communion and fellowship," and the "fostering soil of community life."[133] Although Addams's vision was communal it was certainly not hierarchical. Instead her ideal, as historian Daniel Levine describes it, was "a society in which there were no barriers of class, occupation, or social standing between individuals." Addams apparently made "a conscious effort to cast out the devil of social distinctions in her own mind"; for instance, in a letter to a close friend Mary Rozet Smith, she carelessly referred to Miss Smith as a lady but to the recipients of charity as girls. Catching herself, Addams apologized for "the false social distinction, a remnant of former prejudices."[134]

While Addams tried to create an egalitarian community among the urban poor of inner-city Chicago, John Dewey (a close friend of Addams) looked to the public schools to institutionalize communitarian ideals. Dewey rejected the kind of classroom that was so imbued with competitive ethics that "for one child to help another in his task has become a school crime," calling instead for a new form of cooperative learning.[135] He envisioned this new type of school as a kind of "embryonic community" or "miniature community."[136] Dewey's Hull-House was the Laboratory School, which he directed from 1896 to 1904. Like Hull-House, the Laboratory School was conceived as a means of working for "a new society where cooperation rather than competition should rule." Its teachers were committed to counteracting the pernicious "competitive antisocial spirit and dominant selfishness" that prevailed in society.[137]

Progressives of both a religious and secular stripe would have nothing to do with the notion of a "hidden hand" transforming private vices into public benefits. As Herbert Croly laments in *The Promise of American Life*, "the plain fact is that the individual in freely and energetically pursuing his own private purposes has not been the inevitable public benefactor assumed by the traditional American interpretation of democracy."[138] The good society, progressives reiterated, could result only through the principled attachment of each individual to "the public interest." It was a man's duty, explained the *Outlook*, "to subordinate private to public interest, to seek first of all the common good in which our own is involved."[139] "The most glorious career that love can conceive for its object," lectured George D. Herron, a reform clergyman who was appointed to teach in a new Department of Applied Christianity at Grinnell College, "is one of complete sacrifice in the service of the common life."[140] To "work not for wealth but for Weal" was the "shining goal" that W. E. B. Du Bois held out to the black community.[141] And when Frederic C. Howe, aide to Cleveland's reform mayor Tom Johnson, looked hopefully to an emerging "city consciousness," he intended to denote "that instinct which is willingness to struggle for the common weal, and suffer for the common woe."[142]

Whether "the desire to . . . inculcate in Americans a higher sense of brotherhood and selflessness and a devotion to the public interest" was, as David Danbom argues, "the essence of progressivism" is questionable.[143] But that this was one important cultural strand within progressivism seems undeniable. To characterize progressive reformers as advocates of "moral capitalism," as Robert Wiebe does, is to understate the tension between a way of life based on "fraternal equality" and one based on competitive individualism.[144] For the individualist, "moral capitalism" is redundant; capitalism *is* a moral order. For the egalitarian, the competition and self-aggrandizement of capitalism is inherently immoral because it neglects the community interest that transcends the sum of private interests and wants.

Nor was the progressives' hostility to competitive individualism mere rhetoric. Their attempts to institutionalize alternative patterns of social relations in schools, settlement houses, and community centers attest to that. So, too, do their efforts (sometimes successful, sometimes not) to assert public control over the economy. At the local level, this meant pushing for municipal ownership and operation of essential services like garbage collection and lighting. The attempt to assert a public interest separate from and superior to private interests also manifested itself in citywide planning, zoning and building codes, regulation of noise, smoke and waste disposal, and standards for the treatment of workers and consumers. At the state and national level, progressives backed regulation of trusts, a progressive income tax, and the protection of labor through child labor laws, workman's compensation, and collective bargaining.[145]

Progressivism was more then than just an effort, in Richard Hofstadter's words, to "re-create the old nation of limited and decentralized power, genuine competition, democratic opportunity, and enterprise."[146] If theirs was in many ways a "backward-looking vision,"[147] the past that progressives idealized was more often the small-town virtues of fraternity, equality, and civic participation.[148] Typical, for instance, was William Allen White's praise of the small-town society of the Midwest for imbuing Americans with "the gospel of a fraternal equality."[149] Although progressives often looked back nostalgically to the small town for models of community, many simultaneously looked forward hopefully to a new age that would usher in "a free, full community life."[150] Even a hardened cynic like Thorstein Veblen could cautiously predict that "the Christian principle of brotherhood should logically continue to gain ground at the expense of the pecuniary morals of competitive business."[151] If not all progressive reformers were as optimistic as Frederic Howe that the city was drawing individuals "into an intimacy, a solidarity, which makes the welfare of one the welfare of all," many did share Howe's sense that such public-spiritedness was the ideal that people should strive for and that the self-interested pursuit of individual gain characteristic of capitalism needed to be remedied if not jettisoned.[152]

Progressivism, like abolitionism, civic republicanism, and Puritanism, poses puzzling anomalies for the conventional dichotomy between traditional hierarchy and modern individualism. If one focuses on its commitment to individual conscience and autonomy, its adherents seem to be forerunners of a modern individualistic age. But if attention is turned to its vision of the individual subordinated to the collective will, progressivism seems to point backward to a bygone age of hierarchy. Only if one recognizes that group and grid, integration and regulation, are distinct social dimensions can one begin to make sense of these apparently paradoxical movements that envision an egalitarian community that is neither individualistic nor hierarchical. That the communitarian vision of Puritanism, classical republicanism, abolitionism, and progressivism recurs in many of the social movements of our day tells us that this way of life is not a vestige of the past. Rather it represents a basic choice that all humans, whatever their level of technological advancement, face in organizing their social life. In spite (or perhaps because) of the fact that the United States has lacked the strong hierarchical institutions and values of Europe, the tension between competitive individualism and egalitarian community has taken on even greater importance in America than in Europe.[153]

The Paradox of Crusading Capitalism

Students of American political culture have frequently commented on the tone of moral absolutism that so often animates our political life. Americans, writes Hofstadter, are prone "to fits of moral crusading" and "do not abide very quietly the evils of life." Seymour Martin Lipset concurs that Americans are "particularly inclined to support movements for the elimination of evil." "There have been few more persistent themes in American politics," echoes Grant McConnell, "than that of outraged virtue. Sporadically, but repeatedly and insistently, national indignation has erupted at revelations of corruption in public life." Samuel Huntington coins the label "creedal passion period" to describe those recurrent periods in United States history characterized by "widespread and intense moral indignation" about the gap between American ideals and practice, and by a "rush to moral judgment on the rights and wrongs of politics."[154]

The existence of this moral fervor in American politics presents a puzzle for those who would portray American political culture as pervasively entrepreneurial and capitalistic. Why should a country overwhelmingly committed to bidding and bargaining declare a holy war on the urban boss who buys and sells votes? Why would entrepreneurs accustomed to transacting with all comers erect an impassable boundary separating good from evil, virtue from corruption? Why should a nation unambiguously committed to possessive individualism bewail the spread of selfish materialism? Why should a people "notorious for its worship of success" suddenly turn "savagely upon those who had achieved it"?[155] How is it, in short, that a nation of traders can possess (in the words of G. K. Chesterton) "the soul of a church?"[156]

Perhaps Americans are just a peculiarly paradoxical bunch. This is the answer offered in Michael Kammen's brilliant if elusive *People of Paradox*. Kammen delights in such paradoxical descriptions of America as a place of "godly materialism," "pragmatic idealism," and "practical moralism."[157] America, he notes, couples "boisterous competition" with "self-effacing cooperation," "concern for private, material prosperity with a propensity for periodic public renewal and religious enthusiasm."[158] For Kammen, paradoxes such as these define the American character. What interests Kammen is less the "power struggles between radicals and moderates, conservatives and liberals [than] the larger and *internalized* tensions within the society as a whole, as well as within many of those individuals who comprise it."[159] Cultural conflict, for Kammen, takes place within the American psyche.

I have no wish to take issue with the notion that value systems contain paradoxical contradictions. Egalitarians, for instance, confront such a paradox in trying to combine a collective life with individual autonomy. Paradox and irony are basic to the human condition. But there is a danger of replacing social and political conflict with psychological paradox. "Godly materialism" may be an internalized tension, but it may also be a conflict between the godly and the materialistic. Rather than view paradoxes as defining the American character, perhaps it is better to view such antinomies as part of the sociocultural struggle over how America is to be defined. Paradox may be evidence not of a single national character but of rival cultural systems. In any event, to identify a paradox is the beginning, not the end. It poses the puzzle but does not explain or resolve it.

How then is one to resolve the paradox of crusading capitalists? One might, as Hartz and Boorstin do, downplay or even ignore the importance of the crusading side of American political culture. Hartz's analysis of the American Revolution, for instance, contrasts the "sober temper" of America with "the crusading spirit" and "wild enthusiasms of Europe." For Boorstin, too, the American Revolution constitutes "a prudential decision" free of dogma or religious enthusiasm.[160] But this interpretation cannot do justice to "the fanatical and millennial thinking" that historians have identified as characteristic of American revolutionary rhetoric. The ideas expressed during the American Revolution, historian Gordon Wood tells us, are "remarkably similar" to the ideas articulated during the seventeenth-century Puritan Revolution and the eighteenth-century French Revolution. All three upheavals share "the same general disgust with a chaotic and corrupt world, the same anxious and angry bombast, the same excited fears of conspiracies by depraved men, the same utopian hopes for the construction of a new and virtuous order."[161] Nor can Hartz or Boorstin account for the moral intensity animating the Jacksonian war on "the monster bank," the abolitionist assault on "the slave power," the Populist attack on "the money power," the Progressive crusade against "the interests," or the 1960s "war on poverty."

Several scholars have tried to account for the crusading side of American political culture by pointing to the country's Puritan origins. Huntington, for instance, identifies the Puritan Revolution as "the original source . . . of the moral passion that has powered the engines of political change in America." Puritanism "bequeathed to the American people the belief that they were engaged in a righteous effort to ensure the triumph of good over evil." And from the Puritans came the recurring drive "to cleanse and purify government."[162] If Locke, as Hartz says, led in the direction of "the free and easy play of pressure groups," Puritanism led toward "the politics of movements and causes, of creedal passion and reform." If Americans were capitalists on account of a pervasive Lockean liberalism, they were moral crusaders because the zealous spirit of Puritanism had been "permanently lodged deep in the American consciousness."[163]

Placing the millennialism and moralism of Puritanism at the fount of American culture is an improvement over Hartz's uniform Lockean liberalism or Boorstin's hegemonic "pragmatic spirit." Unexplained, though, is how cultural traditions get "lodged" in a nation's consciousness. What stops such traditions from being dislodged or pried loose from the collective memory? Ideas must not become disembodied, handed down from generation to generation yet untouched by human hands.[164] To explain why some Americans have expressed a fervent commitment to transforming a corrupt and selfish society into a virtuous and cooperative commonwealth, it is not sufficient to invoke our Puritan antecedents. One needs also to ask about the social and political experiences that make these ideas plausible as well as the function the ideas serve.[165] Placing ideas in a social context invites us to probe the process by which ideas are continually negotiated and renegotiated by real human beings.

Mary Douglas's grid–group typology helps us to think systematically about the types of social experiences that are viable, and the kinds of cultural biases that go with these different forms of social experience. Douglas hypothesizes that "it takes a certain kind of social experience to start to worry about the problem of evil." "Where a man is expected to build his own career by transacting with all and sundry as widely as possible, . . . men are not set on either side of a line dividing sheep from goats." Under these competitive conditions, men "are seen to be unequally endowed with talents, but the inequality is random, unpredictable and unconnected with moral judgment." Rigid boundaries separating good from evil would be dysfunctional to the individualistic entrepreneur, as they would serve only to impede individual transactions.[166]

It is in strongly bounded but internally egalitarian collectivities, Douglas suggests, that rigid distinctions between "pure, good men and utterly vile men" will flourish. The central organizational dilemma for people who wish to live a collective life without hierarchical authority is how to keep the group together in the absence of coercive sanctions. Belief in an evil and corrupt outside world serves to knit together the members of the group by reminding them of their precarious position as a "city on the hill." The rigid dichotomy between a good inside and an evil outside thus serves to legitimize their preferred pattern of social relations.[167]

Douglas's theory helps identify what it was about Puritan social organization that generated a heightened concern with the boundaries separating purity and corruption. If Douglas is correct, it is not Puritan ghosts lodged in our national subconscious that account for America's periodic "fits of moral crusading" but rather the recurrent effort among some Americans to organize social life along communitarian lines. Puritanism thus becomes an instance of a more general phenomenon rather than the explanation itself. Radical abolitionists, civic republicans, communitarian progressives, and radical feminists manifest a preoccupation with rooting out corruption and evil and a fervent desire to usher in a new world of purity and virtue not because they inherited a disembodied Puritan vision but because, like the Puritans, they tried to structure their lives in ways that combine a commitment to community with an insistence on individual conscience and autonomy.[168]

The paradox of crusading capitalism begins to dissolve. Crusades against corruption in the name of purity stem not from capitalists suddenly deciding to subordinate bidding and bargaining to moral passion but rather from egalitarians who repudiate the bidding and bargaining of competitive individualism in favor of egalitarian community. From the decision to opt for community over competition, the collective good over private interests, flows other commitments and behaviors. Some of these patterns of behavior and preferences are selected consciously; other bits of behavior, like the concern with purity, are selected "behind the back" of individuals.[169] Those who select a life of competitive individualism will end up with a package of values and beliefs that differ systematically from the values and beliefs of those who choose a life of egalitarian community. In the next four chapters I attempt to spell out in greater detail the conflict between these two political cultures as manifested in their contrasting conceptions of property, equality, democracy, and authority.

2

Radical Lockeanism

American political culture is commonly portrayed, to borrow a phrase from Emerson, as the lengthened shadow of John Locke. This view of American history has been advanced not only by political scientists influenced by Louis Hartz[1] but by generations of frustrated radicals who have tried to explain why socialism failed to take root in the United States.[2] Marxist ideas, these observers maintain, could never flourish in a nation wedded to the Lockean justification of property as a "natural right."

As an explanation for "why no socialism" this interpretation is not without a certain plausibility. A Lockean commitment to private ownership of property does conflict with Marxist doctrines of collective ownership. But the antithesis between socialism and bourgeois liberalism, although useful for certain purposes, is inadequate because it offers an impoverished view of radicalism (equating radicalism with a revolutionary, class conscious working class)[3] and an undifferentiated view of liberalism (conflating individualism and egalitarianism). In part the weakness of this dichotomy reflects the fact that, as Larry Siedentop points out, "socialism is itself . . . parasitic on the norms of liberal theory . . . for its commitment to human equality."[4] Equally important, I contend, is that throughout this nation's history Americans have interpreted Lockean precepts to justify egalitarian as well as individualistic patterns of social relations.

This chapter argues that two distinct belief systems—the first egalitarian and communitarian, the second individualistic and competitive—exist within what is commonly termed "liberalism"[5] and that proponents of both belief systems appeal to Lockean precepts but interpret these precepts to support fundamentally different political cultures.

It may be true, as Richard Hofstadter argues, that in America "the range of vision embraced by the primary contestants in the major parties has always been bounded by the horizons of property,"[6] but what kind of cultural constraint is entailed by a commitment to the private ownership of property? If "genuine" radicalism is equated with socialism, then a belief in private property is of decisive importance. If, on the other hand, radical egalitarianism is conceived of as a commitment to equal conditions, and if egalitarian social relations are compatible with private property, then the significance of a nation being "bounded by the horizons of property" recedes.

The political thought of Jean-Jacques Rousseau strongly suggests that private property is compatible with an egalitarian worldview.[7] Rousseau embraced private property at the same time that he championed a radically egalitarian vision of society and politics. As Judith Shklar observes, "Rousseau did not doubt any more than Locke did that men had a right to their property and that rightful ownership was the origin of all justice. That there could be no liberty and no true security of obligations without the sacred rights of property was perfectly clear to Rousseau."[8] Rousseau's defense of private property did not prevent him from abhorring great inequalities, excoriating the ill effects of competition, or extolling the virtues of fraternity and community. Private property, *as long as it was relatively evenly distributed*, was, in Rousseau's view, perfectly consistent with the just society. The real social evil lay not in private property per se but in the damage inflicted by gross inequalities in the distribution of property. Rousseau's preoccupation with the ill effects of unequal conditions differentiates him, of course, from Locke. My point is not that Locke and Rousseau were agreed on the value of equality. Clearly they were not. Rather I mean to suggest that the distinction between Lockean individualism and Rousseauean egalitarianism turns less on the question of property than on questions of equality, fraternity, and competition.

Interpreting Locke on Property

In recent years a debate over how best to interpret Locke's views on property has raged among students of political philosophy. According to some, Locke was the philosopher of "possessive individualism," who sanctioned unlimited accumulation and competitive capitalism. This is the Marxist's Locke as well as Hartz's Locke. Others have argued that Locke is more accurately understood as an exponent of a radical, indeed revolutionary, politics, and that his attitude to property was "not so far removed" from that of the Diggers, the agrarian communists of the 1640s.[9]

The precise details of these controversies need not concern us here. My interest does not lie in what Locke really said or meant, but rather in what this controversy suggests about the argument that Americans have always been gripped by an "absolute and irrational attachment" to Locke.[10] For if contemporary scholars disagree over what Locke intended, might not we expect to discover that American citizens, too, have differed in their constructions of Locke? If so, a recasting of the Hartzian thesis is in order. For if adherents of different political cultures have appealed to different Lockes, then "the power of Locke in America" seems less constricting than Hartz and others have believed.[11] Rather than posit a Lockeanism that controls and confines American political discourse, I accent the ways in which adherents of rival political cultures have used Lockean language and principles to bolster their preferred patterns of social relations.

Disagreements about how best to understand Locke stem in part from a basic ambiguity in Locke's political thought, an ambiguity that invites both individual-istic and egalitarian readings of Locke.[12] By emphasizing that private property is inviolate, Locke seems to sanction existing inequalities. This is the Locke seized upon by individualistic merchants and entrepreneurs. At the same time, however, Locke suggests that in the state of nature men can rightfully acquire only enough to meet the immediate necessities of life. "Whatever is beyond this," Locke ex-plains, "is more than his share, and belongs to others."[13] Here is a Locke more attractive to those imbued with an egalitarian ethos.

To be sure, Locke also says that the creation of civil society supersedes the limitations that restricted legitimate accumulation in the state of nature. The introduction of money makes it possible for people to accumulate goods beyond immediate use without having them spoil. But if civil laws permit greater accu-mulation than natural law, they do not necessarily justify unlimited accumulation. For the validity of a civil law depends on it being "conformable to the Law of Nature," which decrees that no man can have such a "Portion of the things of this World" as to deprive "his needy Brother a Right to the Surplusage of his Goods."[14] To those who rely on this Locke, redistribution does not seem out of the question.

More important, Locke's notion that laws should secure to each man the fruits of his labor can be harnessed to competing cultural visions. It can be used by individualists to justify the right of each man to keep what he has acquired, but it can also be employed by egalitarians to attack large concentrations of wealth on the grounds that such holdings do not really derive from productive labor. Interpreted literally, Locke's injunction that a person has a right to that property which "he hath mixed his Labour with"[15] can be used to attack the speculator and trader in the name of the "workingman." Locke's labor theory of value is thus potentially subversive of the bidding and bargaining essential to competitive indi-vidualism.

Put more formally, private ownership of property is a necessary but not suffi-cient condition for competitive individualism. Affirming private (as opposed to collective) ownership of property does not preclude an attack on the existing dis-tribution of that property. Nor does accepting the principle of private property necessarily entail rejecting the collectivity's power to regulate the individual's ability to dispose of that property. It is true that in its most extreme manifestations, egalitarianism will incline toward holding property in common. But egalitarian-ism is viable even where property is not shared communally. What is necessary to sustain an egalitarian political culture is not the elimination of private property but rather the preclusion of large concentrations of property and the restriction of an individual's freedom to dispose of that property when it conflicts with the com-mon good.

That different interpretations of Locke's political thought are possible does not mean that all interpretations are equally plausible. Individualists arguably do have a better claim on Locke than egalitarians or hierarchists. The notion of a natural right to property, as Richard Schlatter points out, works against "the claims of all those, whether paternal autocrats or levelling democrats, who want to redistribute property in accordance with some ideal rule of justice."[16] To argue that individual property predates government is to arm individualists with a potent instrument with which to resist the communitarian impulses inherent in the egalitarian vision.

These individualistic tendencies in Locke's thought have led some American egalitarians to reject Lockeanism and private property altogether. More commonly, however, American egalitarians have tried to reconstruct Locke to suit their own cultural ends. In this chapter, I examine four periods in late-eighteenth- and nineteenth-century American history, each of which witnessed important efforts to fashion a radically egalitarian interpretation of Lockean ideas. I begin with Tom Paine.

Tom Paine: Constructing an Egalitarian Locke

Although Paine professed never to have read Locke, there are obvious parallels in their political thought. Indeed so similar were their arguments in places that Paine was accused of having cribbed from Locke's *Treatise of Government*.[17] But if Paine echoed Lockean precepts, he turned these principles to often radically egalitarian ends. Hartz seems to sense the difficulties that Paine poses for the consensus thesis, tagging Paine an "American radical liberal."[18] But if we conclude that Paine is indebted to Lockean liberalism (as most assuredly he was), we must also conclude that strikingly different visions of the good life are consistent with Lockean premises about private property and natural rights. For from the Lockean premise of a natural right to property, Paine launched an uncompromising assault on the existing distribution of wealth.[19]

Paine and other spokesmen for Philadelphia's radical artisans used Locke's labor theory of property to formulate an invidious distinction between productive and unproductive labor. Property derived from trade or speculation did not possess the same moral justification as property derived from "real" work. Artisans, craftsmen, and farmers were entitled to the fruits of their labor, whereas merchants, bankers, and speculators, who were perceived to have gained their property by the labor of others, were not.[20]

Paine's egalitarian construction of Locke reached its fullest expression in a pamphlet entitled *Agrarian Justice*, which he wrote in 1796 while in France. Like Locke, Paine began from the premise that "the earth, in its natural, uncultivated state was, and ever would have continued to be, the common property of the human race." Each man, Paine argued (again echoing Locke), thus had a natural right to property. So too Paine agreed that the origins of private property lay in the individual improving the land by mixing his labor with the earth. But Paine also insisted that since "it is the value of the improvement, only, and not the earth itself, that is individual property, . . . every proprietor . . . of cultivated lands owes to the community a ground-rent . . . for the land which he holds."[21] This payment would create a fund that would pay everyone reaching the age of twenty-one the sum of fifteen pounds "as a compensation in part, for the loss of his or her natural inheritance, by the introduction of the system of landed property."[22]

Paine radicalized Locke still further through a narrow interpretation of the Lockean injunction regarding the personal (as opposed to landed) property that accrues to a man as a result of "the Labour of his Body, and the Work of his Hands."[23] For Paine, "all accumulation . . . of personal property, *beyond what a man's own hands produce*, is derived to him by living in society." One thus had an absolute right only to what one had personally made and not to what one traded or exchanged for. Because personal property was "the effect of society," Paine reasoned, each person "owes . . . a part of that accumulation back to society from whence the whole came."[24]

Whether property is conceived of as prior to society (as in the case of landed property) or as subsequent to society (as in the case of personal property), Paine thus arrives at the same egalitarian conclusions: the community is justified in redistributing wealth from the haves (with "overgrown" shares of property) to the have-nots.[25] In the case of landed property, individuals are obligated as a matter of justice to compensate the less fortunate, who have been dispossessed by the march of civilization.[26] In the case of personal property, the community has a right to make claims upon individual wealth because it is society that makes possible the accumulation of virtually all personal property.

As Paine's thinking moved in an increasingly egalitarian direction, he also inclined increasingly toward system-blame rather than self-blame. In *Common Sense*, Paine had been content to explain economic inequalities primarily in terms of individual differences in talent and industry, but in *Agrarian Justice* he announced that "the fault [for dispossession] is in the system" of landed property. "The present state of civilization," Paine continued, "is as odious as it is unjust," for it deprives people of their natural inheritance without any kind of restitution for that dispossession. It was "next to impossible," Paine believed, for those born into poverty "to get out of that state themselves." Far from being able to pull themselves up through hard work, the poor had become virtually a "hereditary race."[27]

It is true that Paine stopped short of advocating an "agrarian law" dividing existing landed property, reasoning that since in the real world "it is impossible to separate the improvement made by cultivation from the earth itself upon which that improvement is made," it would be unjust to take land from those who had improved that land. Ground rents and taxes were thus a means of doing justice to both those who had been "thrown out of their natural inheritance" and those who had mixed their labor with the land. But if Paine was loathe to redistribute land, so too were the French Jacobins, who in 1792 had made advocacy of agrarian laws a capital offense.[28]

For some Paine's refusal to follow the French socialist Babeuf in attacking the principle of private property indicates that at bottom Paine was "one of the purest ideological spokesmen for the bourgeoisie."[29] But if Paine is a bourgeois liberal, one cannot but marvel at the incredible variety of political values and beliefs that exist within that category. Far from being an ideological straitjacket, Lockeanism begins to seem at best a loose-fitting overcoat. Paine's commitment to private property did not prevent him, for example, from excoriating the existing distribution of that property as "odious and unjust," or from insisting that it was "necessary that a revolution should be made in it." Sounding more like Marx than Locke, Paine declared that "the contrast of affluence and wretchedness continually meeting and offending the eye, is like dead and living bodies chained together."[30] Nor did accepting Locke's labor theory of property prevent Paine from suggesting, again sounding more like Marx than Locke, that "the accumulation of personal property is, in many instances, the effect of paying too little for the labor that produced it; the consequence of which is that the working hand perishes in old age, and the employer abounds in affluence."[31] Whatever Paine's debts to the bourgeois world of private property and John Locke, Paine's beliefs about human nature, blame, authority, and wealth are far more consistent with radical egalitarianism than competitive individualism.[32]

Radicalizing Locke in the Jacksonian Era

The radicalizing of Locke evident in the political thought of Thomas Paine is even more pronounced in the thinking of radical Jacksonians. Like Paine, they interpreted Locke to suit egalitarian ends. When Locke said that "Justice gives every Man a Title to the product of his honest Industry,"[33] radical Jacksonians used it to attack wealthy speculators who turned a quick profit. When Locke said there was a natural right to property, radical Jacksonians took it to mean not only that an individual's property should be protected from government but that every individual should possess property.

If Locke was used by some to justify unlimited appropriation, radical Jacksonians showed that Locke could also be invoked to check the accumulation of wealth. Lemuel D. Evans invoked a radical Locke at Texas's 1845 constitutional convention when he insisted upon invalidating the huge land grants made under the Mexican regime: "The great Locke laid it down that we could rightfully appropriate so much, and only so much as we might need for the plough, or to graze our flock, only so much as we can mix our labor with."[34] Although their commitment to Lockeanism may have prevented radical Jacksonians from renouncing private property, it did not stop them from advocating, in the words of one Pennsylvania Democrat, an equal "possession, distribution, and transmission" of that property.[35]

George Henry Evans, founder of the National Reform Association, had little difficulty constructing an egalitarian vision upon Lockean precepts. If every man had a natural right to property, Evans reasoned, the current distribution of property, which left many laborers landless, must be unnatural and unjust. The availability of unsettled lands in the West meant that Evans and his Reform Association allies felt they did not need to redistribute existing property. Instead they advocated giving away unsettled western land to the urban poor. But their plan was not to give the "haggard, care-worn" laborer an opportunity to become an acquisitive capitalist in his own right. Rather they hoped, as one historian points out, that "the holdings might be arranged so as to form small communities." Evans and his associates planned to have each village be made up of 160 farms, each of 160 acres.[36] That their vision was bounded by the horizons of private property should not obscure its radically egalitarian character. The primary objective of the Reform Association was not to get the urban poor into the game of competitive enterprise but to create a new social system premised on equality of condition.

The potential radicalism of Lockean natural rights doctrines is perhaps nowhere more fully realized than in Thomas Skidmore's *The Rights of Man to Property!*

Skidmore regarded the right to property as so fundamental that he criticized Jefferson for substituting "the pursuit of happiness" for "property" in the Declaration of Independence.[37] Property was an inalienable right, given by "the Creator of the Universe" to every person. "I am; therefore, property is mine."[38] But from these rather orthodox Lockean premises, Skidmore issued a call for social revolution, insisting that the rights of labor and the poor require that "we rip all up, and make a full General Division" of property.[39] Such drastic measures were justified, Skidmore reasoned, because no one could be denied his birthright without his consent and none could have possibly consented to a propertyless condition or one of meager property holdings.

The best means of realizing "man's natural right to an equal portion of property,"[40] Skidmore believed, was to give to each man upon coming of age an equal share of society's property. What is significant is less the mechanism (essentially a 100 percent inheritance tax to be distributed equally to those who had reached the age of majority) than the reasoning used to arrive at this scheme. For Skidmore's argument is not Andrew Carnegie's argument that inheritance dulls the spirit of enterprise (the individualist rationale for inheritance taxes) but that inheritance is unjust because it gives more to some than to others (the egalitarian argument). Carnegie's concern was the effect upon those who inherit; Skidmore's concern was with those who inherit little or nothing.

Another radical Jacksonian who was too much of a Lockean ever to favor European-style socialism was William Leggett. "The protection of property," Leggett agreed, was "among [a good government's] first and principal duties."[41] But Leggett's rationale for limiting the scope of government activity was less a Lockean desire to protect private property from the state than a belief that governmental action inevitably tends to favor the strong few at the expense of the weak multitude. Leggett could not sanction governmental activity because the power of the state "has always been exercised under the influence and for the exclusive benefit of wealth. It was never wielded in behalf of the community."[42] Were this premise to change and government to become seen as an agent of the public interest or a protector of the weak, so too, presumably, would Leggett's conclusion about the legitimacy of governmental action. Indeed this was the basis for Leggett's approval of unions (or, as he called it, "associated effort"), which he believed to be "the great instrument of the rights of the poor."[43] What kind of a republic is it, he asks, "where men cannot unite in one common effort, in one common cause, without rousing the cry of danger to the rights of person and property"?[44] The answer is an individualist republic, and Leggett's rhetorical question makes clear that this is not his preferred way of life.

Basic to the radical Jacksonian worldview was the idea that "all the wealth of the world is the product of labour."[45] By labor these Jacksonians usually meant those people who worked with their hands for wages: "those who do the work and fight the battles; who produce the necessaries and comforts of life; who till the earth or dig for its treasures; who build the houses and the ships; who make the clothes, the books, the machinery, the clocks and watches, the musical instruments, and the thousands of things which are necessary to enable men to live and be happy."[46] The person who risked all in speculative ventures, who extended credit to others, or "who does but exchange,"[47] was seen as parasitic rather than wealth-creating. Given this narrow reading of what constituted labor, the Lockean injunction concerning the fruits of one's labor became subversive rather than supportive of competitive individualism.

Most Democrats would have concurred with Amos Kendall that "the interest of every man in every nation . . . is in the full enjoyment of the fruits of his own labor."[48] But radical and even many moderate Democrats felt that there was often too little correspondence between those who did the labor and those who had the wealth. The editor of the Detroit *Free Press*, John S. Bagg, observed that although "all wealth is the result of labor, yet, strange enough, the man who does the labor never has the wealth."[49] Radicals questioned the justice of a social system in which "those by whose toil all comforts and luxuries are produced or made available enjoy so scanty a share of them."[50] Workers, went the typical radical lament, "produce all the wealth of society without sharing a thousandth part of it."[51] If radical Jacksonians began with Locke, they had ended up closer to Marx than to Adam Smith.

The more intellectual of the radical Jacksonians sometimes recognized the distance they had traversed from Locke. George Bancroft, who served as James Polk's secretary of the navy and unofficial historian/philosopher of Jacksonian democracy, displayed a "strong animus" toward Locke in his multivolume *History of the United States*, explicitly repudiating those aspects of Locke's thought that seemed to impede an egalitarian way of life. By basing society upon contract, Bancroft lamented, Locke tried to build political life on nothing more than self-interest.[52] This, Bancroft insisted, was a terrible mistake. The "abandonment of labor to the unmitigated effects of personal competition," Bancroft argued, "can never be accepted as the rule for the dealings of man to man." Bancroft's vision of the good life was not a world of acquisitive enterprise but a world in which "humanity will recognize all members of its family as alike entitled to its care."[53]

One of the reasons that the debate between consensus and conflict theorists over the ideological character of the Jacksonians has never been adequately resolved is that both schools are correct in identifying what the Jacksonians were not. Hartz is surely right that the Jacksonians' commitment to private property prevented them from embracing socialism's doctrine of collective ownership.[54] But Schlesinger, too, is correct that many Jacksonians were far too hostile to inequalities, untrammeled exchange, "the allurements of credit,"[55] and speculative risk-taking to be just another manifestation of the philosophy of acquisitive enterprise.[56] As long as we insist on the dichotomy between socialism and individualism, the philosophy of the radical Jacksonians, like that of Paine, will remain anomalous. Only by abandoning the dichotomy can one begin to appreciate their preference for egalitarian social relations.

The Radical Reconstruction of Locke

That the Civil War poses difficulties for the consensus thesis has long been recognized. Hartz brilliantly tries to account for the anomaly of the South in terms that are consistent with a Lockean liberal consensus. Oddly, however, he does not appear to feel a need to explain the efforts at Radical Reconstruction that occurred during and immediately after the war. No mention is made of Thaddeus Stevens, George W. Julian, Benjamin F. Butler, or Benjamin Wade; of Wendell Phillips, Hartz observes only that he was "not a creator of political ideas."[57]

Although these architects of Radical Reconstruction were Lockeans, they nonetheless used Lockean ideals in the name of confiscation and redistribution of property. From Locke's precept that people have a natural right to the land they work, Radical Republicans concluded that southern slaves had thus earned a title to their masters' land. Confiscation and redistribution were, in Wendell Phillips's view, merely "naked justice to the former slave," who was entitled to "a share of his inheritance."[58] Confiscated lands, Samuel May insisted, "ought to be given in suitable portions to the colored people, who so long have tilled them without wages."[59] George Julian agreed that slaves had "earned their right to the soil by generations of oppression."[60] When Elizur Wright called for a policy of "the Soil to the Tiller," it was thus a revolutionary appeal for redistribution of land from planters to freed slaves and poor whites.[61]

None of these men, of course, advocated collective ownership of property. They all fervently believed with Thaddeus Stevens that "small, independent landowners are the support and guardians of republican liberty." Only if men were "the owners of the soil which they till" could they be moral and industrious.[62] For Julian it was axiomatic that "the nation will be powerful, prosperous, and happy in proportion to the number of independent cultivators of its soil."[63] Realizing this vision of a South populated by small family-owned farms, however, required them to significantly modify Lockean precepts concerning the inviolate right to private property.

The Radical vision of a society of small landowners should give pause to those who would portray the Radical plan to confiscate slaveholder property as simply a manifestation of the emerging system of industrial capitalism. Radicals did want to break the power of the planter aristocracy but they also worked hard to forestall the concentration of land in the hands of northern capitalists and speculators. When the government scheduled a public auction for plantation lands on the South Carolina Sea Islands in early 1863, for instance, abolitionists and Radicals, fearing that the land would become concentrated in the hands of "sharp-sighted speculators," successfully pushed to have the government reserve most of the land for the future benefit of the freedmen.[64] Their aim, in the words of George Julian, was "an equitable homestead policy," in which plantation lands would be parceled out in small farms rather than being sold in large tracts to the highest bidders. Such a policy was necessary to avoid "laying the foundation for a system of land monopoly in the South scarcely less to be deplored than slavery itself."[65] "Every man," Julian reminded his congressional colleagues, "has by nature . . . a right to a reasonable portion of [land], upon which to subsist."[66]

Speculators were criticized by Radical Republicans for putting profit above humanity. Thomas D. Eliot, a Massachusetts House Republican who sponsored and authored the Freedman's Bureau bill, excoriated northern speculators as "white blood hounds . . . whose pursuit is for gold." Policy, he argued, should be determined by "humanity" not by "the greatest revenue."[67] Northern speculators were depicted by Julian as vultures, "hovering over the public domain, picking and culling large tracts of the best lands, and thus cheating the government out of [its] productive wealth, and the poorman out of the home which else might have been his."[68]

Radicals hoped not only to destroy the Southern slaveholding aristocracy but also to remake the entire nation's institutions in a more equalitarian direction. Thaddeus Stevens expressed the Radical hope that the war against slavery would provide the occasion for "the intelligent, pure and just men of this Republic [to remodel] *all* our institutions" in such a way as to free them from "every vestige of human oppression, of inequality of rights, of the recognized degradation of the poor, and the superior caste of the rich. . . . No distinction would be tolerated in this purified Republic but what arose from merit and conduct."[69] The egalitarian character of the Radical vision is evident, too, in Benjamin Wade's lament that "property is not equally divided, and a more equal distribution of capital must be wrought out."[70]

Equally noteworthy was the Radicals' repudiation of the liberal notion of procedural justice. In Thaddeus Stevens's view, it was "impossible that any practical equality of rights can exist where a few thousand men monopolize the whole landed property."[71] "The nominal freedom of the slaves," agreed Edmund Quincy, "must be actually secured by the possession of land. If the monopoly of land be permitted to remain in the hands of the present rebel proprietors . . . the monopoly of labor might almost as well be given them, too."[72] "To give to these people only freedom, without the land," concurred Thomas Wentworth Higginson, "is to give them only the mockery of freedom which the English or Irish peasant has."[73] Without confiscation and redistribution, reiterated Wendell Phillips, "the negro's freedom [was] a mere sham."[74] "Real liberty," echoed George Julian, "must ever be an outlaw where one man only in three hundred or five hundred is an owner of the soil."[75]

Nor did these Radicals share the liberal faith that securing the suffrage was sufficient to ensure other rights. Most Republicans would have sided with Ohio congressman James Ashley that "if I were a black man with the chains just stricken from my limbs, without home to shelter me or mine, and you should offer me the ballot, or a cabin and forty acres of cotton land, I would take the ballot."[76] But a significant and powerful minority, headed by Stevens, vigorously dissented from this position. Homesteads, Stevens insisted, "are far more valuable [to the freedman] than the immediate right of suffrage, though both are their due."[77] "The nation must recognize," explained Higginson, "that even political power does not confer safety upon a race of landless men."[78] "They who own the real estate of a country," reasoned Elizur Wright, "control its vote."[79]

The Radical argument for confiscation and redistribution ultimately failed. Some who favored confiscation wished to see the land go to the highest bidder and had no interest in ensuring that farms stayed small and relatively equal. Many believed that to give the freedman land would only reinforce the lack of initiative learned in servitude. Self-reliance could be learned, they believed, only when the freedman was thrown into the labor market.[80] Still others had absolutely no interest in offering help to a race of people they believed to be inferior. But it is also important to recall that the issue of confiscation was hotly contested during and in the immediate aftermath of the Civil War.

If Stevens's social vision was bounded by the horizons of property, he (like Phillips, Julian, Wade, and Butler) nonetheless raised fundamental questions about how concentrated that wealth should be. In the hands of these Radicals, the natural right to property became a weapon to take property away from those who had much and to give to those who had little or none. If Radicals did not favor collective ownership of property, nor were they simply apologists for competitive individualism. Their social vision, like the one adhered to by Paine and the radical Jacksonians, presented an egalitarian challenge to the dominant competitive individualist ethos.

Populism: Locked In?

Much is often made of Populism's "propertied consciousness." Chained to the Lockean rock, the Populists were unable to forge a fundamental alternative to capitalism. Populists failed to fully transcend their "confinement within a propertied society" or transform a "social landscape . . . of unrelieved Lockean drabness."[81] But if the Populists often began from Lockean premises, their rendition of Locke undermined many of Locke's most cherished tenets, not the least of which was limited government.

In the hands of the Populists, the Lockean labor theory of value, rather than legitimating existing property, became a bluntly redistributive instrument. The narrow definition that Populists attached to "labor" lent an unmistakable radicalism to such orthodox Lockean sentiments as "labor creates all wealth" and "wealth belongs to him who creates it."[82] The problem, as Populists saw it, was that the common people "have produced but they possess not"; the "wealth producers" were not the "wealth owners."[83] Those who owned wealth had gained their wealth by "robbing" the laborer (in that quintessentially Lockean terminology) of "the fruits of his labor."[84] "The wealth of the Vanderbilts, Rothschilds and Goulds," for instance, was "but the accumulated labor of millions who have received but part of a just reward for their services." "The robber," therefore, "must be made to disgorge."[85]

The Lockean conception of property as the "fruits of labor" was thus transformed into a substantive conception of "just reward." "The true principle," explained W. Scott Morgan, "is that the laborer should be rewarded according to that that he does, and not according to what the employer can get the labor performed for." Although this might seem "debatable ground," Morgan explained that to grasp its validity we need only to "fall back upon that universal natural law, 'The laborer is entitled to all the fruits of his toil.'"[86] This view was echoed by Kansas senator William Peffer, who said that the hired workman is entitled to "a fair, just, equitable reward for his toil. If he does *all* the work, surely he is entitled to a large share of the profits."[87] If these Populists did not directly repudiate the wage labor system, as Marxists might wish, their narrow definition of labor coupled with their adherence to a substantive conception of justice nonetheless constituted a frontal assault on the central organizing principles of competitive individualism.

The Populists' invidious distinction between "wealth owners" and "wealth producers" radicalized many of their seemingly moderate claims. When Kansas governor Lorenzo Lewelling stated, "If it be true that the poor have no right to the property of the rich let it also be declared that the rich have no right to the property of the poor,"[88] he was not defending the status quo but rather slyly attacking property in the name of property. For if the wealth of the rich really belonged to the poor who had produced it, then redistribution did not involve taking property away from the rich but only restoring property to its rightful owner.

Populists also skillfully exploited the Lockean notion that the earth was originally common property. If the earth was, as Locke said, "theirs by a common inheritance," then it "must belong without possible alienation of title to all individuals of all generations." No one could be "disinherited" from a natural birthright, and all were entitled to "share equally the abundant natural provisions for a happy existence." Each person was entitled to "a portion of the earth and its forces" so as to "sustain life and gratify legitimate desires."[89] By denying people "the right to occupy the earth, . . . plutocratic capitalism [denied] to the people the heritage which the Creator gave them 'without money and without price.'"[90]

Working from Lockean premises, the Populists not only derived a right to subsistence but even formulated a defense of public ownership of industry. Not just land but "all the natural sources of wealth" were part of "the heritage of the people."[91] "The terrible elements of physical nature . . . steam, electricity, compressed air," reasoned Frank Doster, a prominent Kansas Populist and later Chief Justice of the Kansas Supreme Court, were "the common property of all [but] have been made the monopoly of the few." The solution, reasoned Doster—along with many other Populists—was to "have the government, that is, the people, assert their rightful dominion over" these natural resources.[92] When Lockean principles are invoked to justify public ownership, one has grounds to question whether Lockeanism is quite the ideological straitjacket it is sometimes alleged to be.

Populist plans for government ownership were, to be sure, largely limited to those spheres where private monopolies seemed endemic, such as railroads, telephones, and the telegraph. When it came to land, private ownership remained the cultural ideal. But here, too, the bourgeois ideal of property ownership was given a radical twist as it became transformed into a demand for a guarantee of housing. When the Populists said that "each citizen shall have . . . his own home secured,"[93] the statement implied not only a rejection of collective ownership of land but also outrage against people "being robbed of their homes" through foreclosure and unemployment.[94] The Populist Locke becomes an advocate for the homeless. No man, preached the Texas Populist James H. Davis, should be without "a home and some portion of the earth from which to produce comforts, and upon which to rest our weary limbs after a day's toil in production." The absence of a home ("around which lingers a halo of endearment to every human being") "tends to make man an alien to his God, an alien to his country, and to convert him into a vagabond, a wanderer and an outcast." So essential is a home to human happiness "that the decay of liberty, the downfall of society and the wreck of happiness in every age and every country have been measured by the homeless numbers within her borders."[95] In the hands of the Populists, the bourgeois idealization of hearth and home became a means to criticize competitive individualism.[96]

Nor did the Populists' defense of private property entail an acceptance of unlimited accumulation. Davis believed in "the sacred rights of home *in reasonable amount of acres.*"[97] Ignatius Donnelly, who penned the preamble to the Omaha Platform, believed that it was "right and wise and proper for men to accumulate sufficient wealth to maintain their age in peace, dignity and plenty" but that it was also desirable to "establish a maximum beyond which no man could own property."[98] Henry George considered it one of the great virtues of his Single Tax on rent that the "equalization in the distribution of wealth that would result . . . must lessen the intensity with which wealth is pursued." Having abolished "the fear of want, . . . whoever would toil to acquire more than he cared to use would be looked upon as we would now look on a man who would thatch his head with half a dozen hats."[99] Unlimited accumulation, far from being the cultural norm, would become the mark of a crazy person.

Measured by a Marxist yardstick, of course, George's vision of "diffused proprietorship"[100] is patently bourgeois. After all, George denied there was "a necessary conflict between capital and labor" and wished to "let the laborer have the full reward of his labor and the capitalist the full return of his capital." Moreover, George clearly begins his analysis from the conventional Lockean assumptions of "possessive individualism." "What constitutes the rightful basis of property?" George asks. "Is it not, primarily the right of a man to himself, to the use of his own powers, to the enjoyment of the fruits of his own exertions? . . . As a man belongs to himself, so his labor when put in concrete form belongs to him."[101] But from these orthodox Lockean assumptions, George drew conclusions that subverted basic tenets of competitive individualism.

If "the exertion of labor in production is the only title to exclusive possession," George reasons that "no one can be rightfully entitled to the ownership of anything which is not the produce of his labor, or the labor of someone else from whom the right has passed to him." Because land is not the product of labor but "a gratuitous offering of nature," it can justly be denied to none. Private property in land denies men's "equal right to the bounty of nature" and was thus "a bold, bare, enormous wrong." Moreover, to "admit the right of property in [nature's bounty] is to deny the right of property in the produce of labor . . . for the right to the produce of labor cannot be enjoyed without the right to the free use of the opportunities offered by nature." By allowing "nonproducers" to "claim as rent a portion of the wealth created by producers," private property in land denies "the right of the producers to the fruits of their labor." Beginning from the axiom of possessive individualism that "there can be to the ownership of anything no rightful title which is not derived from the title of the producer and does not rest upon the natural right of the man to himself," George ends by assailing private property in land as nothing short of "robbery."[102]

George's egalitarian commitments are explicit in his vision of a future society in "which individual interests [will] be subordinated to general interests" and "the fraternity that is born of equality [will take] the place of the jealousy and fear that now array men against each other." His hostility to competitive individualism is evident in his impatience and even contempt for "the higgling of the market," as well as in his searing indictment of "the forms of freedom" that disguise the "virtual slavery . . . of the masses in every civilized country." George sounds far more like Rousseau or even Marx than Locke when he dismisses political liberty in industrial society as "merely the liberty to compete for employment at starvation wages."[103] To portray George and the Populists as just another manifestation of bourgeois liberalism is to miss the ways in which, beginning from Lockean liberal premises, they derived a radically egalitarian critique of competitive individualism.

The End of History?

The spectacular collapse of socialist economies in the Soviet Union and Eastern Europe has prompted competitive individualism's champions to trumpet "the end of history."[104] With socialism having fallen to the wayside, entrepreneurial capitalism will now have the world stage to itself.[105] If socialistic collectivism and competitive capitalism are our only alternatives, then it necessarily follows that socialism's demise (if that is what we are observing) signals the triumphant reign of competitive individualism. The conclusion seems inescapable, however, only so long as one accepts the dichotomy.

Competitive individualism's defenders, ironically, commit much the same error that has long been made by Marxist critics of American capitalism, who, in their perpetual preoccupation with the question of "why no socialism in the United States," greatly exaggerate competitive individualism's "hegemony." These critics overlook, belittle, or explain away the unremitting hostility that American egalitarians like Paine, Skidmore, Julian, and George have shown toward unlimited appropriation, wide disparities in outcomes, and "the race and scramble for wealth."[106] Both individualism's critics and defenders make the untenable assumption that rejection of state ownership of the means of production or collective ownership of property necessarily entails support for competitive individualism. American egalitarians, at least in the nineteenth century, opposed both European-style state socialism *and* competitive individualism.

In many ways this general pattern still holds. Surveys show that few Americans express a preference for collective ownership of property or nationalizing industry,[107] and that socialism remains a suspect ideology.[108] But a shared preference for private ownership of property does not mean that all Americans are satisfied with the existing distribution of wealth or that they are enamored with private enterprise.[109] Surveys have found that 75 percent of Democratic party elites, 68 percent of labor leaders, 82 percent of black elites, 83 percent of nonfundamentalist Protestant religious leaders, and 68 percent of media elites believe that government should reduce the income gap between rich and poor.[110] By more than a two-to-one margin, feminist leaders believe that free enterprise is unfair to workers. Other groups also show a considerable skepticism of the market: only 31 percent of black elites and 44 percent of labor elites feel that free enterprise is fair to workers. When it comes to accounting for poverty, Democratic elites are fourteen times more likely to blame the free enterprise system than the poor, black elites are seventeen times more likely to fault the system, and feminist elites are eight times more likely to blame the system.[111]

Americans continue to believe with Locke that one has a natural right to the fruits of one's labor, but what counts as labor remains subject to vigorous debate. Egalitarians still refuse to accept that those who possess great fortunes could possibly have earned them with "real" labor. "I ain't seen anybody made worth a millionaire by working," reports a respondent in Hochschild's *What's Fair?*, directly echoing Henry George's rhetorical question, "How many men are there who fairly earn a million dollars?" Starting from a narrow definition of labor, it is only a short step to the conclusion that wealthy Americans "manipulate and steal. . . . You're dishonest someplace—that's how you get the big money."[112]

Nineteenth-century American egalitarians insisted that absolute equality was impossible, and that justice demanded that earnings be related to ability.[113] Contemporary American egalitarians take a similar position but, today as yesterday, a belief that the more able should earn more does not prevent egalitarians from rejecting unlimited accumulation, belittling the gains of capitalists as ill-gotten, or pressing for more equal outcomes. In *Equality in America: The View from the Top*, Sidney Verba and Gary Orren found that although American elites largely agreed that earnings should be based on ability, a significant number did not believe that the wealthiest Americans deserved anything like the incomes they were earning. Seniors at ten of the nation's elite colleges, for instance, would like to see a "president of one of the top hundred corporations" earn no more than $70,000 a year. Feminist leaders, consistently the most egalitarian of the groups surveyed by Verba and Orren, share this view of business executives as grossly overpaid and undeserving. In fact, close to one in four feminist elites would like to see the government place a top limit on all income.[114]

In sum, although Americans are overwhelmingly committed to the institution of private property, they are deeply divided about what constitutes a fair distribution of private property and whether the community should decide how the individual disposes of that property. Americans continue to believe with Locke that "Justice gives every Man a title to the product of his honest Industry,"[115] but they differ over what is to count as "honest Industry." A Lockean consensus on private property does not, in short, translate into a consensus on a competitive individualistic way of life. That egalitarians in the United States only infrequently reject private property in favor of collective ownership of property should not obscure the strength of egalitarianism in this country or the often thoroughgoing nature of their critique of competitive capitalism. The egalitarian conception of equality as equalizing results, I show in the next chapter, differs radically from the individualistic definition of equality as equalizing opportunities to get ahead.

3

Rival Visions of Equality: Process Versus Results

In a letter to a young friend, Tocqueville counseled the need for "taking pains to use words in their true sense, and so far as possible in their most limited and certain meaning, so that the reader is always sure what object or image you want to offer him."[1] Precision in the use of words is, as Tocqueville suggests, essential to good scholarship. In politics, as in life, however, the premium is on ambiguity. Avoiding spelling out precisely what one has in mind by a word or phrase enables one to reach agreements and make alliances that might not be possible were everyone to define precisely his or her terms.

Among the vaguest words in political discourse is "equality." Political philosophers have long lamented its elusive, protean quality.[2] Whether equality is, as one nineteenth-century critic complained, "a word so wide and vague as to be by itself almost unmeaning,"[3] there is no doubt that the term possesses many, often contradictory, meanings.[4] This chapter identifies two fundamentally different meanings that Americans have historically attached to the word "equality." The first, the competitive individualist definition of equality, conceives of equality in terms of *process*; the second, the egalitarian definition, conceives of equality in terms of *results*.[5]

A results-oriented conception of equality should not be equated with a preference for complete equality of condition. If absolute equality is to be the standard for equal results, then few if any egalitarians would qualify—not Marx, not Tawney, not even Rousseau. Both the results and the process vision of equality, Thomas Sowell explains, "recognize degrees of equality, so the disagreement between them is not over absolute mathematical equality versus some degree of equalization, but rather over just what it is that is to be equalized."[6] No egalitarian wishes for everyone to be equal in all respects; all make some allowances for differences in interests and talents. What differentiates the egalitarian vision from the individualist vision is that the former focuses on equalizing outcomes whereas the latter focuses on equalizing processes.

Studies of American political culture have often assumed that in this most bourgeois of nations the process-oriented definition of equality has gone largely unchallenged.[7] The lack of a vibrant equal-results tradition is commonly thought to set the United States apart from other Western nations, to form the basis of its "exceptionalism"—its lack of a strong socialist party, its weak labor unions, its position as a welfare state "laggard." Any number of factors have been adduced to explain American exceptionalism, including the frontier, natural abundance, the absence of feudalism, and the timing of universal male suffrage.[8] All of these explanations have merit, but they tend to overstate the consensus on competitive individualism and thus misstate the nature of American exceptionalism. What is exceptional about America is not that it lacked a results-oriented vision of equality but that those who favored equalizing results believed that equal process was a sufficient condition for realizing equal results. Not the absence of egalitarianism but rather the belief among egalitarians that equal results could best be achieved through limited government is what made America exceptional.[9]

Exceptionalism is thus in part responsible for the slighting of egalitarianism in the American past. Many who have viewed the nineteenth century through a twentieth-century frame of reference have equated a preference for laissez-faire with rejection of an equal-results vision. In the twentieth century, by and large, this assumption makes sense, but in the nineteenth century it fails miserably. Many Jeffersonians and Jacksonians favored a limited central government not because they rejected a results-oriented vision of equality but because they believed that the federal government was the primary source of artificial inequality. A faith in the natural equality of men (that they were roughly equal in the "riches of the mind," as Thomas Skidmore put it)[10] together with a belief in the naturally harmonious workings of the self-regulating marketplace allowed even the most radical of egalitarians, like Skidmore, to endorse both laissez-faire and equal results.[11]

My aim in this chapter is twofold: to document the existence of two rival definitions of the meaning of equality in the American past and to analyze the changing relationship between these rival visions of equality that has taken place over the last two centuries. I tell the story, albeit in a highly episodic and sketchy way, of how competing visions of equality, once deemed mutually supportive, have increasingly come to be viewed as incompatible. In America's first century, egalitarians from Paine to Henry George believed that the opportunity to trade and compete as equals would lead to roughly equal and noncoercive social conditions. Today, it is widely believed among egalitarians that the opportunity to compete creates unconscionable inequalities, and that the call for equal treatment is often little more than an excuse to perpetuate existing inequalities.

Merchants and Radicals in Revolutionary Philadelphia

Equality has long been recognized as a core value of the American creed. Participants in the American Revolution believed it to constitute the very "life and soul" of the republican experiment.[12] Yet the republican doctrine of equality, as Gordon Wood observes, "possessed an inherent ambivalence: on one hand it stressed equality of opportunity which implied social differences and distinctions; on the other hand it emphasized equality of condition which denied these same social differences and distinctions."[13] Equality may have been widely agreed upon by patriots, but whether this meant equal process or equal results was sometimes a matter of bitter dispute. Nowhere in Revolutionary America was the conflict between these opposing visions of equality more evident than in Pennsylvania.

Neither "Constitutionalists" (as those who supported the 1776 Pennsylvania state constitution were called) nor "Republicans" (the opponents of that constitution) supported the inequalities of the Old World. Both sides rejected the hierarchical conception of society in which "in due gradation ev'ry rank must be, Some high some low, but all in their degree."[14] Both parties opposed entail, primogeniture, and other legal means by which inequalities were passed down from generation to generation. The individual's position in society, both sides agreed, should not be determined by birth.

A shared opposition to the inequalities of hierarchy, however, did not prevent each side from arriving at its own distinctive conception of equality. For the Philadelphia merchants and entrepreneurs who led the effort to overturn the 1776 constitution, equality meant eliminating the marks of ascriptive status that prevented talented individuals from rising to the top. The social environment these men inhabited was intensely competitive, as Thomas M. Doerflinger documents in *A Vigorous Spirit of Enterprise*. Barriers to entry were low, risks were high, bankruptcies were frequent, and payoffs were tremendous.[15] For these entrepreneurs, equality meant opportunities not guarantees, process not results.

The radical framers of the 1776 constitution, in contrast, insisted that equality meant more than merely the opportunity to become unequal. For radicals like William Findley, the essence of republicanism was equality "of wealth and power." "Enormous wealth, possessed by individuals" always posed a "danger in free states." "Great and over-grown rich men," James Cannon agreed, "will be improper to be trusted." To vest "an enormous Proportion of Property . . . in a few Individuals," believed the radicals, was "dangerous to the Rights, and destructive of the Common happiness of Mankind."[16] From the radicals' point of view, the question of *how* that property was gained was largely irrelevant.

To the radicals, the individualistic merchants were as much an "aristocratic junto" as were the born-and-bred landed gentlemen.[17] Untrammeled competitive individualism seemed to lead as readily to hierarchical stratification as did hierarchy itself. Although radicals concerned with equalizing results could see no significant difference between the merchants' way of life and that of the independently wealthy gentlemen, there were in fact fundamental cultural differences. The merchants were "intense entrepreneurs who were tough, grasping, and willing to take large risks. . . . With few exceptions they did not have a truly genteel education. . . . They tended . . . to be narrow and materialistic, with politics of secondary interest." In contrast, Philadelphia's "true aristocrats . . . were born rich, and their money was invested in relatively safe investments, rather than in trade. As young men they could afford to go to the College of Philadelphia and take the grand tour of Europe. . . . Wealth was a condition of life, not a goal to be achieved through scrambling and gambling in the market place. Such gentlemen in Philadelphia, like their counterparts in Virginia, were likely to turn their attention to politics and civic affairs at an early age."[18] In short, it was the difference between competitive individualism and hierarchy.

In the eyes of individualistic merchants, egalitarian radicals like Cannon, Findley, Timothy Matlack, Thomas Young, and John Smilie were "levelers," perverting the noble republican ideal of equal opportunity. "Different degrees of industry and economy," Republican spokesmen insisted, would "ever create inequality of property, especially in a commercial country." Radicals did not necessarily deny that this was so, nor did they endorse an absolute equality of condition. Theirs was nevertheless a vision based on equalizing results. They believed, with Thomas Young, that it was important to keep "the whole people as much on a level as may be." This meant not simply equal political rights but "equal interests, equal manners, and equal designs."[19]

The policy consequences of these conflicting visions of equality became manifest in the acrimonious debate over how to respond to the sharp rise in prices that accompanied the Revolutionary War. For the radicals who shared a vision of equal results, price controls were a necessary device to ensure a "just price" (i.e., one that did not unduly disadvantage the common people). The merchants, in contrast, opposed radical efforts to regulate procedures in the name of substantive ends. Trade, they insisted, should "be as free as air." Prices should be determined by the natural laws of supply and demand, not by what the community deemed a just result.[20]

That both the Constitutionalists and the Republicans affirmed equality does not mean, as consensus theorists would have it, that their rhetoric was mere shadow boxing. For although few if any Philadelphians defended ascriptive hierarchical inequalities, beneath the shared commitment to equality lay competing visions of what equality meant. Nor can the conflict between Constitutionalists and Republicans be described in Progressive terms as the party of equality versus the party of privilege. For if the consensus interpretation fails to distinguish between individualism and egalitarianism, the Progressive interpretation fails to distinguish between individualism and hierarchy.

Democrats and Republicans in Antebellum America

Outside of Southern slaveholders and New England Whigs, few public figures in antebellum America openly defended inequality (at least among white males). Politicians on both sides of the political spectrum pronounced themselves friends of equality. Equality, Jacksonians reiterated, lay "at the heart" of both democracy and the Democratic party.[21] But Republicans (and, to a somewhat lesser extent, Whigs), refused to let the Democrats have a monopoly on equality. Presidential aspirant William Henry Seward, a former Whig turned Republican, insisted that the new Republican party stood for "one idea . . . the idea of equality."[22] But although both Democrats and Republicans affirmed their commitment to equality, their rhetoric often concealed underlying disagreement about what equality meant. For most Republicans (and Whigs), equality was defined in terms of process; for many (but by no means all) Jacksonians, equality was defined in terms of results.

When Seward claimed that the Republican party stood for "the idea of equality" he meant, as he subsequently explained, "the equality of all men before human tribunals and human laws." Similarly, when Lincoln stated that the United States should be "a society of equals," he was attacking the legal barriers to free competition imposed by slavery. The "leading object" of free government, he announced in an address to the joint session of Congress in July 1861, was "to elevate the condition of men—to lift artificial weights from all shoulders . . . to afford all, an unfettered start, and a fair chance, in the race of life."[23] For Republicans, then, equality was less about sameness than it was about the opportunity to distinguish oneself. The great virtue of Northern society was not that social conditions were roughly equal but that birth was no obstacle to distinction. "The northern laboring man," boasted Cassius Clay, "could, and frequently did, rise above the condition [in] which he was born to the *first rank* of society and wealth." What distinguished America from Europe, explained another Republican, was that "the door is thrown open to all, and even the poorest and humblest in the land, may, by industry and application, attain a position which will entitle him to the respect and confidence of his fellow-men." Equal treatment meant not only the opportunity to succeed but also to fail. The Republicans' objective, explained Iowa senator James Harlan, was to place the laborer "on a platform of equality—let him labor in the same sphere, with the same chances for success and promotion—let the contest be exactly equal between him and others—and if, in the conflict of mind with mind, he should sink beneath the billow, let him perish."[24]

Confident that equal opportunity did exist in the North, Republicans, like the Whigs before them, were relatively sanguine about the inequalities of wealth that existed in the North.[25] Jacksonians, in contrast, worried a great deal about the "vast disparity of condition" that they saw opening up. Banking, tariffs, manufacturing, even government itself—all seemed to conspire "to make the rich richer and the poor poorer." It would not be long, many feared, before society would be scarred by "inordinate wealth, on the one hand" and "squalid poverty on the other."[26]

Without roughly equal conditions, Jacksonians believed that it would be impossible to sustain equal procedures, for the powerful few would inevitably bend laws to advantage themselves at the expense of the powerless multitude. "To secure the enjoyment of equal laws," announced an official resolution endorsed by Vermont and Massachusetts Democrats, it was "essential that the people . . . should be on an equality in their social and political condition." It was agreed that "social equality [was] the legitimate foundation of our institutions, and the destruction of which would render our boasted freedom a mere phantom." Political democracy, echoed Massachusetts Democrat Robert Rantoul, depended on a society in which "inequalities both of property and of power "were" comparatively trifling." Only where equality of condition prevailed could democratic procedures be "real, substantial, and of course, permanent."[27]

Equality of condition was seen by a few Jacksonians as a higher stage of equality than "mere" legal or procedural equality. Orestes Brownson, for instance, remained dissatisfied with "the inequality in wealth, intelligence and social position" that he saw "even in this land of equal rights." "If we have realised political equality," Brownson explained, "we have not yet realised social equality."[28] More commonly, Jacksonians defined "equal rights" so broadly as to include a rough equality of conditions. Equal rights, as Thomas Skidmore defined it, was "the title which each of the inhabitants of this globe has to partake of and enjoy equally with his fellows, its fruits and its productions."[29] Few Jacksonians went as far as Skidmore in the extent of equal conditions advocated, but many mainstream Democrats shared a conception of equal rights that involved some amount of equalizing results. For Amos Kendall, a leading member of Jackson's celebrated "Kitchen Cabinet," equal rights meant rejecting any measures that "subject one man to the power or influence of another." At its core, as historian John Ashworth shows, the Democratic demand for equal rights "was no less than a demand for an equality of social power."[30]

Populists and Entrepreneurs in the Gilded Age

The last decades of the nineteenth century witnessed intense social and ideological conflict even though both "reformers" and "conservatives" claimed the mantle of equality. This common vocabulary has led some observers to downplay the differences between conservatives and reformers as "chiefly a matter of temperament—the former's complacency in contrast with the latter's sense of urgency."[31] What separates Americans, in this view, is not their ideals but rather the willingness to tolerate the discrepancy between real and ideal. Conservatives talk about equality as an alternative to practicing it; reformers talk about equality in order to practice it.[32]

This view underestimates the idealism of late-nineteenth-century conservatism because it misunderstands the nature of their ideal. Gilded Age conservatives seem complacent or hypocritical only if equal results is assumed to be their ideal. Conservatives' relative lack of concern over unequal results reflected less a habitual complacency than a fervent commitment to a different vision of equality—equality of process—than that which animated their radical critics.

The fundamental difference between these two visions of equality was evident to many conservatives of the day. In 1883, Yale political and social scientist William Graham Sumner criticized those who feel that "they have a right, not only to *pursue* happiness, but to *get* it." "The State," he explained, "gives equal rights and equal chances just because it does not mean to give anything else. It sets each man on his feet, and gives him leave to run, just because it does not mean to carry him."[33] Two decades later, William Allen White criticized the Populists for having "preached not merely equality before the law, but tacitly demanded that there should be an equality of the enjoyment of the things of this life."[34]

Among those who vigorously upheld the ideal of equality as process was the industrialist Andrew Carnegie, who, in *Triumphant Democracy*, celebrated "the equality of the citizen" that prevailed in the United States. There was, he insisted, "not one shred of privilege to be met with anywhere in all the laws."[35] If there was complacency among businessmen like Carnegie, it stemmed from their shared belief that equality of process had been fully realized in America. They believed, with

Senator Chauncey M. Depew, that "we live under just and equal laws and all avenues for a career are open." That Depew could simultaneously celebrate equality and point to the multimillionaire Cornelius Vanderbilt as "a conspicuous example of the product and possibilities of our free and elastic conditions" reflected not hypocrisy but rather the conviction that "the same . . . open avenues, the same opportunities which [Vanderbilt] had before him are equally before every other man."[36] Given greatly unequal talents and energies, vastly unequal results were a just outcome of equal processes.

The Populists issued a direct challenge to this process-oriented understanding of equality. At a minimum, they demanded a social order in which there would be "neither millionaires nor paupers."[37] The more radical among them dreamed with Edward Bellamy of a time of "equality in the distribution of the fruits of labor."[38] If Populists differed over how equal social conditions should be, there was little disagreement among them that it was results and not merely process that needed equalizing.

For individualists like Sumner the accent is on the equal right to the *pursuit* of happiness, but for the Populists this constituted a stingy conception of equality. "The equality of man," explained one, "means something more than equal privileges in money getting."[39] Individuals deserved the right not just to pursue happiness but to realize it in some measure. All men, one Populist spokesman insisted, were entitled to "share equally the abundant natural provisions for a happy existence." Others looked to a social order in which each member would be "equally nourished, equally exercised, and receive equal honor for equal exertion." "The products of labor must be for all; services must be rendered for all; the fruits of genius must be shared by all. Who does the best he can has done all any man can do, and should have his equal share of the social fund of happiness." "What one individual or class . . . secures by virtue of superior capacity and opportunities, above the general average of the comforts of life, belongs of right to other individuals or class of individuals failing by lack of capacity and opportunity to secure an average."[40]

Equal process without equal results was at best a meaningless abstraction and at worst a cruel fraud.[41] "Shame on your boasted institutions of liberty," Populists proclaimed. Unless America changed in a more egalitarian direction, they warned, the common people "will have only three rights . . . to work, to starve, and to die."[42] America would be "free, only in name."[43] "How," asked Kansas governor Lorenzo Lewelling rhetorically, "is liberty to be enjoyed, how is happiness to be pursued under such adverse conditions?"[44] "What is life and so-called liberty," echoed the *Farmer's Alliance*, "if the means of subsistence are monopolized?"[45] "Real" freedom, in short, required not just equal process but equal results.

The Origins of American Exceptionalism:
From Paine to Jackson

That these two conflicting visions of equality—equal results and equal process— have coexisted throughout American history does not mean that the relationship between them has remained unchanged. Far from it. Perhaps the most significant development in American political culture has been the growing tendency for results and process to be seen as incompatible. One of the most distinctive features of American political culture in its first century was the widespread belief among egalitarians that equal processes *would* produce more equal results. The market-place, far from being seen as a source of inequality, tended to be seen as a great equalizer. Great disparities in wealth, egalitarians of the early republic believed, could result only from government intervening to benefit one group at the expense of the many.

The dispute over price controls during the Revolutionary era brought to the surface the underlying conflict between the process-oriented vision and results-oriented visions of equality, but in the main conflicts between these two visions remained relatively muted in the early republic. Common ground was possible because, as Gordon Wood writes, at the time of the American Revolution "it was widely believed that equality of opportunity would necessarily result in a rough equality of station, that as long as the channels of ascent and descent were kept open it would be impossible for artificial aristocrats or overgrown rich men to maintain themselves for long."[46] An unregulated marketplace increasingly came to be seen as a force for greater equality. Equal process would produce relatively equal results.

This fusion of the results-oriented vision of equality with a faith in equal process is most clearly embodied in the person of Thomas Paine, particularly the Paine who wrote *Common Sense* and part 1 of *The Rights of Man*. That Paine's vision of equality was results-oriented is evident even in his earliest writings. Before coming to America from Britain, Paine scored "the rich, [who lived] in ease and affluence," and whose wealth became "the misfortune of others."[47] Throughout his life Paine deplored the "extent of riches" and "extreme of poverty," and he sought more equal outcomes.[48] At the same time, as Eric Foner notes, Paine "accept[ed] the self-regulating market—in labor as well as in goods—as an instrument of progress." Although Paine originally backed the idea of price controls, he soon soured on the idea as ineffective and coercive. In *The Rights of Man* Paine rejected the idea of government setting workers' wages. "Why not leave them as free to make their own bargains," Paine asked, "as the law-makers are to let their farms and houses?"[49]

To the modern reader, Paine's simultaneous endorsement of the free market and equal results, capitalistic enterprise and social justice, and commerce and community seems incongruous. For Paine, however, there was no conflict because inequality stemmed from political rather than economic causes. It was government and not the market, he believed, that was largely responsible for the oppression and misery of the poor. Economic inequalities could be remedied by doing away with the oppressive and corrupt monarchies of the New World and by establishing the procedural equality of republican governments in their place.[50]

"The natural propensity of society," Paine believed, was to create relatively equal and harmonious social relations. Taxation, instead of being a means of redressing inequalities, was itself a principal cause of inequality. Because "the real burden of taxes" invariably fell on the poorer and laboring classes, the result of taxation was that "a great mass of the community are thrown thereby into poverty and discontent." The wretchedness of the poor was due to government's "greedy hand" thrusting itself "into every corner and crevice of industry" and grabbing "the spoil of the multitude."[51] Paine allowed that differences in industry, frugality, talents, and luck meant that even under the most propitious circumstances "property will ever be unequal,"[52] but he insisted that conditions would be considerably less unequal if government let individuals regulate their own affairs.

For those who would understand "Why no socialism in America?" the political thought of Paine is critical. Louis Hartz's thesis of a liberal consensus correctly focuses attention on the weakness of Old World hierarchy in America, but it does not adequately specify how or why those like Paine who adhered to a results-oriented conception of equality also accepted the individualistic doctrine of self-regulation. The Paine of *Common Sense* could embrace both self-regulation and equal outcomes because he believed both that society in its natural state tended toward abundance and a rough equality of condition and that America roughly resembled society in its natural state. It was only when Paine turned his attention to conditions in Europe that he lost faith in the self-regulating market and procedural equality as a sufficient condition for greater equality of condition and came to the conclusion that government intervention would be necessary to achieve an approximate equality of results. Paine continued to believe, though, that for America at least the surest road to equality of result lay not in government redistribution but in avoiding the European path of development—of big government, big business, big banks, and big cities.[53]

Paine's belief that self-regulation would produce relatively equal results (a belief he modified considerably in part 2 of *Rights of Man* and in *Agrarian Justice*)[54] became an article of faith for most Jacksonians. Under the leadership of Andrew Jackson, the Democratic party fused a strong egalitarian impulse with an equally fervent commitment to the unregulated marketplace.[55] This fusion of equal results with equal process has left historians deeply divided over how best to characterize the ideology of the Jacksonian party. For those historians who focus on the party's laissez-faire beliefs ("The world is governed too much" was the lodestar of the Jackson administration mouthpiece, the Washington *Globe*)[56], Jacksonian ideology seems a classic expression of entrepreneurial capitalism.[57] For those who focus on its commitment to equal results, the Jacksonian party—or at least its radical wing—seems to be a forerunner of welfare state liberalism if not social democracy.[58]

To focus on one side of Jacksonian ideology at the expense of the other is to miss what is most distinctive about Jacksonian ideology, namely, the effort to combine equal results with equal process. Interpreting Jacksonian ideology as simply a manifestation of bourgeois capitalism slights the results-oriented conception of equality that animated the Jacksonian language of "equal rights." But to say that Jacksonians endorsed a results-oriented conception of equality is not to accept the view of Jacksonianism as a precursor of New Deal liberalism. For Jacksonians rejected the central premise of the welfare state, namely, that government intervention could redress inequality. Government, in the Jacksonian view, was the problem, not the solution. It was government intervening to benefit the well-organized few at the expense of the many that was largely responsible for inequalities in wealth.

Equality between individuals, Jacksonians believed, was rooted in nature.

"Every law of nature," Ohio senator William Allen asserted, was "a law of equality." Ely Moore, a prominent New York Democratic labor leader and politician, agreed that "nature has set no difference between her children."[59] Nature had made people fundamentally "alike in their hopes and fears, their joys and sorrows, their passions, appetites, and aspirations."[60] What differences between people in talent or industry that did occur naturally were insufficient to explain the vast disparities in conditions that they witnessed all around them. "There is nothing in the actual difference of the powers of individuals," Orestes Brownson insisted, "which accounts for the striking inequalities we everywhere discover in their condition."[61]

Inequalities in condition were attributed not to the natural workings of the marketplace but to the distorting influence of government intervention. Left to themselves, the natural laws of supply and demand "would tend to equalize the distribution of wealth." "Among a free and enterprising people," explained the *Democratic Review*, "the rates of profit, realised by individuals engaged in the various employments of life, have a constant tendency to the same level." There was, echoed another, a "natural tendency of capital to an equal distribution among the people."[62] If society would only trust to "*nature's own levelling process*," America would realize "that happy mediocrity of fortune which is so favorable to the practice of Christian and republican virtues."[63]

Government intervention upset this natural equality and introduced artificial inequalities. By favoring some groups at the expense of others, the invariable result of governmental activity was "in the end, to introduce ranks and classes and family distinctions."[64] "Four fifths of the action of all legislation," reported the *Democratic Review*, "promote[d] the accumulation of prosperity in a few hands."[65] Elsewhere it estimated that "legislation has been the fruitful parent of nine-tenths of all the evil, moral and physical, by which mankind has been afflicted since the creation of the world, and by which human nature has been self-degraded, fettered, and oppressed."[66] Inequality, echoed the *Globe*, is "created by legislative enactments, and sustained by the partiality of the law."[67]

Equalizing outcomes thus required not expanding government's role but contracting it. It was an unquestioned axiom of Democratic political thought that "if all were left without any special aid from government, both land and the products of industry would be far more equally distributed than they are." Allow no citizen to be the beneficiary of special legal privileges and "the strange contrasts created by overgrown affluence and wretched poverty would give place to apportionments of property more equitably adjusted to the degrees of personal capacity and merit."[68] Because men were born equal, formal or legal equality was not only necessary but sufficient to ensure roughly equal outcomes.

Radical Jacksonians rejected socialism not because they rejected a results-oriented vision of equality but because they believed equal results could better be achieved by relying on competitive markets than by enlisting government authority. Socialism could make little headway so long as American egalitarians continued to believe in "nature's own leveling process." Only as egalitarians began to lose faith in the ability of equal process to bring about equal results did the exceptionalist alliance of individualism and egalitarianism, equal process and equal results, begin to unravel.

The Populist Challenge to American Exceptionalism

The egalitarian belief that the free market was a source of equality and that government was a cause of inequality did not collapse overnight. Its demise was instead piecemeal. The belief was challenged in the aftermath of the Civil War by those, like Wendell Phillips and Thaddeus Stevens, who came to feel that procedural inequality was insufficient to help the freed slave. Given only the ballot and a chance to sell his services on the open market, the black man would become a slave once more to the economically powerful. Only through governmental activity could more equitable outcomes be realized. In the contemporary period, it has been the attempt to help blacks and other minority groups that has perhaps contributed most to the collapse of the belief that securing equal process is sufficient to achieve more equal results.

The most sustained of the nineteenth-century attacks upon the exceptionalist belief, however, had little connection with the issue of race. It stemmed instead from the Populist crusade against the railroads. The Populists were much like the Jacksonians in their zeal for redressing inequalities, their denunciations of corporations and the "money power," as well as their defense of the "producing classes" against parasitic speculators and money-changers. But they differed in one important respect. Where the Jacksonians viewed governmental legislation as a major source of inequality, many Populists came to see government as a means of increasing equality.

Few if any Jacksonians had seen any conflict between increasing individual freedom and redressing inequalities. Many Populists, in contrast, became persuaded that individual freedom to do as one pleased created gross inequalities. "The plutocracy of to-day," the editors of the *Farmer's Alliance* explained, "is the logical result of the individual freedom which we have always considered the pride of our system." It was because "individual enterprise was allowed unlimited scope" that the country was plagued by gross inequalities, selfishness, and corporate abuse. The corporation, after all, "in its nature and development . . . is only the original and cherished principle of individual liberty."[69] Untrammeled competition, many Populists came to believe, led inevitably to concentration and monopoly.

W. Scott Morgan, a prominent southern Populist, rejected the notion "that competition will correct all inequalities arising in the various conditions of labor." Far from being self-correcting, as Adam Smith (and Paine) had it, the unregulated marketplace invariably tended to concentrate wealth in the hands of the few: "Competition in wages, when based upon necessity, is decidedly injurious and signifies an unhealthy condition of the industrial interests of the country. Competition in commerce, trade and transportation fails 'at the moment something is expected of it,' because it leads to combination." An equitable distribution of wealth thus could not trust to allegedly natural laws of supply and demand but instead required active governmental intervention.[70]

The view that monopoly was the logical result of competitive capitalism was given widespread currency in Henry Demarest Lloyd's *Wealth Against Commonwealth*, which was published in 1894. "What we call Monopoly," Lloyd observed, "is Business at the end of its journey." Indeed in Lloyd's view the central paradox of modern society was that "liberty produces wealth, and wealth destroys liberty."[71] More equitable conditions thus required placing governmental restrictions on economic liberty. A more respectable and systematic expression of this view was outlined by the sociologist Lester Ward. "All the evils of society," Ward explained, "are the result of the free flow of natural propensities." "Unbridled competition," he argued, invariably "destroys itself" by concentrating power in the hands of the few. It was thus incumbent upon government to regulate society to prevent the occurrence of injustice (i.e., greatly unequal results).[72]

Populists tended not to share the Jacksonian trust in "nature's levelling process." Frank Doster, who in 1896 would become chief justice of the Kansas Supreme Court, expressed this skeptical view of nature in a Fourth of July (1893) oration in Marion, Kansas. The state of nature, Doster believed, was characterized by ruthless selfishness and competitive strife, governed as it was by the doctrine of the survival of the strongest. It was these inequalities of nature that made government necessary. "All government and all necessity for government," Doster maintained, "grows out of the fact of inequalities and that government which does not provide for the leveling and equalizing of the conditions which grow out of the unrestricted exercise of the natural powers of its citizens has failed in the purpose of its creation." What "we call equality," Doster reiterated, "must be realized through the process of human government, and it is the business of government to discover and enforce those laws of harmony which raise men above the barbarous antagonisms of the natural state into relationships of unity and fraternity."[73]

Absent, too, was Paine's faith in commerce as a source of spontaneous sociality and cooperation. The "fierce competitive strife" of bidding and bargaining, in the view of many Populists, led only to the destruction of "brotherly love." Human fellowship, for Paine, flowed out of the competitive marketplace; for Populists, in contrast, fellowship and cooperation were more likely to be seen as directly antithetical to competition and the accompanying "selfish, struggling individualism."[74]

Having begun to lose faith in the self-regulating market as an instrument of equal results, egalitarians in the late nineteenth century began to view government in a new, more positive light. Rejecting Paine's premise that "government even in its best state is but a necessary evil"[75] as well as the Democratic axiom that "the best government is that which governs least,"[76] the Populists declared (in the 1892 Omaha Platform) that "the powers of government—in other words, of the people—should be expanded . . . as rapidly and as far as the good sense of an intelligent people and the teachings of experience shall justify, to the end that oppression, injustice, and poverty shall eventually cease in the land."[77] The people "have come to believe," explained the Topeka *Advocate*, "that many of the abuses to which they are subject might be remedied, and their condition be bettered by a proper exercise of the power of the government."[78] Government could "by enactments procure and secure to the workers [their] equal, rightful share of the labor-saving, wealth-making power of steam, electricity and machinery."[79] Government could also be used to "take away the power of men to keep what they do not need and do not use themselves from people who do need and wish to use what heaven designed should be used."[80] Once government came to be seen as an egalitarian instrument ("a machine to insure justice and help the people," in the words of Ignatius Donnelly),[81] Populists tended not to concern themselves with placing limits on its activity.[82] "The public," explained Henry Demarest Lloyd, "have the right to use public powers for the public welfare to any extent the public demands."[83]

Some Populists, particularly in the South, still clung to the Jacksonian axiom that government intervention inevitably exacerbated inequalities. This was the view of James H. Davis, a Texas Populist, who warned his audiences that "just in proportion as the organic structure or practices of a government, permit or extend paternal aid in helping the individual man, that aid will in some way be appropriated for the benefit of the few and not the many."[84] Similarly, W. A. McKeighan, a Nebraska Populist, spoke in recognizably Jacksonian tones when he called for government to "undo its mischievous legislation protecting capital employed in manufacturing, and leave the distribution of wealth to follow natural laws free from government meddling and interference."[85]

The persistence of American exceptionalism is also very much in evidence in the political thought of Henry George. In *Progress and Poverty* George set out "to show that laissez faire (in its full true meaning) opens the way to a realization of the noble dreams of socialism." Liberty would produce equality; indeed, for George, "freedom [is] the synonym of equality." George chided socialists for attempting "to secure by restriction what can better be secured by freedom." The cause of inequality, in George's view, was not competition but monopoly, specifically monopoly in land. Destroy land monopoly, he promised, and "competition [will] accomplish the end which cooperation aims at—to give to each what he fairly earns." Destroy monopoly and "industry must become the cooperation of equals." By "throwing open natural opportunities now monopolized," "a single tax levied on the value of land" would dispense with "an immense and complicated network of governmental machinery" and "produce greater and greater equality."[86]

In the Single Tax, George found a wonderful device that enabled him to reconcile egalitarianism with individualism, cooperation with competition, an equal distribution of wealth with limited government. And yet George's diagnosis of existing conditions—that there was a "constant tendency to greater and greater inequality as material progress goes on"—led him away from exceptionalist thinking in important ways as well. George's view that inequality was caused by the monopolization of land led him to reject the conventional Jacksonian view that inequality was largely caused by "the immense burdens which . . . governments impose." "No reduction in the expenses of government," George insisted, "can of itself cure or mitigate the evils that arise from a constant tendency to the unequal distribution of wealth." It is not that "governmental economy is not desirable," he reiterated, but that reducing government expenses "can have no direct effect in extirpating poverty and increasing wages, so long as land is monopolized."[87] Given George's diagnosis of the root cause of inequality, only his utopian faith in the efficacy of the Single Tax enabled him to preserve, in his own mind at least, the exceptionalist alliance of limited government and greater equality of results.[88]

No Populist doubted that government as currently constituted benefited the few at the expense of the many. It was the wealthy few, after all, who controlled the state.[89] Increasingly, though, Populists began to believe that the answer was not to minimize the role of government, but rather to have the people take control of government and put it to work for them. Government, after all, was not "a foreign entity," but simply "the agent of the people" designed "for the purpose of executing the popular will."[90] "The State," wrote Lloyd, was "the organ of self-consciousness of peoples. . . . Dominating all as the noblest, strongest, best passion and force is that which acts through the state, through government, to curb the strong, and protect the weak to preserve the common weal."[91] For many Populists, government had become, in the words of historian Norman Pollack, "the locus of regenerative political and social relations."[92]

Although Populism collapsed as an effective political force after McKinley's smashing victory in the 1896 election, Populist ideas about government proved more hardy. The railroads were not taken over by government, but they were regulated in the public interest. The graduated income tax that they had pushed became the law of the land in 1913. The Populists' belief that government was a potential instrument of equalizing results reached fruition in the New Deal. By the close of Franklin Roosevelt's administration, few if any who adhered to an equal-results vision accepted the Jacksonian notion that self-regulation would generate relatively equal results.

The New Left: A New American Exceptionalism?

What was distinctive about American egalitarianism in the eighteenth and nineteenth centuries was its hostility to government as a source of inequality. Populism, Progressivism, and especially the New Deal contributed to bringing the United States more in line with the European model of using government to redress inequalities. Although the United States continued to lack a sizable Socialist party, the Democratic party after the New Deal became in many ways a surrogate Social Democratic party.[93] The post–New Deal Democratic party was committed to using government to remedy the inequalities generated by capitalism, and significant proportions within the party even favored some measure of public planning.

The rise of the New Left in the 1960s presented a challenge to the Social Democratic model of using government to redress the inequalities of competitive individualism. The New Left shared the Old Left's suspicion of capitalism, but it also shared the Jacksonian suspicion of government as a source of inequality. This simultaneous hostility to both capitalism and government, the competitive marketplace as well as authority, was the defining characteristic of the New Left. It was not their egalitarianism ("it is a crime," declared Students for a Democratic Society president Carl Oglesby, "that so few of us should have so much at the expense of so many")[94] that was new; what was novel was their refusal to ally with either hierarchy or individualism to achieve egalitarian ends. The Old Left fought against the inequalities of capitalism by enlisting bureaucracy. Jacksonians resisted bureaucracy and placed their hope for an egalitarian order in a liberated capitalism. The strategy of the New Left was different; their aim, in the words of Tom Hayden, was "to fight against capitalism *and* bureaucracy."[95]

The egalitarian ideology of the New Left is summed up in Kirkpatrick Sale's notion of politics on a "human scale." Like the Jacksonians, Sale believes that government intervention is inevitably a source of economic inequality. "U.S. government regulations and policies have worked," he explains, "both deliberately and accidentally, to create and sustain the large corporations that are the under-pinnings of our economic system." If Sale sounds typically Jacksonian in his denunciations of the "discriminatory, privilege-creating actions of Big Government," his jeremiads about the inegalitarian effects of the competitive marketplace are quite un-Jacksonian. "Capitalism," according to Sale, "is a system that depends, quite simply, on growth," and it is this dependence on "continual growth" that is the motor force generating "the burden of bigness."[96] For Sale, capitalism and government are co-conspirators in the modern onslaught against small-scale egalitarian community.

The New Left's antipathy to both competitive individualism and hierarchical authority was explicitly spelled out by former SDS president Todd Gitlin in his 1966 essay "Power and the Myth of Progress." In it Gitlin attacks "the unchallenged domination of the values of the marketplace" and denounces free enterprise as nothing more than a fraud. But he does not stop with an attack on the inequities of capitalism. For government, too, is at fault. "It would be a mistake," Gitlin explains, "to cast government in the role of hapless bystander, for it too is a mechanism of irresponsible power, almost as free of democratic control as are the corporations, and as comfortably respectful of business as business is of it." Although "New Deal institutions were intended to circumscribe [corporate] power," the fact was that they did "so marginally if at all." In urban renewal, for instance, "public power does not countervail against 'private' power—instead the two combine to exclude from their plans the people most abruptly affected by their decisions." The welfare state, "intended to remedy the excesses of unbridled capitalism," in fact feeds the "powerlessness of the poor." Nor has the welfare state made significant progress in removing inequalities in income: equality, Gitlin says, "is nearly a literal millennium away." The war on poverty is simply more of the same: "irrelevant at best and inimical at worst to the standard of democracy." "Top-down social agencies," Gitlin insists, are "no help." Nor are "traditional liberal instruments of change," such as the ballot box, social mobility, or education.[97]

A dilemma faced the New Left. How could the world be made a more equal place if government and the marketplace were both forces for inequality? If neither governmental authority nor capitalism could be harnessed to egalitarian ends, then how was equality to be realized? Must equal results remain "a literal millennium away"? The New Left's answer was that equality could be achieved only by building egalitarian "counter-institutions" that remained uncorrupted by the inequities of the larger society. Transforming society in an egalitarian direction, explained Hayden, entailed "building institutions outside the established order which seek to become the genuine institutions of the total society."[98] These counter-institutions included community unions, freedom schools, experimental universities, community-formed police review boards, and antipoverty organizations controlled by the poor.

Within these new institutions, power and resources would be shared equally among its members. A shared sense of sacrifice, purpose, and belonging would unite members of the collectivity. Every person would be able to participate equally in the decisions that affected them. None would be treated as better than another. Cooperation would replace competition, and equality would replace authority.

But the New Left's idea of building "moral communities within an amoral society"[99] begged many of the questions it was supposed to answer. To establish small communities of like-minded equals was one thing; to make the outside world a more egalitarian place was quite another. To hope that "these islands" of egalitarian morality would "spread by force of example or by guerrilla warfare until they control the whole country"[100] was utopian. The tension between means and ends was left unresolved. New Left organizations like the Student Non-Violent Coordinating Committee (SNCC) and SDS that operated on strictly egalitarian premises—each individual must consent to every decision, no individual should exercise authority over another—found themselves often unable to reach decisions, particularly as the organizations grew into something resembling a mass movement. One former SDS member recalls a twenty-four-hour discussion over whether group members could take a day off to go to the beach![101] The result was that these organizations often moved painfully slowly, failing to seize opportunities or promptly rebut attacks. Staughton Lynd justified this on the grounds that "the spirit of a community, as opposed to an organization, is not, We are together to accomplish this or that end, but, We are together to face together whatever life brings."[102] But the "building of a brotherly way of life even in the jaws of Leviathan"[103] was, as Richard Flacks recognized, necessarily in tension with the goal of "a redistribution of wealth and power in the society."[104] By refusing to ally with (or be coopted by) established institutions, New Left organizations like SDS and SNCC consigned themselves to the borders of society. This allowed them their purity but often at the cost of effecting the "social change" they so fervently desired.

For allies the New Left looked not to capitalism or to government but to the oppressed. "Our place," they insisted, "is at the bottom."[105] The poor of the rural South and northern ghettos held out the most hope for a radical reconstruction of society because they were least corrupted by involvement with the dominant values and institutions of society.[106] It was "in the culture of poverty," the New Left insisted, that was to be found "a culture of resistance."[107] "Students and poor people," explained Tom Hayden, "make each other feel real." Students could give the poor a sense of democratic possibilities, and the poor could "demonstrate to the students that their upbringing has been based on a framework of lies."[108] The downtrodden's firsthand experience with the brutality of the system made them an untapped "source and reservoir of opposition to the . . . American system."[109]

Unfortunately for the New Left, "the radical potential of the poor"[110] remained largely that: potential. Projecting their own values onto those of the poor (black sharecroppers, for instance, were said to possess "a closeness with the earth, . . . a closeness with each other in the sense of community developed out of dependence, [and] the strength of being poor"),[111] they underestimated the passivity and distrust that marked the "culture of poverty." Try as they might, they could not avoid the dilemma of reconciling their commitment to "redistribute the common wealth"[112]

with their distrust of bureaucracy and governmental power. Maximizing the participation of the poor, as the War on Poverty showed (and the Old Left never tired of
pointing out), was not necessarily the best means of reducing societal inequalities.

The "new exceptionalism" of the New Left should not be exaggerated. The
New Left had more in common with the Populists, Progressives, and New Dealers
than they did with the Jacksonians or Revolutionaries. Asked to "name the system" responsible for injustice in America, most in the New Left answered "capitalism."[113] Despite their suspicion and denigration of government, their desire to
remedy unequal outcomes generated by the market led to policies that required an
expansion of the scope of governmental power. Still, the New Left's distrust of
bureaucracy, authority, and hierarchy remained an important obstacle to a European-style social democratic vision that would combine support for authority with
a commitment to equalizing results. Thus despite its hostility to capitalism, the
New Left's brand of egalitarianism ironically contributed in some ways to furthering American exceptionalism.

At the same time, however, the New Left and the policies it helped generate
(particularly in the areas of affirmative action and environmentalism) have accel-
erated the deterioration of the nineteenth-century exceptionalist alliance of individualism and egalitarianism, equal process and equal results. For the New Left is
even more suspicious than the Old Left of the unregulated market and "mere"
equality of process. "Legal equality and equality of opportunity" came to be seen
by many in the New Left as just a "part of the liberal ideology" that "gets the
Negro in the South no more than a Harlem."[114] Issues of race and gender, more
than economics, have perhaps contributed most to contemporary egalitarian disillusionment with the process-oriented vision of equality.

The Demise of American Exceptionalism:
Redefining Discrimination

Nothing so clearly indicates the total collapse of American exceptionalism as the
changing definition of discrimination that has taken place over the past quarter-
century. The test of discrimination has moved from intent to effect, from process
to results. In the past, most Americans understood discrimination to mean a violation of equal process, as when an individual must sit at the back of a bus because
of his race or is denied a job because of her sex. But since the 1960s, a new egalitarian results-oriented definition of discrimination has arisen. Practices devoid of
discrimination in terms of process are now deemed "discriminatory in effect" if
they increase unequal results. One can now discriminate without intending to, or
be a racist while treating every racial group the same.

In early American history, egalitarians believed that equal process was sufficient to produce equal results. In contemporary America, the presumption among
egalitarians has become entirely different. If the outcomes are unequal, the process
must be unequal. If minorities and the poor are more likely to end up in prison,
then incarceration is discriminatory. If blacks are more likely to be sentenced to
death, then the death penalty must be racist. If women drop out of graduate school

at a higher rate than men, then the process must be sexist. If whites do better than minorities on standardized tests, then the standards are discriminatory. If minorities are underrepresented on university faculties, then the process must be racist. And so on, ad infinitum.

This redefinition of discrimination makes an alliance of individualists and egalitarians extremely difficult. For individualists can hardly be expected to take kindly to having their conception of equality labeled as discriminatory, racist, and sexist. Individualists insist that terms such as discrimination make sense only when applied to the realm of process and intent. Such phrases as "discriminatory in effect" are, from the individualist point of view, transparent attempts to cloak the equal-results vision in the highly charged language of discrimination.

Perhaps no issue better dramatizes the eclipse of American exceptionalism than the controversy over comparable worth. Egalitarian proponents of comparable worth cite the "wage gap" between women and men as prima facie evidence of "wage discrimination." Individualists regard this as patently absurd. How can wages set by gender-blind laws of supply and demand be discriminatory? They may generate unequal results, individualists maintain, but that is as it should be. If men earn more than women, it is not because of discrimination but because, for instance, women are more likely to opt for jobs permitting easy exit from and reentry into the labor force. The problem is not in the process but in the people. If a woman wants to earn more, the answer is not to have government mandate higher wages for traditional female jobs but rather for women to enter those male-dominated labor markets where wages are higher. If a woman tries to enter a traditionally male labor market and is not hired or not promoted *because she is a woman*, only then do you have discrimination. The process is presumed equal until proven otherwise.

Egalitarians have tried to reverse the burden of proof: the process is presumed to be unequal until the results are equal. Women who occupy "traditionally segregated jobs" are said to have a prima facie case for wage discrimination, with the burden of proof falling on the employer. By importing the rhetoric of segregation and discrimination used to win civil rights for blacks in the 1950s and 1960s, proponents of comparable worth have attempted to link their cause with the fight against segregated schools and disfranchisement. But the rhetoric obscures a fundamental distinction: segregation and disfranchisement in the South were inequities in the process; the "wage gap" is an unequal result.

In response to individualists' concerns with inequities in the process, those who argue for salaries based on comparable worth have sometimes claimed that the process itself is iniquitous. The argument made by some feminists that employers have actively colluded to hold down women's wages has won few converts. More common is the contention that the "herding" or "crowding" of women into a "pink-collar ghetto" produces an oversupply of labor and a consequent decline in wages. But in the absence of evidence of somebody doing the herding, this theory (stripped of its loaded metaphors) concedes that wages are determined by market forces of supply and demand.[115] What is discriminatory or iniquitous about that, asks the individualist. What is discriminatory and iniquitous, answers the egalitarian, is that it produces unequal results. Between these two visions there is an impassable gap.

American exceptionalism cannot endure when the laws of supply and demand are seen as inherently discriminatory. In their call for government to correct the inequities of the marketplace, those who support comparable worth can be seen as simply furthering the gradual erosion of American exceptionalism in the twentieth century. But, in many ways, comparable worth is qualitatively different from past policies. Much of the New Deal was conceived as saving the free market from itself, but comparable worth is designed not to check but to override the market. As Michael Levin points out, comparable worth does not even "pretend to facilitate the best tendencies of the free market; rather, it is explicit about seeking to flout the market."[116] The minimum wage, for instance, is designed to provide a "decent" wage for workers, but without any claim that this is what the worker's labor is "really" worth. Comparable worth, in contrast, rests on the notion that the value of one's work can be determined outside of the market. Individualists might be persuaded to put up with the former notion, but they can never countenance the latter, for it is an invitation to endless government intervention.

The same redefinition of discrimination has occurred, as Abigail Thernstrom shows, in the area of voting rights. The Voting Rights Act of 1965 had one purpose only: to ensure that every American citizen has an equal opportunity to vote. Its aim was to do away with those techniques—literacy and "understanding" tests (how many bubbles in a bar of soap?), poll taxes, intimidation, and violence—by which southern whites effectively disfranchised blacks. The premise was that eliminating discriminatory processes would create more nearly equal results. Give blacks the opportunity to vote and the rest would take care of itself. "From participation in elections," Attorney General Robert Kennedy explained in 1961, would flow "all of what [blacks] wanted to accomplish in education, housing, jobs, and public accommodation."[117] Here was the American exceptionalist belief reasserting itself.

The judicial reinterpretation of the Voting Rights Act that occurred in subsequent decades is testimony to the unraveling of the American exceptionalist belief that equal process would promote equal results. In its place has arisen the belief that unequal outcomes are prima facie evidence of an unfair process. Whereas the Voting Rights Act defined discrimination in terms of process (removing obstacles to voting), subsequent interpretations have defined discrimination in terms of results (ensuring that the impact of minority votes is not "diluted," which in practice means maximizing the number of elected minorities). As understood today, the Voting Rights Act is held to mean that if the results are unequal (for instance, if an annexation of surrounding areas decreases the proportion of blacks within the city), then the process can be assumed to be racially discriminatory, no matter what the intent. Originally a voting rights violation was conceived of in terms of giving blacks a test that whites were not required to take. Today, as Thernstrom points out, "an alleged voting rights violation . . . is a districting plan that contains nine majority-black districts when a tenth could be drawn." In short, voting rights has been "reshaped into an instrument for affirmative action in the electoral sphere." It has "evolved from a vehicle for enfranchisement to a means by which political power is redistributed among blacks, whites, and (since 1975) Hispanics." As its focus has shifted from securing equal process to securing equal results, the Voting

Rights Act has become a justification for an "unprecedented federal involvement in local electoral matters."[118]

Affirmative action programs, in all their guises, have provided the final nails in the coffin of American exceptionalism.[119] Finding common ground between individualists and egalitarians on issues of affirmative action is next to impossible, although sometimes the differences between these two visions have been muted by adopting the individualist rhetoric of ensuring equality of opportunity. But equality of opportunity, as philosopher Charles Frankel suggests, "is a highly stretchable [and] ambiguous concept, which cloaks strongly divergent ideas over which people do in fact disagree."[120] Agreement on the proposition that individuals ought to have an equal opportunity to develop their talents masks disagreement about what constitutes an equal opportunity as well as what counts as "natural talents."

Consider the frequently used metaphor of a "fair race." How would one determine whether a race was fair? The notion of equality of opportunity suggests that a fair race is one in which all enjoy an equal start. If individualists and egalitarians can accept this formulation,[121] it is only because this answer begs more questions than it answers. It does not tell us what constitutes an equal start or even when the race starts. Egalitarians take the position that "the race should begin anew every day and that society should do everything in its power to put all competitors on an equal footing at the start of the day."[122] For individualists, this is an open invitation to constant government intervention and tantamount to equal results.

Many egalitarians insist that the only real equality of opportunity is a situation in which one cannot predict an individual's future income, occupation, or status on the basis of that individual's race, sex, religion, ethnicity, or family background. In a truly egalitarian society, ability alone would determine one's position. But what is meant by "ability"? What is "natural ability" and what is a product of one's upbringing? Was I denied an equal opportunity to compete for a position on a professional baseball team because I grew up in a family that placed a higher priority on schooling than sports?[123] To wipe the slate clean, to make everyone's chances identical, as James Coleman points out, requires that "the equalizing institutions must invade the home, pluck the child from his unequalizing environment, and subject him to a common equalizing environment."[124] Interpreted in this way, equal opportunity becomes indistinguishable from equality of results.

The contemporary debate over how to remedy racial and gender inequalities has repeatedly brought to the surface the dissensus that lies beneath the consensus on equal opportunity. As long as affirmative action (which privileges equal results over equal process) remains at the forefront of the public policy agenda, there is little chance that individualists and egalitarians will be able to find common ground. In place of this exceptionalist alliance, Americans will continue to witness a divisive battle between egalitarians, who insist that the belief that equal process will translate into more equal results is naive if not racist, and individualists, who accuse advocates of affirmative action of practicing "affirmative discrimination" and setting America on the road to socialism.

4

Competing Conceptions of Democracy

A central contention of the consensus interpretation of American history is that broad and deep agreement has existed in the United States on the form of government, namely, popular, representative government. As a corrective to the Progressive tendency to counterpoise democrats and aristocrats, the point is well-taken. But as a statement of the political cultural makeup of the United States it misleads by understating the different meanings that Americans have attached to terms such as "democracy" or "republic." This chapter illustrates this point by contrasting the different meanings that Anti-Federalists and Federalists, as well as antebellum Democrats and Whigs, attached to the notions of representation and democracy. Far from being a peculiar feature of our distant past, disagreements over the meaning of democracy remain an integral part of contemporary American political life.

Anti-Federalists and Federalists: Mirrors and Filters

Anti-Federalists routinely portrayed the Federalists as aristocrats and monarchists. Federalists fueled the rhetorical fire by railing against levelers and the "riotous mob."[1] Neither side's description of its opponents, as Louis Hartz correctly points out, was particularly accurate. Hartz is right that the rhetoric generated by the debate over the Constitution often obscured the shared commitment to popular, representative government. But that same rhetoric also reveals sharply contrasting conceptions of democracy and representation.

When Anti-Federalists attacked Federalists as aristocrats, they were suggesting that the proposed government ran counter to their egalitarian vision of democracy "in which the citizen stands in a close relation to the state, in which active and responsible participation is a serious concern, and in which equality holds a very high place."[2] A strengthened central government, Anti-Federalists believed, would make representation unrepresentative as well as inhibit the civic involvement essential to sustaining popular government. Only in relatively small communities was it possible for representatives to resemble those they represented and for citizens to directly participate in governmental affairs, whether in town meetings, juries, political clubs, constitutional conventions, or elections.

Federalists, in contrast, tended to see popular government in terms that foreshadowed Joseph Schumpeter's "competition of elites."[3] Democracy, in this view, is less about giving people a direct say in the formulation of public policy than it is about giving people an opportunity to choose between rival elites. Federalists tended to associate local, face-to-face politics with momentary passion and short-sightedness, while associating a broader, more refined view of the public interest with national representatives.

For the Federalists, representation was a positive good. They considered it to be "one of the greatest improvements which has been made in the science of civil government."[4] The effect of representation, as Madison explained in *Federalist No. 10*, was "to refine and enlarge the public views, by passing them through the medium of a chosen body of citizens, whose wisdom may best discern the true interest of their country."[5] If the Federalists viewed representation as an opportunity to ensure that the wisest and most able would govern, the Anti-Federalists, explains Herbert Storing, "accepted representation reluctantly, as a necessary device in a community where the people cannot assemble to do their common business."[6] As the dissenters from the Pennsylvania ratifying convention expressed it: "Representation ought to be fair, equal, and sufficiently numerous, to possess the same interests, feelings, opinions, and views, which the people themselves would possess, *were they all assembled.*"[7] Instead of the Federalists' metaphor of a *filter*, Anti-Federalists offered the metaphor of a *mirror* to describe the proper relation between constituents and representatives.[8]

The legislature, in the view of many Anti-Federalists, should mirror not only the views but the social condition of their constituents. "The idea that naturally suggests itself to our minds, when we speak of representatives," explained Melancton Smith, "is that they resemble those they represent; they should be a true picture of the people; possess the knowledge of their circumstances and their wants, sympathize in all their distresses, and be disposed to seek their true interests."[9] "The very term, representative," Brutus maintained, "implies, that the person or body chosen for this purpose, should resemble those who appoint them—a representation of the people of America, if it be a true one, must be like the people."[10] A representative system, the "Federal Farmer" agreed, should seek not "brilliant talents" but "a sameness, as to residence and interests, between the representative and his constituents."[11]

The state legislatures, in the minds of most Anti-Federalists, were the model of democratic government. The beauty of the state legislatures, as the Federal Farmer expressed it, was that they "are so numerous as almost to be the people themselves."[12] The House of Representatives, in contrast, was dismissed as only "the mere shadow of representation."[13] "The members of our State legislature," observed "Cornelius," "are annually elected—they are subject to instructions—they are chosen within small circles—they are sent but a small distance from their respective homes: Their conduct is constantly known to their constituents. They frequently see, and are seen, by the men whose servants they are."[14] Under the proposed government, echoed a "[Pennsylvania] Farmer," it would be impossible for representatives to "know and be known by the citizens, [and] have a common interest with them."[15] Many shared the concern of "Brutus" that "in a large extended country, it is impossible to have a representation, possessing the sentiments . . . to declare the minds of the people."[16] Only in a small republic could representation be "real and actual," because only there could representatives "mix with the people, think as they think, feel as they feel."[17]

The citizen's duty, Anti-Federalists taught, did not end with the election of representatives. Rather than leave the elected free to deliberate, citizens were urged to closely monitor the legislative proceedings. Avoiding legislative abuses, as John Francis Mercer explained, entailed keeping representatives under "the constant inspection, immediate superintendance, and frequent interference and control of the People themselves."[18] The proposed constitution, Anti-Federalists feared, would make governing the exclusive preserve of the few. They rejected "the fashionable language" that the "common people have no business to trouble themselves about government."[19] Anticipating themes later elaborated by the Jacksonians, Anti-Federalists argued that no great skills were required for governing. Whereas Federalists maintained that government was "a complicated science," requiring "abilities and knowledge, of a variety of other subjects,"[20] Anti-Federalists insisted that "there are not such mighty talents requisite for government as some who pretend to them . . . would make us believe—Honest affection for the general good and common qualifications are sufficient."[21] In matters of liberty, "the mechanic and the philosopher, the farmer and the scholar, are all upon [an equal] footing."[22] A large territory, moreover, would make it impossible for people to understand what government was doing and how it operated. In such a "vast extent of . . . territory," "Cato" warned, "the science of government will become intricate and perplexed, and too mysterious for you to understand."[23] Only in a small territory, agreed another, will the people "possess a competent knowledge of the resources and expenditures of their . . . government."[24]

Not only were superior abilities and specialized knowledge not needed in government, they were in fact a positive danger to a republic.[25] "Great abilities," many Anti-Federalists believed, "have for the most part been employed to mislead the honest but unwary multitude, and to draw them out of the open and plain paths of public virtue and public good."[26] "The love of domination," warned another, "is generally in proportion to talents, abilities, and superior acquirements; and . . . men of the greatest purity of intention may be made instruments of despotism in the hands of the artful and designing."[27] To avoid such tyranny, it was necessary to promote those "means by which the people are let into the knowledge of public affairs," such as jury trials.[28]

Anti-Federalists aimed to undermine deference and to discredit notions of noblesse oblige. The people, "Centinel" bemoaned, "are too apt to yield an implicit assent to the opinions of those characters, whose abilities are held in the highest esteem," and are thus taken advantage of by "the wealthy and ambitious, who in every community think they have a right to lord it over their fellow creatures."[29] "It is the character of freemen," William Findley insisted, "to examine and judge for themselves; they know that implicit faith respecting politics is the handmaid to slavery, and that the greatness of those names who frame a government cannot sanctify its faults."[30] People should not defer to the opinions of their alleged betters, because the "great . . . do not feel for the poor and middling class. . . . They cannot have that sympathy with their constituents which is necessary to connect them closely to their interest: Being in the habit of profuse living, they will be profuse in the public expenses. They find no difficulty in paying their taxes, and therefore do not feel public burthens."[31] No elite, not even an elected one, they insisted, could "feel sympathetically the wants of the people."[32]

The Anti-Federalist vision of republican government depended not only on an active, public-spirited citizenry, but also on a relatively homogeneous citizenry. "In a republic," Brutus explained, "the manners, sentiments, and interests of the people should be similar."[33] Centinel, too, insisted that "a republican, or free government, can only exist where . . . property is pretty equally divided."[34] "Equality among the citizens," agreed another, "is the only permanent basis of a republic."[35] Disparities in wealth, influence, knowledge, or education were believed to be inconsistent with republican government, for such inequality made it impossible to achieve the public-spirited consensus necessary to a republic. It was only by keeping politics on a "human scale" that "the people may have a common interest," and thereby avoid the sorry spectacle of "a constant clashing of opinions; and the representatives of one part . . . continually striving against those of the other."[36] To "enlarge the circle," as Cato put it, was to "lose the ties of acquaintance, habits, and fortunes" that make democratic government possible.[37]

In addition to emphasizing the ways in which republican government was dependent on the egalitarian makeup of the citizenry, Anti-Federalists also stressed that the habits of the populace depended on the form of government. As Melancton Smith put it at the New York ratifying convention, "Government operates upon the spirit of the people, as well as the spirit of the people operates upon it."[38] The proposed constitution, they believed, paid insufficient attention to the need for the education of citizens. "Our duty," explained Smith, "is to frame a government friendly to liberty and the rights of mankind, which will tend to cherish and cultivate a love of liberty among our citizens."[39] A distant central government, by making participation impossible for the average person, would create an apathetic populace that would turn inward to private pursuits and thus be ripe for despotism. Echoes of this Anti-Federalist argument can be heard in present-day egalitarian charges that the American political system creates feelings of apathy, powerlessness, and helplessness in its citizens.[40]

Those scholars who have attempted to determine whether the Federalists or Anti-Federalists were "really" more democratic inevitably end up rehashing the cultural biases of the day. Joshua Miller, for instance, comes down decisively on the side of the Anti-Federalists when he insists that "democracy without community and small-scale is just a name."[41] Such assertions, rather than lifting us out of bias, only plunge us in deeper. Instead of attempting to award the label "democratic" to one side or the other, we would do better to allow both sides their democratic credentials. The admission that both Federalists and Anti-Federalists were democrats does not, however, commit us to the proposition that there were no significant ideological differences between the two sides, or even that their similarities outweighed their differences. For what is to *count* as democratic is the arena in which adherents of these competing political cultures have fought. The argument over how to define democracy has been part and parcel of the argument between competing political cultures in the United States.

Jacksonians and Whigs: Delegates and Trustees

Much the same lesson can be gleaned from the ideological controversy generated by the second party system of Whigs and Democrats. Although the Jacksonians were not the wild-eyed levelers that the Whigs made them out to be, any more than the Whigs were the aristocrats of rank and privilege that the Jacksonians portrayed them as, there were nonetheless fundamental cultural differences between the two parties. Both Whigs and Democrats fervently proclaimed their adherence to representative government, but each party had significantly different visions of the proper relationship between elected and elector.

Democratic charges to the contrary, Whigs did not reject representative government. What they did oppose was the doctrine of instruction, as well as the egalitarianism that underlay that doctrine. The practice of instruction, many Whigs believed, was "totally irreconcilable to the first principles of Representative Governments," because it compromised "that lofty independence and integrity of mind which should characterize the representatives of the people."[42] Daniel Webster expressed the view of a number of leading Whigs when he told the Senate that "all the world out of doors is not so wise and patriotic as all the world within these walls."[43] Because the representative was possessed of superior knowledge and/or morality, he should be left "entirely free to act after thorough discussion and mature deliberation, as his best judgement shall dictate."[44] In a public letter, congressional candidate Millard Fillmore explained that he was "opposed to giving any pledges that shall deprive me hereafter of all discretionary power. My own character must be the guaranty for the general correctness of my legislative deportment."[45]

Like the Anti-Federalists before them, Democrats rejected what they regarded as their opponents' cramped conception of democracy. The legislative system, explained the *Democratic Review*, was "but a . . . convenient labor-saving machinery, to supersede the necessity for the assemblage of the great masses of the people themselves." Ideally, the political representative should be only a "delegate," with "no will of his own which is independent of that of his constituents." The right of constituents to instruct their representatives, as well as the propriety of leaders making pledges to their constituents, was "the only safeguard of the people against the encroachments of power."[46]

Jacksonians, again like the Anti-Federalists, looked for leaders who would resemble those they represented.[47] The successful Democratic leader had to have the ability to lead people and yet simultaneously appear as one of the people. In Jacksonian thought, as John William Ward explains, "the vertical distance that separates the leader from the led must be denied." Nicknames were one of the devices Jacksonians used to shrink the distance between leader and follower.[48] Indeed, Andrew Jackson—"Old Hickory"—was the first president to be accorded a nickname.

Jacksonians echoed the Anti-Federalist distrust of great talents. A campaign biography of Van Buren, for instance, proudly broadcast that the Democrats—in contrast to the Whigs, who "cast around for great men"—"want, for public office, servants and not masters; agents who will execute their will and not dictators to control it."[49] The Cincinnati *Daily Enquirer* agreed that "it is not great talents that we admire most in public men, but sterling public integrity, disinterestedness of purpose, and purity of moral character."[50] The Democratic theorist George Sidney Camp freely acknowledged that in the United States, men "of the most shining abilities" were not chosen for office. But he regarded this as "a subject for congratulation rather than for chagrin to the friends of popular government."[51] Indeed Democrats denounced Whigs for believing that "the common people could be better ruled by those that are a little uncommon than by themselves."[52]

The Whigs, in turn, ridiculed the Democratic view that a political leader should be a mere "specimen of his constituents."[53] To them it was incomprehensible that a Democratic editor could praise presidential candidate James Polk for being "a man whose virtues are more conspicuous than his talents."[54] National leaders, most Whigs believed, should be men whose special talent and wisdom set them apart from the common crowd. Whigs did not feel the same need as the Democrats did to lessen the distance between leader and follower. For Whigs believed that the further the leader was from the pressure of the follower the more "disinterested" would be the leadership. Only the leader unfettered by commitments and pledges could be trusted to deliberate in the national interest. Political leaders, in the Whig mind, were to be "trustees," not delegates.[55]

The Jacksonians propagated the view that governing took no special skills, and so should be open to all. As Andrew Jackson explained to Congress in his First Annual Message: the duties of officeholders can be made "so plain and simple that men of intelligence may readily qualify themselves for their performance."[56] The same theme was sounded by the Democratic governor of Michigan, John Barry, who argued that "plain men of sound heads and honest hearts are found adequate to the highest and most responsible duties of government."[57] Walt Whitman only echoed Jacksonian orthodoxy when he poured scorn upon the Whig preoccupation with the "intricacy and profundity of the science of government." Whitman ridiculed the Whigs' belief that "the most elaborate study and education are required in any one who would comprehend the deep mysteries, the hidden wonders, of the ruling of a nation and the controlling of a people," when in fact "the principles that lie at the root of true government are not hard of comprehension."[58]

The Whigs vehemently took issue with this claim. "Years of study," insisted one Whig, were "indispensable to the formation of the sound politician in mixed governments." Those elected to office must be men of great distinction for the qualifications required to govern wisely were "of the highest order and the most arduous attainment." Another Whig agreed that not only "high political morals" but "proper qualifications" were essential to governing. "The legislature should be composed of our wisest, best, and most experienced men." "It is doing the people no injustice," echoed another, "to say, that the average capacity and intelligence of those whom they elect to high offices are, or should be, greater than their own."[59]

Beneath the consensus on democracy was thus dissensus about the meaning of democracy. Neither Whigs nor Jacksonians rejected popular government. But adherents of the two parties had very different conceptions of the appropriate relation between representative and represented. Whigs defended deference toward popular representatives; Jacksonians insisted that elected officials "should be considered [as] having nothing of dignity, or power, or splendor about them."[60] For Jacksonians, democracy was defined in terms of the extent of popular participation; for Whigs, democracy meant elite competition.

The Contemporary Debate Over Participation

This culturally grounded debate over the meaning of democracy did not end with the Jacksonians and Whigs. The same debate over what constitutes democracy, grounded in the same fundamental cultural impulses, remains central to the political disputes of today. Egalitarians today, like those of yesteryear, insist that democracy is primarily about popular participation in the governing process. Individualists, in contrast, are more likely to stress the importance of electoral competition between elites.

From the individualist point of view, the United States is democratic because it is competitive. This usage follows Schumpeter, who defined the democratic method as "that institutional arrangement for arriving at political decisions in which individuals acquire the power to decide by means of a competitive struggle for the people's vote."[61] Competition gives the people an opportunity to participate in the selection of its representatives and governors.

From the point of view of contemporary egalitarians, the representative system of the United States barely qualifies as a "real" democracy. People in twentieth-century America, according to Lawrence Goodwyn, "instead of participating in democratic cultures, live in hierarchical cultures . . . that merely call themselves democratic."[62] Benjamin Barber echoes this theme, contending that "the simple fact is that party government and the representative system to which it belongs are both deeply inimical to *real* democracy."[63] The ideological fount of the New Left, the Port Huron Statement, set itself the task of searching out "*truly* democratic alternatives to the present."[64] America, explained Tom Hayden, was a "false democracy."[65]

What, from the egalitarian point of view, is "truly democratic"? "The only true democracy," Kirkpatrick Sale reports, "is direct democracy." Representative government, though it may be "a desirable expedient in a government of great size," is simply "an oligarchy of the elite" and "has nothing to do with citizen participation, popular decision-making, or democracy."[66] Benjamin Barber repeats this view that "real democracy" (or, as Barber also calls it, "strong democracy") "is defined by politics in the participatory mode: literally, it is self-government by citizens rather than representative government in the name of citizens."[67] "True democracy," egalitarians reiterate, is "democracy of participation."[68]

Voting in regular elections is less an expression of democratic participation than it is an abdication of democratic citizenship. Periodic voting, suggests Hayden, is "very weak, a declining form of democracy."[69] Voters, Barber reasons, are "as far from citizens as spectators are from participants or patients are from the doctors they select to heal them." "To exercise the franchise," Barber continues, "is unfortunately at the same time to renounce it." For the franchise "is used only to select the few who exercise every other duty of civic importance in the nation."[70] Following Rousseau, here as elsewhere, egalitarians argue that "the instant a people allows itself to be represented it loses its freedom."[71] Representative democracy, from this point of view, is an oxymoron: representation is the death of democracy.

Representative government undermines democracy, egalitarians insist, because it breeds an apathetic, passive, and ill-informed populace. "The representative principle," Barber argues, "steals from individuals all significant responsibility for their values, beliefs, and actions; ultimately it turns them into passive clients of party bosses ... or active pawns of public opinion manipulators."[72] Politics without participation, writes Robert Pranger, produces not citizens but a "curiously childlike but hardly exemplary creature."[73] Representative government, from this point of view, breeds fatalistic resignation and dependency. "On the underbelly of America's vaunted freedoms," Barber concludes, is "apathy and alienation and anomie."[74]

The commitment to participation flows from (and, in turn, reinforces) egalitarianism. If all people are of equal worth, then all should have an equal say. If equality is to be the master principle, then what gives one person the right to decide for another? As explained by one SDS member, "'participatory' meant 'involved in decisions that were going to affect me.' How could I let anyone make a decision about me that I wasn't involved in?"[75] How indeed, if one is committed to egalitarian social relations?

"Our Practice of Your Principles"

Samuel Huntington's *American Politics: The Promise of Disharmony* opens with a student speech at Harvard's 1969 commencement. The student, Meldon E. Levine, tells the assembled audience of alumni, faculty, and parents that the student protest is not an effort "to subvert institutions or an attempt to challenge values which have been affirmed for centuries." Rather than "conspiring to destroy America," Levine insists, "we are attempting to do precisely the reverse: we are affirming the values which you have instilled in us and which you have taught us to respect." All we ask is that "you allow us to realize the very values which you have held forth." In Huntington's view, Levine's address "precisely caught the spirit of that decade." The 1960s (like the Revolutionary era, the Jacksonian period, and the Progressive era), in Huntington's view, involved "a reaffirmation of traditional American ideals and values; they were a time of comparing practice to principle, of reality to ideal, of behavior to belief."[76]

Certainly this is the manifest form that much New Left rhetoric took. New Left leaders portrayed themselves as the true champions of the democratic ideal. "We are anxious to advance the cause of democracy," declared Paul Booth in the "Build, Not Burn" speech.[77] Too often, explained Tom Hayden, students have "encountered social institutions that . . . betrayed the democratic ideals they were taught" by their parents.[78] What they were asking for, according to Richard Flacks, was "a restoration of democratic consciousness and democratic action," a return to the "first principles of democratic action."[79] "Democracy," reiterated an SDS leaflet, "is an integral part of the original American ideal." Many of those who admired the New Left noted their fervent commitment to democracy. "More than anything," wrote Andrew Kopkind in the *New Republic*, "the new generation of students cares about democracy." So absolute was this commitment that one SDS member could write, "What thinking person cannot subscribe to democracy?"[80]

Is this not conclusive evidence of the liberal absolutism that Hartz decries? Does it not vindicate Huntington's claim that SDS, at least in its pre-Weatherman years, "was a radical reform organization [that only] . . . wanted to realize American liberal values"[81] and support Michael Harrington's contention that the students at Port Huron were "nonsocialists who took the formal promises of American democracy with innocent and deep seriousness"?[82] Was the only difference between the students and "the establishment" they so despised that the students took the democratic promise of America seriously? I think not. The differences between SDS and the establishment were more fundamental than Huntington or Harrington or Hartz would have us believe. For what SDS meant by democracy—equalizing power relationships—differed fundamentally from what the establishment meant by democracy—competitive elections. Students were not simply practicing their elder's democratic principles but rather were attempting to put into practice a different set of democratic principles.

The Port Huron Statement, drafted in the summer of 1962, is often presented as a classic illustration of the "deep and innocent seriousness" of the New Left's commitment to traditional American democratic ideals. But in calling for a "democracy of individual participation," the Port Huron Statement was doing more than just calling attention to the gap between promise and performance. It was trying to redefine the character of American ideals in an egalitarian direction. Championing the individual's right to "share in those social decisions affecting the quality and direction of his life"[83] did not seem particularly subversive until its implications were spelled out. As James Miller points out, "it was easy to fudge the radical implication . . . and speak of the democratic ideal as if it marked the restoration of a lost American dream."[84] But beneath the vague rhetoric of participatory democracy lay the New Left's deep distrust of those mechanisms that most Americans took to be the essence of the democratic ideal: voting in competitive elections, representative government, accountable leadership, and the freedom to pursue one's own private interests. On each of these issues, the New Left offered a vision of democracy that conflicted fundamentally with the democratic ideals of much of the rest of American society.

When individualists spoke of democratic freedoms, they had in mind freedom from governmental restraint, the freedom to buy, sell, consume, or vote as one

pleases. Egalitarians in the New Left put forth a very different definition of democratic freedom that centered on the freedom to participate in the decisions that
affect one's life.[85] So a pamphlet of the Economic Research and Action Project
(ERAP), an SDS spinoff, could proclaim that "freedom is an endless meeting."[86]
Freedom, in this conception, did not mean being left alone by government, but
rather having an equal say in the decisions that governed one's life. The New Left's
proclivity to define freedom as participation is evident in the claim that "the fundamental issue is not whether the few rule in the interest of the many or in their
own interest. It is rather that they rule and *thereby* deprive the many of their
freedom."[87] That the New Left and the establishment both used the vocabulary of
freedom should not obscure the fundamental differences in meaning.

When the New Left charged that the American political system "shuts [the
individual] out of meaningful participation,"[88] they were redefining participation.
Organizing in pursuit of one's interests, contributing money to favorite candidates
or causes, writing letters to congressmen, even voting itself were not real or meaningful participation. The registration of black voters in the South was an instance
of "our practice of your principles"; the denunciation of voting as "undemocratic"
or "irrelevant" was not.[89] Tom Hayden's idea that "parliamentary democracy" was
"a contrivance of nineteenth-century imperialism and merely a tool of enslavement"[90] was derived not from the American Creed but from Herbert Marcuse, who
dismissed "elections and representatives" as "institutions of domination."[91]

The SDS motto, "Let the People Decide," is a telling instance of vague rhetoric papering over fundamental value disagreement. Few Americans would reject
this slogan. Agreement would quickly evaporate, however, as different people began
to spell out what they meant by "letting the people decide." To individualists, voting in competitive elections is letting the people decide. To egalitarians in the New
Left, "letting the people decide" meant consensual decision making in small groups
and an opposition to all forms of authority. Many in the New Left looked hopefully toward a future world in which people would "not even have to do such things
as vote or have leaders or officers." "Everybody is a leader," Tom Hayden exulted.
"We don't believe in leadership," echoed James Foreman of the Student Non-
Violent Coordinating Committee (SNCC). "The people on the bottom don't need
leaders at all," explained SNCC spokesman Robert Moses. "What they need is the
confidence in their own worth and identity to make decisions about their own
lives."[92] After a demonstration on Washington in 1967, one participant reflected,
"There was no leadership; that was what was so beautiful." Leadership was quite
simply undemocratic. As one SDS member formulated it: "Leaders mean organization, organization means hierarchy, and hierarchy is undemocratic."[93]

SDS did not simply embody, as the Harvard undergraduate Meldon Levine had
it, "our practice of your principles." Rather this movement, like others that arose
in the 1960s and 1970s, put into practice values that posed a fundamental challenge to the dominant individualistic political culture. From civil rights groups to
the women's movement to antinuclear groups, social movements of the 1960s and
1970s denigrated authority and "mere" procedural democracy, expressed a profound hostility to the inequalities generated by capitalism (surely the sine qua non

of competitive individualism), and set out to equalize power relations—between leaders and led, rich and poor, men and women, whites and blacks, gays and straights, parents and children, even animals and humans. The anti-authority animus of the 1960s and 1970s, like the anti-authority sentiment that existed during the revolutionary and Jacksonian periods, resulted not because the individualistic consensus "came alive," as Huntington would have it,[94] but rather because of the rise of egalitarianism as a potent cultural force.

5

An Anti-Authority Consensus?

American political culture, Samuel Huntington argues, is characterized by a pervasive "antipower ethic." All the basic tenets of the American Creed—equality, liberty, individualism, constitutionalism, democracy—are "basically antigovernment and antiauthority in character." Each of these values, Huntington maintains, places "limits on power and on the institutions of government." The American Creed is thus "a much more fruitful source of reasons for questioning and resisting government than for obedience to government."[1]

The notion that Americans are naturally disposed to distrust political authority is not new. Over two hundred years ago, Luther Martin explained that "the genius and habits of the people of America were opposed to government."[2] But this formulation is incomplete. It fails to distinguish between two distinct anti-authority visions that coexist in America. The first is individualistic and stresses freedom from governmental control. The second is egalitarian and rejects outright the idea that one person has the right to exercise control over another.[3] To conflate these two distinct cultural visions is to misunderstand the nature of American attitudes to political authority. In addition, it neglects altogether those pro-authority voices who have from time to time been important players in American political life.

The Competing Antipower Ethics of Tom Paine and James Madison

The difference between the egalitarian impulse to question authority and the individualist instinct to limit or escape authority can be seen in the persons of Tom Paine and James Madison. Paine glories in berating authority, ridiculing its pomp, stripping away its secrecy, and exposing its hypocrisy. Madison's forte, in contrast, is in devising ways to check and limit power. Paine's ideas are fashioned to bring down a government; Madison's to establish a limited one.

Paine is most in his element when directly challenging and criticizing authority. He meets aristocratic pretensions with withering scorn: "When I reflect on the pompous titles bestowed on unworthy men, I feel an indignity that instructs me to despise the absurdity." "High sounding names" like *"My Lord"* serve only to "over-awe the superstitious vulgar" and make them "admire in the great, the vices they would honestly condemn in themselves."[4] Monarchical authority is unmercifully ridiculed, its most revered traditions and customs made to seem foolish. But it is not alone Old World monarchy and nobility that are made to feel the acid lash of Paine's tongue. Churches of every type are indiscriminately denounced as "engines of power" and allies of despotism.[5] Even the revered George Washington (to whom only a few years previously Paine had dedicated part 1 of *The Rights of Man*) could not escape Paine's blistering attacks. In a public letter, Paine denounced President Washington as "treacherous in private friendship" and "a hypocrite in public life." "The world will be puzzled to decide," mused Paine, "whether you are an apostate or an impostor; whether you have abandoned good principles, or whether you ever had any."[6]

A commitment to individualism led Madison to design a constitutional structure that would check and limit governmental authority. Paine's egalitarianism, in contrast, led him to question the right of one individual to exercise authority over another. Paine never reconciled himself to the office of the presidency, for instance, because he opposed concentrating executive authority in a single person. A plural executive was far better, he explained, because "it is necessary to the manly mind of a republic that it loses the debasing idea of obeying an individual."[7] Paine agreed with his fellow Pennsylvania radical John Smilie that "a democratic government like ours admits of no superiority."[8]

For Paine and other Pennsylvania radicals, the Pennsylvania constitution of 1776 represented a close approximation of the ideal institutional arrangements.[9] The radical framers of the 1776 constitution replaced the governor under the colonial charter with a twelve-person executive council. This plural executive was designed to do little more than carry out the will of the people's representatives in the assembly. To ensure strict popular accountability, the executive council was to be elected directly by the people, and one-third of the council would leave office each year. The legislature selected annually a president of the council, who did exactly what the title suggests: preside.

A comparison with the Virginia constitution of 1776 reveals something of the distinction between individualist and egalitarian attitudes to authority. Madison and the other framers of the Virginia constitution wished to limit executive authority but, unlike the Pennsylvania radicals, they did not reject the notion of vesting executive power in a single person. The governor was to be elected annually (by a joint vote of the two legislative bodies) and was limited to three consecutive terms. The few, limited powers granted to the governor—such as granting reprieves and pardons, making interim administrative and judicial appointments when the legislature was not in session, and calling out and directing the militia—depended on the consent of an eight-member Council of State (also selected by the legislature, with two members replaced every three years). Executive power in revolutionary Virginia was thus tightly tethered, but the framers did not reject the notion of a unitary executive.

The Pennsylvania and Virginia constitutions also differed in the structure of the legislature. The Virginia constitution established a bicameral legislature, thus institutionalizing the checking and balancing of authority. Pennsylvania's framers, on the other hand, followed Paine in repudiating mixed government, preferring instead a unicameral legislature. Where Madison sought to erect a complex institutional system of checks and balances, Paine (like Rousseau and Condorcet)[10] was attracted to a simple structure of government. As Paine explained in *Common Sense*: "I draw my idea of the form of government from a principle in nature, which no art can overturn, viz. that the more simple any thing is, the less liable it is to be disordered."[11]

How could Paine, in view of his anti-authority leanings, oppose checks and balances? Much of the answer lies in his optimistic view of human nature. Although he and his fellow Pennsylvania radicals distrusted governmental authority, they put great faith in the potential goodness of people. Madison, in contrast, trusted neither government nor the people. Madison crafted institutional checks to guard against the natural self-interest and passions of man; Paine believed that it was institutions that had corrupted man's innate moral sense. "Human nature," Paine insisted, "is not of itself vicious. . . . Man, were he not corrupted by governments, is naturally the friend of man."[12] Where Madison took people much as he found them, Paine had great faith that the inherent goodness of human nature could be reclaimed by eliminating oppressive and slavish institutions. For Madison the individualist, human nature is a constant; for Paine the egalitarian, it is a variable to be changed.[13]

Paine's view of human nature and governmental authority is characteristic of egalitarian thinkers from Rousseau to Charles Reich.[14] The belief that human beings are born good and corrupted by evil institutions sustains the egalitarian commitment to a noncoercive social order. An egalitarian who had become persuaded that human nature was irretrievably bad, could hardly resist hierarchical arguments for increasing institutional restraints upon individuals. Nor could that egalitarian deny the individualist claim that there was no sense in trying to remake human nature. If people are naturally good, egalitarians can persuade each other that antisocial behavior is a product of the false consciousness which coercive institutions have imposed on individuals and that a noncoercive social environment is a viable way of organizing life.

By the same token, the individualistic way of life, whether articulated by Madison or Adam Smith, is invariably grounded in a pessimistic assessment of human nature. Madison's preference for a limited government is sustained by his distrust of human nature. In *Federalist* No. 10, Madison posits that the causes of faction are "sown in the nature of man." Human nature, in the individualist view, is both self-seeking and unmalleable. Because no institutional arrangement (other than despotism) can prevent men from pursuing their self-interest at the expense of the larger community, Madison deduces that the political system should be structured so as to take advantage of this inevitable conflict of self-interested individuals and groups. Pitting interest against interest, Madison reasoned, would create a political system that would generate a beneficial collective outcome (limiting the scope of political power) that was no part of the intention of any of the participants.

Huntington suggests that what is most striking about the American case is the way in which both optimistic and pessimistic evaluations of human nature have led to favoring weak government over strong government.[15] There is validity to this view, but it neglects important differences in attitudes toward authority that result from these rival visions of human nature. It matters a great deal whether opposition to authority is animated by a skepticism about human nature or a hostility to institutions as inherently oppressive. Paine's roseate conception of human nature leads him to denigrate institutions as inherently corrupting and to condemn authority as an obstacle to fulfilling human potential. Madison's more sober evaluation of human nature leads him to support institutional structures that harness individual self-interest. Paine's attitude to authority is instinctively adversarial; Madison, in contrast, is inclined to support authority as long as its reach and scope are sharply circumscribed.

This difference between the anti-authority bias of Paine and that of Madison was not lost on Paine's contemporaries. Many of them recognized that Paine's version of the antipower ethos would have very different consequences for authority from the version put forth by Madison and his individualist allies. Paine, observed a caustic John Adams, has "a better hand in pulling down than building up."[16] A similar judgment of Paine was rendered by Madame Roland, who found "him more fit . . . to scatter . . . kindling sparks than to lay the foundation or prepare the formation of a government."[17]

Adams and Madame Roland identify one of the important limitations of an egalitarian bias. Well-suited to criticizing authority from the fringes of society, egalitarianism is less well adapted to providing a governing creed for political institutions on the scale of the modern nation-state. Moreover, by constantly berating and questioning authority egalitarians may delegitimate existing, albeit imperfect, democratic authority. Having thrown over an imperfect democratic order in quest of the perfectly noncoercive order, a nation may be disappointed to find that the alternatives are more rather than less coercive. And yet for all its deficiencies, egalitarianism plays an indispensable role in sustaining a democratic society. A willingness to question and confront authority is essential to keeping authority accountable. Egalitarianism punctures the pomposity of authority and strips away gratuitous secrecy. It exposes authority's cover-ups and corruption, and thereby prevents governmental power from growing arrogant or domineering.[18]

In the life of Thomas Paine, one can clearly see both the strengths and weaknesses of the egalitarian ethos. No American leveled more telling blows against British monarchy than Tom Paine. No pamphlet was more influential in turning American sentiment against the British rulers than *Common Sense.* Yet having assisted in the tearing down of a government, Paine was no longer so sure afoot. He played no role at all in forging the Constitution or setting up a new government, and went instead to Europe where he could continue to draw applause playing the gadfly. When he ventured back into American politics, he misjudged badly, lashing out wildly and irreverently against Washington. His courageous and unyielding hostility to authority no longer gained the same widespread appreciation it had during the Revolution, and he spent his final years in relative isolation on a farm in New Rochelle. At his funeral there were, appropriately enough, no dignitaries, only a handful of neighbors and friends.

The Two Federalists

If Madison expressed a fundamentally different disposition toward authority than did Paine, his views differed just as significantly from those of his coauthor of *The Federalist*, Alexander Hamilton. Whereas Paine's credo was "question authority" and Madison's was "check authority," Hamilton's was "support authority." Paine's cultural bias was egalitarian, Madison's individualistic, and Hamilton's hierarchical.

Ironically it is those scholars who have been most attentive to the alternative, communitarian vision of the Anti-Federalists who have been most inclined to treat the Federalists, and their constitutional handiwork, as if they, and it, were all of a cultural piece. Much like the Anti-Federalists whom they so admire, these scholars tend to view the Federalists as advocates of centralized power, administrative efficiency, elitism, and inequality. Hamilton's vision is made into the whole. Political scientists—who for the most part ignore the Anti-Federalists as irrelevant, wrongheaded, or worse—have often made the opposite error of inflating Madison into the whole. From this perspective, the architects of the Constitution were concerned with limited government, checks and balances, and the safeguarding of minority rights. A balanced assessment of the framers must account for their commitment both to limited government and to the centralization of power.

The task of persuading opponents to adopt the Constitution naturally led the Constitution's supporters to downplay areas of disagreement. These areas of divergence would become obvious to all in the 1790s as Madison moved into a position of leadership in the Republican party and Hamilton assumed a preeminent position within the Federalist party.[19] But if the difference between the Madisonian and Hamiltonian visions of authority was sometimes muted in 1787 and 1788, it was there nonetheless. Madison indicated as much to Jefferson, explaining that although *The Federalist* was "carried on in concert, the writers are not mutually answerable for all the sides of each other."[20]

Madison's basic concern was to establish a system of checks and balances that would limit governmental authority; Hamilton's primary interest, in contrast, was in securing efficient and effective administration. It is impossible to imagine Madison writing, as Hamilton did in *Federalist* No. 68, that "the true test of a good government is its aptitude and tendency to produce a good administration."[21] Hamilton, as Gerald Stourzh correctly points out, was relatively uninterested in theories of checks and balances.[22] Good government, for Hamilton, was a function not of checks and balances but rather of an "energetic" administration.

Limiting authority was a secondary concern for Hamilton and his hierarchical allies. Much more important to them was increasing support for national authority. From this point of view, the preeminent question was not how to check government but rather, as Fisher Ames put it, "how our government is to be supported."[23] These Federalists believed, with John Adams, that the "one thing" absolutely essential in a new republic was to establish "a decency, and respect, and veneration . . . for persons in authority."[24] Hamilton maintained that this public support for authority could best be gained by establishing a strong and effective government. Rejecting the Anti-Federalist view that "a numerous representation was necessary to obtain the confidence of the people," Hamilton argued that "the confidence of the people" would be "gained by a good administration."[25]

Hamilton gave voice to what Isaac Kramnick identifies as "the state-centered language of power." His vision of politics was animated by classical concepts of *imperium, potestas, gubernaculum,* prerogative, and sovereignty.[26] In *The Federalist,* he spoke of "reasons of state"[27] and did not disguise his admiration for European style state-building. In *Federalist* No. 17 Hamilton likened America under the Articles of Confederation to the "feudal anarchy" of medieval Europe, and he approvingly cited England's successful drive to subdue the "fierce and ungovernable spirit" of localism and reduce "it within those rules of subordination" that characterize "a more rational and a more energetic system of civil polity."[28] The Constitution would thus enable the United States to follow the European pattern of state-building.

In *Federalist* No. 26, when Hamilton linked "energy in government" with "the security of private rights," he sounds much like Locke and the liberal theorists.[29] But as Kramnick points out, "Hamilton was interested less in the limited liberal state than in the heroic state."[30] A strong state was, in Hamilton's view, essential for America to assume a respected and active place in the world of nations. "A nation without a national government," he informed his readers at the close of his final contribution to *The Federalist,* is "an awful spectacle."[31] Hamilton's mind turned readily to phrases such as "the splendour of the national government" or "the majesty of the national authority."[32] And when he observed the weakness of the present American state, he was moved to despair at "the imbecility of our Government"; the United States seemed to Hamilton to have "reached almost the last stage of national humiliation."[33]

Madison sometimes voiced similar concerns that the existing national government "is held in no respect by her friends" and "is the derision of her enemies."[34] But whereas Hamilton wanted an expanded American state that would rival in power the great states of Europe, Madison wanted only to give the federal government the means necessary to meet such limited responsibilities as protecting property rights, upholding contracts, and honoring treaties. Scholars have misunderstood Madison's position at the Constitutional Convention and in *The Federalist,* as Lance Banning persuasively argues, because of "a failure to distinguish Madison's determination to secure a general government whose authority would be *effective* or *complete* from his opinion of the *quantity of power,* the nature of the duties, that ought to be confided to federal hands." Madison, Banning continues, can be "legitimately described as a determined 'nationalist' only in his quest for a structure and mode of operation that would make the general government effective and supreme *within its proper sphere,* which he consistently conceived as relatively small. He was not a 'nationalist' in his conception of the duties or responsibilities that should be placed in federal hands."[35] Madison never shared Hamilton's hierarchical vision of an activist state achieving glory abroad and intervening vigorously to promote economic development at home. Instead Madison envisioned a relatively inactive and limited government that would leave as much scope as possible to individual initiative.

To be sure, in *Federalist* No. 10 Madison envisioned a governing elite with "enlightened views and virtuous sentiments" who would pursue the public interest. But, as Kramnick accurately observes,

> Madison's enlightened leaders would demonstrate their wisdom and virtue more by what they did not do than by what they did. Being men of cool and deliberate judgment, they would not pass unjust laws that interfered with private rights. They would respect liberty, justice, and property, and run a limited government that did little else than preside over and adjudicate conflicts in a basically self-regulating social order.[36]

Madison's distrust of state governments stemmed in large part from his perception that they tended to an "excess of law-making." The voluminousness and mutability of state laws, Madison believed, hindered private initiative. "What prudent merchant," Madison asked rhetorically, "will hazard his fortunes in any new branch of commerce, when he knows not but that his plans may be rendered unlawful before they can be executed?"[37] Government was to provide a neutral and orderly arena in which varying societal interests could flourish and compete.

It is against this background that Madison's support for a federal veto over state legislation must be understood. Far from representing support for a more expansive government, it was actually part of an effort to reduce governmental intervention in individual lives. Madison repeatedly identified the veto as a "defensive power." The federal veto, Banning explains, "was not to be an instrument with which the central government could set a positive direction. It was to be a negative on acts that breached a solemn compact or contradicted the most basic principles of a republican revolution."[38]

The *positive* powers of the new federal government, Madison insisted in *Federalist* No. 45, should be "few and defined." Moreover, Banning tells us that during the Constitutional Convention, Madison acted as "one of the convention's most consistent advocates of strict, though full, enumeration."[39] For Hamilton, in contrast, explains Alpheus T. Mason, "the powers granted for achieving [the objects of the national government] were undefined—indeed, undefinable."[40]

The cultural difference between Madison and Hamilton is particularly evident in their very different conceptions of a desirable executive. Madison's preference for a "checked, limited, and precisely defined" executive[41] is evident throughout the convention and contrasts sharply with the kind of president preferred by Hamilton. In a famous five-hour speech to the convention, Hamilton proposed a life term for the chief executive, an absolute veto over legislation, and power to appoint the "chief officers" of the executive branch without congressional approval. Most revealing of all, Hamilton vigorously defended executive patronage (known to contemporaries as "corruption") as essential to securing effective governance. "Support of Government," Hamilton argued, depended on executive "influence," that is, on "a dispensation of those regular honors and emoluments, which produce an attachment to the Government."[42] Madison, on the other hand, desired a considerably weaker, more tightly tethered executive. Madison even preferred vesting the power to appoint judges in the Senate, changing his mind only after the Senate was reconstituted to represent states rather than individuals. Moreover, Madison repeatedly defended the Virginia plan's provision for having the president share veto power with a "council of revision." And in the closing days of the convention, we find Madison supporting George Mason's proposal for a six-member Executive Council to be appointed by the Senate. Even within the executive branch then Madison wanted to establish a system of checks and balances.

Although both Madison and Hamilton shared a pessimistic evaluation of human nature, they differed sharply in their willingness to trust those in positions of power. Hamilton believed that the chief executive in particular could be trusted to act justly because "the sole and undivided responsibility of one man will naturally beget a livelier sense of duty and a more exact regard to reputation."[43] Regard for his reputation as an honorable and virtuous man will prevent the ruler from acting in ignoble ways. In Madison's contributions to *The Federalist*, as Mason points out, "one discovers . . . no such confidence in the purifying effect of power."[44] Instead Madison argues that government must be made "to controul itself" through a policy of supplying "by opposite and rival interests, the defect of better motives."[45] Hamilton is engaged in a constant balancing act: denigrating human nature to combat what he perceives as utopian dreams of "perfect wisdom and perfect virtue"[46] while simultaneously trying to forestall a uniformly bleak picture that would unduly restrict the latitude of those in authority. To portray "human nature as it is," Hamilton argues in *Federalist* No. 76, one must do so "without either flattering its virtues or exaggerating its vices." The "supposition of universal venality in human nature," Hamilton explains, "is little less an error in political reasoning, than the supposition of universal rectitude."[47] No such ambivalence is evident in Madison's writings. Madison consistently accents the negative in human nature: men are not angels is what he says; he never mentions that neither are men demons. The essential difference between the two men is that Madison's vision is focused on limiting power while Hamilton's gaze is fixed on augmenting power.[48]

This is not to deny there were important areas of agreement between Madison and Hamilton. The two men, and the two ways of life they represented, were able to cooperate during this period because they shared certain beliefs. They both believed the Articles of Confederation were inadequate to guarantee public order and private liberties. Both men also believed that "inequality of property" was inevitable, and both adhered to a relatively unflattering view of human nature.[49]

In addition, sometimes Madison and Hamilton agreed on a position but for different reasons, as was the case with the question of whether to make the executive independent from the legislature. For Hamilton and his hierarchical allies, an independent executive was an essential prerequisite to a powerful chief executive. Individualists like Madison, on the other hand, came to feel that a system of competing powers required the executive to have an independent power base. Selection by the legislature would result in the executive being swallowed up by the legislature, and thus reduce competition. In the one case independence was a prerequisite to the exercise of executive prerogatives; in the other independence was necessary to maintain competition within the government.

Ambiguity, too, was instrumental in securing agreement among individualists like Madison and hierarchists like Hamilton. Article II of the Constitution was only the most ambiguous section of a document that left so much unsaid. The opening words of Article II, for instance, read: "[T]he executive power shall be vested in a President." Was this a grant of power or merely a gratuitous prologue indicating the unitary nature of the executive and his title? Agreement on the proposition that the executive had "the executive power" thinly masked sharp disagreements about what that power included. Was a declaration of neutrality, for example, an executive power? Within a few short years precisely this question would dredge up the submerged differences between Hamilton and Madison as to what was "the executive power."

Anti-Federalists were correct in interpreting the Constitution as a blow against local egalitarianism but erred in their portrayal of the Constitution as an unambiguous victory for hierarchy. The view is in error because it ignores the coalitional nature of support for the Constitution and the consequent impediments put in the way of government action. This is not to deny that the new Constitution, compared to the Articles of Confederation, greatly strengthened the central government and executive power. Nor is it to deny that the hierarchical advocates were an important element in the coalition that created and ratified the new government. What the Anti-Federalists failed to appreciate, however, was the vital role of individualism in crafting, and gaining support for, the Constitution. The results were not as bad as their opponents predicted because, as Kramnick correctly concludes, "the new American state created by that 'triple headed monster' of a Constitution was much closer to Madison's state than to Hamilton's."[50]

But if the Anti-Federalists erred in underestimating the antipower component of the Constitution, it is no less a mistake to portray the Constitution as purely a product of an antipower ethos. Key figures at the convention, including Hamilton, Gouverneur Morris, and Rufus King (all of whom were on the five-person Committee of Style that put the critical finishing touches on the Constitution) were inclined to a pro-authority position. The importance of the hierarchical political culture in American history will be elaborated at greater length in the next chapter, but before pursuing this theme I will try to clarify the differences between individualistic and egalitarian attitudes to authority.

The "Splendid Venom" of Wendell Phillips

The important distinction between individualistic and egalitarian dispositions toward authority can be further illustrated by examining the contrasting careers of the Illinois Democrat Stephen Douglas and the Boston abolitionist Wendell Phillips. Although both men adhered to a political ethic that can reasonably be described as "antipower," to leave the issue there ignores their radically divergent views of political authority. Douglas aimed to limit central power, leaving localities and individuals free to regulate their own lives; Phillips believed he had a moral duty to attack "everything and everybody until it secures something no longer capable of being found fault with."[51] Between these two visions of authority lies "a world of difference."[52]

Like Paine, Phillips's stock in trade was castigating those in positions of authority, holding their deceptions and half-measures up to the harsh light of truth and justice. Both men were, as Richard Hofstadter wrote approvingly of Phillips, "a thorn in the side of complacency."[53] Paine's preferred weapon was the pen; Phillips, the spoken word. Phillips's splendid oratory earned him the title "abolitionism's golden trumpet."[54] For politicians, however, the music was anything but sweet. To his friends he was the "master of invective"; to his enemies he was a purveyor of "blackguardism and shameless abuse."[55]

There was hardly a prominent politician in the country who did not feel the sting of Phillips's "splendid venom."[56] Rufus Choate was "like a monkey in convulsions"; Daniel Webster was "a great mass of dough"; Edward Everett, "a whining spaniel"; Robert Winthrop, "a bastard who had stolen the name of Winthrop"; Massachusetts governor Henry Gardner, a "miserable reptile"; James Buchanan, "the lees of a wornout politician"; Ulysses Grant, "an empty soldier"; Andrew Johnson, a "usurping traitor"; Chief Justice Salmon Chase, "diseased"; and Abraham Lincoln, a "slave hound," "a huckster," "a spaniel by nature," and a "first rate second rate man."[57] Phillips felt it was his duty to rip off the mask and reveal the villainy beneath. "If the community is in love with some monster," he once explained, "we must point him truly. . . . It is [our] right and duty to use . . . searching criticism, pitiless sarcasm, bitter invective, [and] rigid analysis of motives . . . to save the people from delusion."[58] His highly personal attacks upon politicians stemmed not from spite but from a conviction that politics was inherently corrupting and from a calculated political strategy that deflating the reputations of "great men" would undermine people's unthinking trust in their leaders and force people to judge questions independently.

When Phillips stated that "every government is always growing corrupt," he had in mind much more than the conventional American fear of power encroaching upon liberty. Individualists like Madison feared that power unchecked would become tyrannical (that, as Lord Acton had it, "power corrupts, and absolute power corrupts absolutely") and so devised a complex scheme of checks and balances and separation of powers to divide power among competing institutions and individuals. But for Phillips, such institutional checks were beside the point because the activity of political compromise was itself corrupting. Every governmental officer, Phillips explained, is "by the necessity of his position, an *apostate* . . . an enemy of the people."[59] For Madison, political give-and-take served to check the corruption of power; for Phillips, politics itself corrupted principles of truth and justice.

To speak truly required the individual to stay out of the political realm. Phillips, like many other abolitionists, refused to vote, join a political party, or hold office because to do so would be to compromise one's moral integrity. The true reformer, Phillips believed, must stand "outside of organizations, . . . outside [of] the press, of politics, and of . . . church, . . . with no object but truth, no purpose but to tear the question open and let the light through it."[60] "The real progress of our cause," he reiterated, "is to be looked for from those who keep aloof—who have rid themselves not only of old parties—but of parties themselves."[61] It was the "duty of every Christian to withdraw from a government which upholds iniquity."[62]

The abolitionists' withdrawal from established institutions was partly strategic (as Ann Phillips told her husband, you "cannot stir their waters unless you are outside")[63] and partly motivated by a dread of the "moral taint" that would ensue from contact with impure institutions. The sense that American political institutions were polluting pervaded abolitionist thinking. Upon returning to Boston after five years in Europe, Henry Wright confided in his journal, "A strange feeling comes over me. I feel that I am once more breathing the pestiferous atmosphere of chattel slavery."[64] Wendell Phillips agreed that slavery "sends out poisonous branches over the fair land, and corrupts the very air we breathe."[65] To a Boston gathering, Phillips warned against the "pollution of our native city" that results from allowing escaped slaves to be hunted down and returned to their masters. Massachusetts should secede from the Union, he believed, rather than "allow her soil to be polluted with the footprints of slavery."[66] If slavery could not be eliminated in the South, the purity of the North required "no union with slaveholders" and the expulsion of "pestiferous leaders" like Daniel Webster from the body politic.[67]

If purity demanded that Phillips remain outside the political system, his antipathy to that system as well as his position as outsider ensured that his voice remained strident and uncompromising. His advice to others was "never to be silent, never to be appeased, never to be anything but implacable,"[68] and he practiced what he preached. He indicted the entire Republican party for being "shuffling, evasive, unprincipled, corrupt, cowardly and mean—mad for office, cankered with gold, poisoned with spite, cowards from the first."[69] He condemned officeholders and populace alike for their "paltry moral vision" and their "timid silence."[70] He was appalled both by the perfidious deceit of the nation's political leaders and by the unquestioning obedience of the people.

By attacking those in positions of authority, Phillips hoped to undermine people's willingness to defer to their leaders. Finding their political leaders to be nothing more than "peddling hucksters,"[71] people would learn to judge each question for themselves. Upon hearing of Lincoln's death, Phillips admitted that although the news made him "very sad for Lincoln," he thought "the removal of men too great and too trusted is often a national gain in times like these and it lifts off a weight and lets people develop and think for themselves."[72] It gladdened him to see people withdraw from "rotten" churches because it showed an admirable "independent purpose of mind."[73] Phillips felt that "we are bullied by institutions," and he called upon citizens to "stand on the pedestal of your own individual independence, summon these institutions about you, and judge them."[74]

Phillips's antipathy toward institutions, authority, and compromise was typical of Garrisonian abolitionists. Some, like Stephen Foster and Parker Pillsbury, were more radical; others, like Lydia Maria Child, somewhat less so.[75] But these differences were differences in degree, not in kind. It was not an unhinged or mean-spirited personality but a shared egalitarian culture that explains "the ferocity of [Phillips's] onslaughts on public men and public measures" that contemporaries so often remarked upon.[76] Phillips was in fact a gentle person, who showed immense kindness in his personal life toward his fellow abolitionists and toward his invalid wife.[77] His sense of humor and tact helped him to defuse potential conflicts and patch up differences within the abolitionist group.[78] Those who saw him lecture, expecting a "ferocious ranter," were often surprised by what they saw: "a quiet, dignified and polished gentleman and scholar, calm and logical in his argument."[79] Phillips described himself as "a quiet moderate halfway sort of sim sam fellow."[80] Not personality but an egalitarian cultural bias accounts for Phillips's unrelenting criticism of those in positions of authority, his harsh rejection of the obfuscation, half-truths, and "timid silence" of the politician, and his overriding concern with personal purity.

Stephen Douglas and Popular Sovereignty

If Wendell Phillips embodied the egalitarian spirit of the abolitionist movement, Stephen Douglas was the political embodiment of frontier individualism.[81] To say that Douglas and the frontier constituency he represented had little love for authority is true but incomplete. Just how incomplete can be seen by contrasting Douglas's political views with those of Phillips. Although separated in age by only two years, Douglas's individualistic bias is worlds apart from Phillips's egalitarian bias.

Phillips and the other Garrisonian abolitionists tried to build an impermeable wall separating the pure inside from the impure outside. Only by walling themselves off from a corrupt and odious system could abolitionists sustain their moral integrity and provide the country a shining example of the possibilities of a harmonious fellowship that knew no artificial distinctions of gender, religion, or politics. No abolitionist could cross the boundary separating the good inside from the evil outside without at once compromising his inner purity and incurring the taint of sin. Those who tried to build bridges to the system were reviled as "apostates" and "traitors" and expelled from the redeemed community of saints.[82]

Among the individualistic farmers of the Illinois frontier, the boundaries between members and nonmembers, purity and corruption, were much more ill-defined and permeable.[83] Notions of pollution and purity, omnipresent in the writings of Phillips and other abolitionists, are notably absent from Douglas's public speeches and private correspondence. Douglas was unwilling to label even slavery wrong for fear of encouraging a moral absolutism that would impede political compromise. There was no "tribunal on earth," Douglas felt, "that can decide the question of the morality of slavery or any other institution. I deal with slavery as a political question involving questions of public policy."[84] Douglas had no patience for Phillips's scorn of those who would "compromise with the world."[85] For Douglas, everything should be open to negotiation and compromise. This, as Douglas saw it, was exactly the beauty of American democracy as well as of the doctrine of "popular sovereignty" that he tirelessly championed throughout the 1850s.

"Popular sovereignty," as Douglas understood and articulated "that great Democratic principle," meant that "it is wiser and better to leave each community to determine and regulate its own local and domestic affairs in its own way."[86] Applied to slavery, popular sovereignty meant that the decision as to whether slavery would be permitted in a particular territory would be left to those who resided in the territory and would thus be kept out of the halls of Congress. The great virtue of this doctrine, in Douglas's view, was that agreement on policy could paper over differences in objectives. Antislavery Republicans like Horace Greeley could support the policy in the belief that those living in the western territories would exclude slavery, while many Southerners could support the policy either because they were confident that new territories would embrace slavery or because they were content to have the federal government refrain from judging slavery to be wrong.[87]

The same ambiguity that recommended popular sovereignty to Douglas made the doctrine abhorrent to Phillips. Phillips and his fellow abolitionists did not want support from those who did not believe in "the cause." During the Civil War, for instance, abolitionists were deeply troubled that the increasing support for emancipation stemmed not from a commitment to abolishing slavery but from a selfish calculation that abolition was the most effective means to win the war and preserve the Union.[88] Douglas, who sought to build broad-based political coalitions, cared little about the motives of those attracted to his policy position; Phillips, who distrusted compromise as inherently corrupting, could not accept policy agreement that arose from selfish or impure motives.

Critics of Douglas's notion of popular sovereignty, like President James Buchanan, dismissed the doctrine as nothing more than "a self-evident political maxim [that] the majority shall rule."[89] But this criticism is inaccurate, for it neglects Douglas's emphasis on local knowledge as a restraint on national majorities. Central authority should be limited, Douglas argued, because local authorities were better informed about local circumstances and thus could make more intelligent decisions than a distant, central authority. Douglas repeated this argument time after time in the decade prior to the Civil War. In 1850, he insisted that the pioneers who emigrated to the western territories, after becoming "familiar with [the land's] topography, climate, productions and resources, . . . were as fully competent to judge for themselves what kind of laws and institutions were best adapted to their conditions and interests, as we were who never saw the country, and knew very little about it."[90] It was unwise, he reiterated in 1854, for those in Washington "to violate the great principle of self-government . . . by constituting ourselves the officious guardians of a people we do not know, and of a country we never saw."[91] "Those of us who penetrated into the wilderness," Douglas explained to a North Carolina audience in the 1860 presidential campaign, "think that we know what kind of laws and institutions will suit our interests quite as well as you who never saw the country."[92]

Douglas placed a premium on what the Austrian economist Friedrich Hayek has celebrated as the "knowledge . . . of local conditions, and of special circumstances." Douglas's principle of popular sovereignty follows the Hayekian precept that "the ultimate decisions must be left to the people who are familiar with . . . the particular circumstances of time and place."[93] Douglas shares Hayek's skepticism of general rules; the primary difference is that whereas Hayek cautions against the theoretical knowledge of the scientist, Douglas is concerned to thwart the moral absolutes of the likes of Wendell Phillips.

Phillips worked from a fundamentally different conception of knowledge than Douglas. For Phillips, "anything right in principle had to be right in practice." Action that was right in Boston must be right in Charleston. If slavery was wrong in the North it must be wrong in the South. A man should look not to local circumstances but should keep "his eyes fixed on God, careless of effects or success." "The truth," Phillips believed, would be "the lever" to "uprear the universe." He judged his fellow men not by special circumstances or local conditions but by "an absolute rule of right."[94] Anything less was moral cowardice.

Douglas did not deny that such a thing as morality existed, although his much-quoted statement that he did not care whether slavery was voted up or down has often been interpreted this way. He did not dispute that there was a difference between good and evil; what he rejected was the notion that a central government was capable of making that determination. "When God created Man," Douglas told a gathering in Philadelphia, "he placed before him good and evil and endowed him with the capacity to decide for himself and held him responsible for the consequences of the choice he might make." Implicit in Douglas's creed, as historian Jean Baker observes, "was the belief that the people had been given the freedom to make mistakes."[95] People would learn to act morally neither through deductive reasoning nor through the tutelage of others, but rather through the trial-and-error experience of making decisions and constructing institutions.

Douglas's advocacy of leaving moral decisions to local communities has led some scholars to portray Douglas as a champion of untrammeled majoritarianism, heedless of individual rights.[96] But this view ignores the thoroughgoing libertarianism of Douglas's political views. He was, for instance, violently opposed to nativism in all its forms. "To proscribe a man in this country on account of his birthplace or religious faith," Douglas preached, "is subversive of all our ideas and principles of civil and religious freedom. . . . It is revolting to our sense of justice and right."[97] He consistently opposed all state legislation that aimed at reforming the private habits of the citizens of Illinois, whether relating to church-going, drinking, or schooling.[98] Government, in Douglas's view, had no business regulating individual morality.

It is true that when Douglas denied government's "right to force a good thing upon a people who are unwilling to receive it,"[99] it was Congress and the issue of slavery that he had first and foremost in mind. His insistence that "communities must be perfectly free to form and regulate their domestic institutions in their own way subject only to the constitution" did beg the question of what happened when a community tyrannized a minority in ways that were consistent with the constitution. And at no time, as Jean Baker points out, "did he discuss the possibility of conflict between repressive local government and individual rights."[100] But if Douglas never faced the contradictions within the doctrine of popular sovereignty, his laissez-faire politics reflected an acceptance of the Democratic axiom, expressed by Orestes Brownson, that "we may not control a man's natural liberty even for the man's good."[101] As the less philosophical Douglas preferred to put the point, "if a man chooses to make darnation fool of himself I suppose there is no law against it."[102]

Douglas worried more about the "meddling" of the central government than the meddling of local and state government because he believed, in a reversal of

Madison's position in *Federalist* No. 10, that national majorities were more prone to arbitrary and tyrannical action than local majorities.[103] Although their different historical experiences led them to different estimations of state and local power, Madison's and Douglas's visions of the proper scope of authority were ultimately quite similar. Central authority, both men believed, should be tightly limited. In *The Federalist*, Madison sought to limit central power through checks and balances within government; in the historic debates with Lincoln, Douglas strove to limit central power by decentralizing power to states and localities. Like Madison, Douglas felt government tended to an excess of legislation. Both men also shared a conviction that "expanding the sphere" would "widen the area in which white individuals could freely pursue their preferences."[104]

Throughout his life, Douglas harbored what one biographer described as an "inveterate suspicion of too much centralized authority."[105] At an early age he was attracted to the frontier, where "no man acknowledges another his superior unless his talents, his principles and his good conduct entitle him to that distinction."[106] But Douglas's distrust of authority differed fundamentally from the antipathy to authority expressed by Phillips. Douglas could never share Phillips's joy that parties and churches had "broken in pieces."[107] Douglas, like Madison, wished to minimize authority to maximize self-regulation. If authority left them alone, they would leave it alone. Phillips, on the other hand, like Paine, felt that oppressive institutions must be "broken" and torn down. It was not enough to limit authority to an agreed-upon sphere; rather authority must be continually assailed and exposed to the harsh light of truth and justice.

When civil war broke out, Douglas and Phillips reacted in radically different ways. Their differing responses are suggestive of their underlying cultural differences. Douglas, Lincoln's longtime political rival, lent the new president immediate and unqualified support. Douglas made it clear to all that although he was "unalterably opposed to the administration on all its political issues, he was prepared to sustain the President in the exercise of all his constitutional functions to preserve the Union, and maintain the government, and defend the Federal Capital." The South's secession meant that it was no longer "a party question nor a question involving partisan policy; it was a question of Government or no Government."[108]

But for Phillips, no government was much preferable to a morally corrupt government.[109] Phillips would not support Lincoln and the war effort unless he was persuaded that the government's cause was righteous.[110] In a speech announcing his support for the war, Phillips seemed ready to believe that such a reformation had occurred. "Yesterday the government had been an agreement with Hell," he told a cheering crowd, "today it was 'the Thermopylae of Liberty and Justice.'" Following the bloodletting, "the world will see under our banner all tongues, all creeds, all races,—one brotherhood,—and on the banks of the Potomac, the Genius of Liberty, robed in light, four and thirty stars for her diadem, broken chains under her feet, and an olive branch in her right hand."[111]

But his attitude to authority had not changed. If it was not perfect, it was perfidious. Phillips demanded leaders who would "dare all and do all," who had "wills hot enough to fuse the purpose of nineteen millions of people into one decisive blow," who would push for "justice—absolute, immediate, unmixed justice to the negro."[112] And when they failed to measure up to these exacting standards, he quickly assailed them as feeble, small-minded, even traitorous. In a private letter to Senator Charles Sumner, Phillips attacked Lincoln as "a timid and ignorant President," who "is doing twice as much today to break this Union as [Jefferson] Davis is." Publicly, he declared that if Lincoln had been a "traitor," he could not have been of more aid to the South. So strident did Phillips's attacks on Lincoln become that by the end of the war even William Lloyd Garrison was accusing Phillips of being "a bayoneter of presidents."[113]

For any political leader, there is a world of difference between the egalitarian like Phillips who calls for the government to undertake a massive crusade on behalf of some moral cause—whether to abolish slavery, poverty, disease, discrimination—while at the same time harshly criticizing all governmental actions that fail or stray from the path of justice, and the individualist like Douglas who is reluctant to ask the government to solve societal problems but is willing to support governmental authority within its legitimate sphere. The individualist pattern balances limited support with low demand; the egalitarian pattern creates a mismatch between high demand and low support.[114] The former pattern tends to result in limited government, the latter pattern is more likely to produce government that lacks legitimacy. Individualism and egalitarianism may both be antipower, as Huntington claims, but that should not obscure the basic and consequential differences between the cultural imperative to contain authority and the cultural imperative to condemn authority.

Exit, Voice, and Loyalty

Albert Hirschman's categories of exit, voice, and loyalty provide a helpful way of elaborating the distinction between egalitarian, individualistic, and hierarchical attitudes to authority.[115] The categories of exit, voice, and loyalty are intended to describe modes of responding to decline, disappointment, or failure.

Hirschman defines "exit" as "withdrawal from a relationship with a person or organization." Dissatisfaction with a product or a decision is dealt with by switching to another firm's products or by leaving an organization. Exit reflects a preference for individual over collective solutions, for "flight rather than fight."[116] Escaping authority is preferred to confronting authority.

"Voice" is defined as "any attempt at all to change, rather than to escape from, an objectionable state of affairs." Voice is asserted when a firm's customers or an organization's members express their dissatisfaction "directly to management or to some other authority to which management is subordinate or through general protest addressed to anyone who cares to listen."[117] Grievances are met with collective action that directly confronts those in authority, challenges their credentials, and/or exposes their duplicity.

Finally, some organizational failures are met with neither individual flight nor collective resistance but with "loyalty." In the case of loyalty, members stick it out through the bad times, trusting that those in positions of authority will be able to turn things around. Voice is regarded as disrespectful, uppity, or at least bad form; exit as dishonorable if not treasonous. "Our country, right or wrong," Hirschman explains, is the paradigmatic expression of this pattern.[118]

It does not take a great leap of imagination to see that these three modes of response described by Hirschman can be made to correspond roughly with the political cultures of individualism, egalitarianism, and hierarchy. Exit is the option of choice for individualists, egalitarians incline toward voice, and hierarchists opt for loyalty. The cultural connection seems clearest in the case of the affinity between competitive individualism and exit. Hirschman himself explicitly associates exit with American individualism in applying his categories to the realm of American ideology. Building upon Louis Hartz and Frederick Jackson Turner, Hirschman portrays exit as the paradigmatic American response to decline. "The United States," Hirschman suggests, "owes its very existence and growth to millions of decisions favoring exit over voice." It was "founded on exit" and has "thrived on [exit]." Echoing Hartz's indictment of "irrational Lockeanism," Hirschman laments our "national infatuation with exit," and he characterizes "the [American] belief in exit as a fundamental and beneficial social mechanism" as "unquestioning."[119]

Hirschman sees the western frontier (the "gate of escape," in the language of Frederick Jackson Turner) as providing "a paradigm of problem-solving" for Americans, even for those social groups for whom the opportunity to "go West" was more myth than reality. Rather than making do or fighting back, Americans more than most people were able to think about solving problems through physical flight.[120] The pioneer manifests the individualist's typically exuberant optimism; as one British visitor observed, the pioneer "has always something better in his eye, further west; he therefore lives and dies on hope."[121] Exit westward offered opportunities for self-advancement—a better farm, greater wealth, higher status—as well as freedom from intrusive authority.

Stephen Douglas's westward move from Vermont to Illinois exemplifies the exit of the individualist. Douglas, explains historian Robert Johannsen, was one of many young men who left Vermont in the 1830s "in the face of declining opportunities for professional advancement." Finding himself in "straitened pecuniary circumstances" and frustrated by slow advancement, the twenty-year-old Douglas determined upon "removing to the western country" and seeing "what I could do for myself in the wide world among strangers." Douglas soon settled in Illinois, where "the lure of quick, easy profit" captured his imagination. He wrote home to his family,

> If you have an opportunity of selling your farm immediately, be sure and embrace it, and all come on at once. . . . There are greater bargains to be had now and better locations than there will be at any future time. Land far superior to yours in beauty, in fertility, in location and in every other respect may be had at one dollar & twenty five cents per acre. And these very lands five or ten years hence will be worth from five to fifty dollars per acre.

Exit created the opportunity to get rich. As he informed his stepfather, "With the capital you can bring with you, you will be able to make a fortune in Lands without laboring any yourself."[122]

Douglas's move westward is typical of the individualistic exit of the mid-nineteenth century pioneer. But all exits are not alike. Exit has a quite different meaning for the radical egalitarian who seeks to "opt out of [the competitive marketplace] by seeking to hedge off a commune of like-minded souls."[123] If the individualist uses exit as a "status escalator," the egalitarian's exit follows the injunction, "Come out of [Babylon], my people, that ye be not partakers of her sins, and that ye receive not of her plagues."[124] There is a world of difference between the exit of the individualistic pioneer and the exit of the egalitarian "come-outer," between exiting one product for another and rejecting one system for another. The individualist exits to seek out more advantageous market conditions; the egalitarian exits to preserve individual purity.

The difference between individualistic exit and egalitarian exit is the difference between "freedom of choice" plans backed by the Reagan and Bush administrations and the "free schools" established in the 1960s and early 1970s. Individualistic supporters of "free choice" believe exit will allow students and parents to choose the product they most prefer and will improve the performance of public schools by exposing them to market forces. Egalitarian proponents of "free schools" favor exit from authoritarian schools and the formation of alternative, "countercultural" schools that reject formal authority structures and encourage greater student participation in governance and choice of curriculum. Both schemes are animated by an antipathy to hierarchical bureaucracy, but the difference is that individualist supporters of the freedom to choose focus their criticism on the inefficiency of school bureaucracies, whereas free school proponents believe that all formal authority is coercive, including the teacher's authority over the student. Both sides agree that bureaucracy stifles initiative, but whereas individualists wish only to limit the size and scope of bureaucratic authority, egalitarians prefer to abolish bureaucratic roles and expertise altogether.[125]

Individualists do not share the egalitarian hostility toward expertise. Rather than condemn expertise as unjustifiable inequality, individualists prefer to cultivate expertise as a means to limit the reach and scope of formal authority (if they don't know how you're doing it, they can't tell you what to do) and thereby extend the sphere of self-regulation.[126] Egalitarians, in contrast, are engaged in a continuous battle with the division of labor and make every effort to eliminate differences in position, status, or knowledge. Job rotation, task sharing, and internal education are among the methods used by egalitarian groups to "demystify" and redistribute on an equal basis the possession of information and power.[127]

Radical egalitarians typically combine exit from the system with a stridently critical voice.[128] Among abolitionists, for instance, the renunciation of evil did not absolve them from the duty of denouncing evil. Even after withdrawing from corrupt proslavery institutions, abolitionists continued to harshly criticize those institutions. The Garrisonian abolitionist Stephen Foster combined exit and voice in a particularly dramatic and combative fashion. Like most other Garrisonians, Foster denounced the "corrupt and profligate" churches—"the American church and clergy," he insisted, "were thieves, adulterers, manstealers, pirates and murderers"—and "came out" from what he deemed an oppressive institution. His renunciation of church membership did not stop him, however, from attending church services and giving the unsuspecting congregation an uninvited lecture on antislavery. To silence Foster's disruptive voice, members of the congregation were forced to physically remove (carry, drag, hurl) him from the church.[129]

Most abolitionists did not go as far as Foster, but all continued to harshly criticize the proslavery institutions from which they had withdrawn. Some observers felt that to be consistent, the abolitionist should leave the country altogether; "he had no right to stay and criticize, for by his very presence he was participating in the social sin he found so monstrous." Wendell Phillips defended the abolitionist position by arguing, "mark the difference between *speaking* and voting or taking office—I speak for the changing of laws, all the time washing my hands of them."[130] Voice would remain pure and strong only if it originated from outside the system. Only then, as Foster wrote to Phillips, can we carry out our business, which "is to cry 'unclean, unclean—thief, robber, pirate, murderer'—to put the brand of Cain on every [political leader]."[131] Only the voices that exited from a corrosive system could speak truth to power.

Phillips and Foster stood outside of established organizations but they did not stand alone. They found support and fellowship amid other outsiders devoted to the cause of abolition. Within this redeemed community of like-minded souls, Phillips experienced the "sublime" feeling of clasping hands with other human beings and feeling their "hot blood" in his hands. Abolitionists, Phillips felt, had "their hearts grappled together as if by hooks of steel." Another Garrisonian referred to his fellow abolitionists as a "battalion of the sacramental host of God's elect." The abolitionists viewed themselves, explains historian James Brewer Stewart, as "a harmonious body of saints, a churchly assembly that conducted its internal relationships by the same [noncoercive] values that it sought to enforce throughout America."[132]

The egalitarian voice remains harshly critical of established authority because its position outside the system means that it need not assume responsibility for decisions reached and because reminding those within the collectivity of the system's corrosive evils helps to keep together a group that rejects authority.[133] At the same time, the uncompromising stridency of the radical egalitarian voice provokes the hostility of the larger society and thus keeps egalitarian groups on the fringes of established society. The result is a self-sustaining loop in which the harsh egalitarian voice is both cause and consequence of exit from the system.

If the radical egalitarian pattern is characterized by the simultaneous use of exit and voice, the individualist is more likely to regard exit and voice as alternatives. Individualists either "vote with their feet" (exit) or vote at the ballot box (voice). The individualistic movers cease to concern themselves with the deficiencies of the community or organization once they have left it,[134] and individualistic voice is tempered by virtue of its coming from within the system rather than from without. If voice fails to produce the desired changes, exit can be used, and if exit is used, voice is no longer necessary. If voice does produce the desired changes, exit is forestalled. For the individualist, then, voice does not fuel exit, as it does for the egalitarian, but rather forestalls exit.

All voice, it seems, is not the same.[135] There is a world of difference between voice that moves along established channels of communication (voting, petitions, campaign contributions, etc.) and voice that emanates from outside of those channels. The harshly critical tone of voice used by abolitionists, who rejected voting and officeholding as corrupting, was radically different from the more compromising tone of voice used by members of the established nineteenth-century political parties. What distinguishes individualism from egalitarianism then is not the presence or absence of voice but rather the form that voice takes.

There is also an important difference between individual voice and collective voice. For the individualist, voice, like exit, is a means of expressing individual preferences. Voting, petitioning, contacting government officials, writing letters, contributing to campaigns—all are means of relaying the preferences of individuals to governmental officials. For the egalitarian, voice is a collective activity. If voting is not corrupting, then it is certainly insufficient to sustain a truly participatory democracy. The collective voice is not to be the sum of individual voices, as individualists seem to posit, but is to represent a public whole that transcends the individual parts. A system based on individual voice, egalitarians insist, allows the triumph of the strong. To hear the voice of the public and the voice of the weak and inarticulate requires collective mobilization.

When Hirschman posits that the recourse to exit "will often diminish the volume of voice," it is collective voice he seems to have in mind. As examples of the "voice-weakening effect of exit," Hirschman cites Frederick Jackson Turner's thesis that the possibility of westward migration explains the lack of a militant working-class movement in the United States as well as Hartz's comment that "physical flight is the American substitute for the European experience of social revolution."[136] The availability of individual exit, according to this interpretation, served as a "safety-valve" that enabled the discontented to move beyond the reach of the intrusive hand of central authority instead of directly challenging authority through collective action. The frontier thus undermined the possibilities for collective mobilization to redress grievances. Escaping authority substituted for confronting authority.

The defiance of collective voice is not the only alternative to the escape of individualistic exit. In the early republic, many of the most conservative and hierarchical elements regarded the western exodus with great suspicion. Loyalty was their preferred mode of response. Members of the Federalist party, for instance, although supportive of the purchase of New Orleans and guaranteed American navigation of the Mississippi that resulted from the Louisiana Purchase, did not generally share the Republican enthusiasm for westward expansion. Pioneers, complained Timothy Dwight, "are impatient of the restraints of law, religion, and morality; [and] grumble about the taxes, by which Rulers, Ministers, and Schoolmasters, are supported." This "vast wilderness world" was widely perceived by conservatives as a threat to established order and authority. "By adding an unmeasured world beyond [the Mississippi River]," Fisher Ames worried, "we rush like a comet into infinite space."[137] It was essential, Federalists agreed, that the government's means of coercion keep abreast with any increases in territory.[138]

Conservative members of the Whig party also voiced uneasiness with rapid territorial expansion. Governmental authority, Daniel Webster reasoned, was "very likely to be endangered . . . by a further [territorial] enlargement," and the Georgia

Whig Alexander Stephens voiced his fear that rapid expansion would "sweep over all law, all order."[139] Whigs did not wish to halt westward migration altogether but they did oppose land policies designed to attract people to the subsistence frontier. "Believing that law and social order were agents of civilization," explains historian Daniel Feller, Whigs "opposed policies that drew men beyond their restraining influence. Whig rhetoric described the frontier as the seat of lawlessness and disorder, and the squatter not as a heroic pathfinder but as a renegade and outcast."[140]

Hierarchically inclined Federalists and Whigs opposed the settlement of western territories for the same reason that individualistic Democrats and Republicans endorsed it: to further their notion of desirable authority relations. Both cultures believed that extending the base of the pyramid horizontally would prevent it from growing vertically. For individualists this was a positive social good; for hierarchists it was a self-evident social ill. Jeffersonians and Jacksonians believed in the "dispersion of authority over wide spaces"; Federalists and Whigs adhered to a philosophy of "concentration of national authority in a limited area."[141] Thus Hamilton could warn ominously that territorial expansion would inevitably be "attended with all the injuries of a too widely dispersed population," while President James Polk, a Democrat, could praise expansion for allowing America to avoid "the tendencies to centralization and consolidation" that plague other less fortunate nations.[142] The divergent evaluations of territorial expansion offered by these two political leaders reflect the radically different conceptions of authority harbored by adherents of individualistic and hierarchical political cultures. It is to an elaboration of the hierarchical cultural bias that I turn next.

6

Hierarchy in America

This chapter takes issue with the Hartzian thesis that there has never been a significant hierarchical political culture in the United States. My argument, it is worth stressing at the outset, is not that hierarchy is powerful in the United States. Nothing so perverse is intended. Relative to other nations of the world, America is distinguished by the weakness of hierarchy. But relative weakness is not the same thing as nonexistence. Skipping the feudal stage, as Hartz suggests, did weaken the social basis for hierarchical values. But hierarchy as a way of life is not limited to feudal social relations. Hierarchy finds support wherever social relations are highly stratified and the group has sanctions over the individual, whether this be in the army, the patriarchal family, the modern corporation, or on a large plantation.[1] If such hierarchical values as deference, authority, noblesse oblige, obedience, and sacrifice of the parts for the whole have never had the same level of appeal in the United States that they have had in India, Japan, Great Britain, and other countries, neither have they been wholly without advocates.

Recent scholarship detailing the structures of everyday life in American colonial history has made it increasingly evident that America was born neither free nor modern.[2] "The old story of individualism and free enterprise coming with the first boatload of English colonists," as Joyce Appleby notes, "is no longer credible."[3] The story has turned out to be much more complex than Tocqueville or Hartz would have had us believe. Historians have documented at length the ways in which "early American society was profoundly conservative, even profoundly European in character."[4] Moreover, New World conditions did not always and ineluctably erode Old World hierarchical relationships. Instead, in the century before the Revolution, American communities increasingly came to resemble Europe in terms of population density, stratification of economic classes, and specialization of labor. Some historians have gone so far as to see in these developments evidence of a "feudal revival";[5] others have talked more modestly of a phase of "social replication" in which "strategically placed elites" attempted to "recreate British society in America."[6] In this interpretation, the American Revolution, far from being the logical outcome of internal colonial development, is better understood as a "social accident."[7]

The cumulative effect of this historiographical revisionism has been to present a picture in which "the original colonists, their children, and grandchildren look more and more at home in the Old World."[8] This revisionism is an important corrective to the Tocqueville-Hartz view, although it is sometimes in danger of replacing a monolithic New World individualism with a monolithic Old World traditionalism. Seventeenth-century England was more than just hierarchy, stratification, deference, and patriarchy. According to Richard Ashcraft, "the late seventeenth century was a high-water mark of democratic participation, not achieved again in England until the mid-nineteenth century."[9] Indeed if we are to believe Alan Macfarlane's account, "the majority of people in England from at least the thirteenth century were rampant individualists, highly mobile both geographically and socially, economically 'rational,' market-oriented and acquisitive."[10] Important, too, as Michael Walzer shows, were the Puritan saints who invoked collective discipline and revolutionary zeal to attack traditional hierarchy.[11] Americans who desired to recreate British society still had to decide which of the rival British ways of life they wished to emulate.

If not all "English ways" were supportive of hierarchy, many were and many of these were successfully transplanted to the New World.[12] This transplanting of hierarchical values and social relations occurred throughout colonial America, but nowhere was the transmission and reinforcement of a hierarchical culture more evident than among Virginia's gentlemen-planters.

Virginia's Gentlemen-Planters

The argument of *The Liberal Tradition in America*, Hartz tells us on the opening page, "is based on on what might be called the storybook truth about American history: that America was settled by men who fled from the feudal and clerical oppressions of the Old World."[13] Although this "storybook truth" may be valid with respect to such groups as the Puritans and Quakers,[14] it ill fits the experience of the Anglican gentry recruited to Virginia by Governor William Berkeley beginning in the 1640s.[15] The names that dominated Virginian society and politics in the colonial era (Bland, Burwell, Byrd, Carter, Lee, Mason) were largely from eminent and well-connected English families. Like Berkeley himself, they tended to be younger sons who stood little chance of inheriting an estate in England. Their aim was less to escape from traditional hierarchy than to recreate on American soil the hierarchical system from which they had been excluded at home. They sought to emulate English ideals, not to reject them.[16]

Hartz, to be sure, does address the anomaly posed by the South to his thesis of liberal consensus. He picks up the South in the decades immediately prior to the Civil War when the likes of George Fitzhugh, Beverly Tucker, Edmund Ruffin, and Albert Bledsoe were developing a self-conscious, patriarchal defense of slavery as a "positive good." The "feudal dream" of these thinkers, Hartz argues, is best understood as a "fraud," a desperate pose tried on by bourgeois individualists to defend the indefensible institution of slavery against the abolitionists' assault. The effort to graft a feudal superstructure upon a Lockean base was, in Hartz's view, a recipe for "philosophic pain" as well as failure.[17]

Granted that the Southern paternal defense of slavery was often forced and frequently resorted to transparent subterfuge, none more transparent than when the young Mississippi sociologist Henry Hughes dressed up slavery as "waranteeism."[18] But hierarchical ideas were not something that Virginians discovered only in the nineteenth century. Seventeenth-century Virginia may have lacked feudal institutions, but its social relations had been at least as close to Sir Robert Filmer's *Patriarcha* as to John Locke's *Two Treatises*. Viewed against the backdrop of Virginia's seventeenth century past, the hierarchical language used by Virginians in the nineteenth century appears less fraudulent than Hartz would lead us to believe.

In contrast to Puritan Massachusetts, which attracted immigrants mostly from the middle strata of English society, Virginia tended to draw people more from both higher and lower ranks. Servants had been a small minority (under 25 percent) of the Puritan migration, but over three-fourths of those who migrated to Virginia in the mid-seventeenth century came as indentured servants. Where the early Puritan leaders actively discouraged the migration of noblemen (as well as servants), Berkeley did everything he could in his thirty-five-year tenure as royal governor to attract men of "good families." And when these notables arrived, Berkeley promptly promoted them to high office and granted them large estates in the (largely successful) hope of creating in Virginia "a Royalist utopia dominated by ideals of honor and hierarchy." The greater stratification of Virginian society compared to the Puritan colonies was not just a demographic or economic accident but a conscious product of a cultural vision brought over from England.[19]

Virginian elites and masses came largely from the south and west of England, traditionally the most conservative areas of England. Land distribution in southwest England was considerably more unequal than in the east. Where East Anglia had a high proportion of independent yeoman, the countryside of southwestern England was "dominated by the great estates of the gentry, with their environing clusters of small houses inhabited by tenants and subtenants." In a study of the links between New World and Old World cultures, David Hackett Fischer draws attention to the "many strong links between the character of the south and west of England and the culture of Virginia." Both regions, he finds, "were marked by deep and pervasive inequalities, by a staple agriculture and rural settlement patterns, by powerful oligarchies of large landowners with Royalist politics and an Anglican faith."[20]

The strong connection with Stuart England was evident in Virginia's religious practices and beliefs. From the middle of the sixteenth to the middle of the seventeenth century, Virginia's religious life was dominated by an Anglican orthodoxy that was "ceremonial, liturgical, hierarchical, [and] ritualist."[21] The architectural plan of Virginia's leading churches was designed to dramatize hierarchy:

> Great oak-walled pews reserved for magistrates and leading families stood at the front of each arm of the church, delimiting the central area. High within the space thus defined stood the pulpit under its grand, ornately canopied sounding board. . . . In tiered hierarchy beneath the pulpit were the desk from which the scriptural

lessons were read, and the clerk's desk from which the parson's lay assistant "lined out" the psalms for communal intonation. Behind pulpit and desks, and hence symbolically and dramaturgically in the lowest position, was the gallery that a small number of slaves would enter by a steep narrow stairway just inside the south door.[22]

The liturgy and church plan, Rhys Isaac concludes, "combined to offer a powerful representation of a structured, hierarchical community."[23]

The principle of hierarchy suffused every aspect of Virginian colonial society. Even sports followed a deliberately created and actively enforced hierarchy. Those of a "lower estate" were forbidden by law and custom to compete in sports that were considered the exclusive privilege of gentlemen, such as horse racing. Heavy fines were meted out to those who failed to keep their station. Even hunting had a distinct hierarchy. "Virtually every male in Virginia," Fischer explains, "could be ranked according to the size of animals that he was allowed to kill for his pleasure." At the top of the hierarchy was stag hunting; lesser gentry chased fox; yeoman and parish clergy went after hares, rabbits, and small vermin; husbandmen, laborers, and apprentices amused themselves by murdering defenseless geese, cocks, and chickens.[24]

Hierarchy also shaped Virginians' ideas of punishment. The punishment a person received for an offense was dependent upon one's social status. If a convicted felon was poor and illiterate, he was condemned to the gallows. Literate felons, in contrast, were permitted (except for the most serious crimes) to plead "benefit of clergy," which allowed them to escape with a branding on the thumb. Gentlemen-felons were sometimes branded with a "cold iron" that would leave no mark. Unequal treatment for unequal stations was sanctioned not just in written law but also in unwritten custom. According to these unwritten rules, "violence was thought to be the legitimate instrument of masters against servants, husbands against wives, parents against children, and gentlemen against ordinary folk. But violent acts by servants against masters, or common folk against gentle folk was followed by savage punishment." This type of discrimination by social rank was much less pronounced in the Puritan colonies, where the aim of punishment was primarily maintaining group solidarity and not, as it was in Virginia, maintaining hierarchical distinctions.[25]

Hierarchical ideals intruded in more subtle ways, such as in styles of architecture. The "great houses" of the wealthy planters were characteristically made up of an imposing centerpiece flanked by subordinate symmetrical wings. This design, Isaac explains, expressed "a strong sense of gradations of dominance and submission. . . . The prominence of an elevated center, or 'head,' to which all other parts, or 'members,' were subordinate silently reinforced the dignity and claims to obedience of the gentleman who was styled 'the head' of the household." That these buildings gave expression to their hierarchical social values was not lost on contemporaries, one of whom praised Colonel Landon Carter's country seat, Sabine Hall, as "an edifice [with] every part . . . in its due place and fit station, neither above nor below its dignity."[26]

Notions of place and station were deeply imbedded in Virginian society. "One sat in church according to one's status," write the historians Darrett and Anita Rutman, "and someone pushing into a place to which he or she was not entitled sent reverberations through society."[27] "There is no truer Emblem of Confusion," explained a Royalist writer in neighboring Maryland, than an attempt by servant or subject "to be equal with him, from whom he receives his present subsistence." "Ranging in contrary and improper spheres" constituted the very definition of disorder.[28] Virginian institutions, whether the church, court, muster, or election, "all displayed principles of descending authority—from those whose rank and accomplishment fitted them for rule, to those whose circumstances and limited understanding ordained that they should be ruled."[29]

Dress in colonial Virginia, as in England, was carefully calculated to signify rank. "The social distinction between gentlemen and 'simple men,'" Fischer finds, "expressed itself in almost every imaginable piece of apparel: hats versus caps, coats versus jackets, breeches versus trousers, silk stockings versus worsted, red heels versus black heels." The gentry pierced their ears for pearls or elegant black earstrings and adorned their persons with silver buckles, snakeskin garters, gold buttons, lace cross clothes, and silver hatbands; men of lower station wore none of these things.[30] The gentry's finery (e.g., a periwig, lace-ruffled cuffs, or fur muffs) conspicuously attested not only to their wealth but to the freedom from manual labor that was the sine qua non of a gentleman, just as the plain clothes of the common people advertised their dependence upon such labor.[31]

Clothing functioned as a constant reminder to Virginians of the relative status of society's members. Ostentatious display on the part of the gentry was designed to instill deference—even awe—in the lesser folk. Recalling his days as a farmer's son in the 1730s, the Reverend Devereaux Jarratt wrote that the sight of a periwig, a "distinguishing badge of gentle folk," would "so alarm my fears [that] I would run off, as for my life." People of his sort, he explained, were "accustomed to look upon, what were called gentle folks, as beings of a superior order. For my part, I was quite shy of them, and kept off at a humble distance."[32] The sight of a gentleman upon his fine horse or riding in his coach or chariot symbolized and reinforced a sense of the gentleman's superiority both on the part of the laborer or farmer, who had to look up to greet the impressive passerby, and on the part of the rider, who might condescend to greet the traveler beneath his foot.[33]

Deference on the part of inferiors was expected to be matched by a sense of noblesse oblige on the part of superiors. When lightning killed two horses belonging to a horse trader, the gentlemen present showed the "liberality" proper to their station by immediately compensating the unfortunate man for his losses.[34] Gentlemen were taught to "condescend" to their inferiors, which meant to treat them with kindness, decency, and respect.[35] A willingness to assume public responsibilities, whether political, civic, judicial, or military, with only scant remuneration, was a defining mark of the gentleman.[36]

These complex relations—deference and condescension, place and station, obligation and responsibility—were instilled at an early age. Among the rules of conduct that a Virginian youth might learn were "Reverence thy Parents," "Submit to thy Superiors," "Despise not thy Inferiors," "Be courteous to thy Equals,"

"Approach near thy parents at no time without a bow," "Never speak to thy parents without some title of respect—viz.—Sir, madam, etc. according to their Quality," "Always give the wall to thy superiors that thou meetest, or if thy walkest with thy elder give him the upper hand," "Give always place to him that excelleth thee in quality, age or learning," "Be not selfish altogether; but kindly, free and generous to others." At the age of thirteen, George Washington had been required to copy out 110 of such "rules of civility and decent behavior," among which were "In pulling off your Hat to Persons of Distinction, as Noblemen, Justices, Churchmen &c make a reverence, bowing more or less according to the custom of the better bred," "If any one far surpasses others, whither in age, Estate, or merit [yet] would give place to [one] meaner than himself the one ought not to accept it," "In speaking to men of Quality do not lean or Look them full in the Face, nor approach too near them," "In writing or Speaking, give to every person his due Title According to his Degree & the Custom of the Place," "Strive not with your superiors in argument, but always submit your judgment to others with modesty," "Let the ceremonies in Courtesie be proper to the Dignity of his place with whom thou conversest for it is absurd to act the same with a Clown and a prince."[37]

The patriarchal family, with its unequal but reciprocal obligations, was viewed as a model for all social relations. The relation between master and servant was seen to be analogous to the relation between husband and wife or, even better, parent and child.[38] In deriving the social and political from the familial, Virginian colonials showed themselves to be closer in many ways to Filmer's patriarchal vision than Locke's contractual vision. Even genealogically, the Virginian ruling elite was closer to Filmer. By the mid-eighteenth century, most of Virginia's leading families were related in some way to Mary Horsmanden Filmer Byrd, whose first marriage had been to the son of Robert Filmer.[39]

Many among the Virginian elite proudly referred to themselves as patriarchs. So, for instance, William Byrd II (son of the aforementioned Mary Byrd) could write: "Like one of the Patriarchs, I have my Flocks and my Herds, my Bond-men and Bond-women. . . . I must take care to keep all my people to their Duty."[40] Among those usually counted as part of the patriarch's family, and hence under the patriarch's charge, were not only members of the nuclear family but all those persons who were dependent upon him, including clerks, tutors, impecunious friends, visiting relatives, overseers, servants, and house slaves. So, for example, General George Washington, while at Valley Forge, referred to his wife, servants, aides, staff, and visitors as his "family."[41] The patriarchal ideal was not only in the minds of the patriarchs but was supported in law. The murder of a father by a son, a husband by his wife, or a master by his servant was treated as treason, not homicide. Masters were required by law, moreover, to "detain and keep within their hands and custody the crops and shares of all freemen within their families," so as to ensure the payment of taxes.[42]

The gentry's patriarchal ethos was accompanied by a disdain, albeit ambivalent, for the transactions of the marketplace. Money, writes Fischer, was something "which Virginians liked to have, but hated to handle." "Half a crown," boasted William Byrd II, "will rest undisturbed in my pocket for many moons together." Virginia's gentlemen-planters, Fischer finds, "repeatedly expressed an intense contempt for trade, even as they were compelled to engage in it." In their ambivalence toward the market they were little different from the English gentleman, who could "scorn base getting and unworthy penurious saving" and yet express a "desire . . . to lay up somewhat for my poor children."[43]

The cultural similarities between the gentry of colonial Virginia and England is attested to by the affinity each felt for the other. "English aristocrats who came to the New World," observes Fischer, "instantly recognized a cultural kinship with the great planters of Virginia." Lord Adam Gordon, for instance, the first son of the second Duke of Gordon, remarked approvingly that the "topping families" of Virginia had all been founded by "younger brothers of good families of England." Were he to live in America, he concluded, Virginia "in point of company and climate would be my choice."[44] Virginians, for their part, did all they could to emulate the styles and customs of the British.

Accenting the similarities in kind between the cultural patterns of English society and colonial Virginia is not to deny the significant differences in degree that existed between the mother country and the colony. Virginia was less stratified than Britain, particularly at the top end. The social distance between gentry and commonalty was comparable in the two countries, but Virginia lacked the finely graded distinctions among the gentry that characterized British society. There was no distinction, for instance, between gentlemen (*nobilitas minor*) and lords (*nobilitas major*); indeed there were no lords (dukes, marquesses, earls, viscounts, barons).[45] The truncated nature of hierarchy in Virginia was evident, too, in the colony's resistance to having a colonial bishop.[46]

The relatively weaker grip of hierarchy in Virginia also manifested itself in clothing. No Virginian gentleman could match the elegant wardrobe of a Sir Walter Raleigh, who wore jewels on one occasion that were said to be worth 30,000 pounds, a sum that exceeded the capital assets of some American colonies. And while the gentry did their best to emulate the style of their British counterparts, lower class Virginians were, to the dismay of many gentlemen, less inclined to accept clothing deemed suitable for their station. One Virginia planter complained, for instance, of a cowkeeper in Jamestown who went to church in "fresh flaming silk" and of a collier's wife who wore a "rough beaver hat with fair pearl hatband, and a silken suit." Climate, too, played a role in modifying Old World fashions. In winter, according to one English gentleman who visited Virginia in 1732, the Virginian gentry dressed "mostly as in England and affected London Dress and wayes." But during the hot, humid summers, he found that planters often jettisoned the full formal regalia in favor of cooler and simpler clothing.[47]

Colonial Virginia, in short, was a more fluid, more disorderly, and less settled society than was England.[48] Although by the mid-eighteenth century eastern Virginia was no longer physically on the frontier, it was still the case, as Bernard Bailyn points out, that "every section of the land, no matter how long settled and sophisticated, had direct and continuous contact with the wilderness."[49] Moreover, the swamps and hot summers of Virginia meant exceptionally high death rates, particularly for immigrants who lacked immunity against fatal diseases such as malaria.[50] In addition, Virginia's major crop, tobacco, exhausted all but the richest soil, thus creating a continual pressure for westward movement to clear new areas.[51] In attracting farmers to the New World, promoters stressed the availability of cheap, plentiful land and the opportunity for economic advancement. After serving out their term of indentured servitude, these men quickly sought to realize the promised opportunities by moving westward.[52] The geographic mobility of the American population troubled many of the established slaveholders along the eastern seaboard. One planter, for instance, condemned the "miserable, selfish, avaricious and dastardly spirit of emigration." "Every virtuous and patriotic citizen," he insisted, "should feel himself bound to the soil which gave him birth, which has been the home of his father, and which contains the bones of his ancestors."[53] Maintaining hierarchy in the face of such mobility was a difficult task.

Nonetheless, by the middle of the eighteenth century eastern Virginia had become, by American standards at least, a settled, stratified community. Mobility was possible, but the planter elite's increasing stranglehold on the most productive lands (along with extensive and calculated intermarriage within the elite) along the Chesapeake meant that dramatic reversals in fortune were uncommon. Between 1650 and 1775, writes Fischer, "few men . . . succeeded in rising above the social order in which they were born." Something of the continuity over time in the dominance of the leading planters is indicated by the fact that as late as 1775 every member of Virginia's Royal Council was descended from a councilor who had served in 1660.[54]

Virginia's cultural development was, of course, never static. If social relations became increasingly hierarchical throughout the seventeenth century, the eighteenth century was witness to a number of challenges to the traditional establishment that left hierarchy significantly weaker at the close of the century than it had been at the opening of the century. These challenges to hierarchical orthodoxy came from both communitarian and individualistic sources. The effect in either case was similar: to undermine the deference to authority upon which hierarchy rested.

Among the first critical challenges to the traditional hierarchical order was the rise of the "New Light" Separate Baptists in the 1760s. In place of a stratified order of unequals, the Baptists offered a tightly bound community of equals "called out of the world . . . to live together, and execute gospel discipline among them." "In sharp contrast to the explicit preoccupation with rank and precedence that characterized the world from which they had been called," explains Isaac, the Baptists "conducted their affairs on a footing of equality." Their emphasis on equality was nowhere more evident than in the Baptist requirement of unanimity in important church elections. Displays of finery, which to the ruling gentry were the hallmarks of position, were to the Baptist a mark of sinful pride. Baptists preferred that their clothing, like their chapels, be simple and unadorned. The evangelical assault on deference was evident, too, in their "readiness to send out the humblest of men, including slaves, to expound Scripture, declaring them qualified by a 'gift' of the Holy Spirit."[55] This explains in part why the gentry at the top of the social ladder tended to resist the Baptists while those at the bottom, including slaves, were more receptive to the evangelical message.

An alternative, secular impulse, best exemplified in the person of James Madison, also helped to delegitimate the claims of hierarchy. In contrast to the Baptists, who offered "a community within and apart from the community,"

Madison offered a more individualistic alternative to hierarchy. Madison sided with the evangelical separationists in the fight against using public monies to provide for support of the Episcopal church because he believed, with the Baptists, that religion should be based on voluntary association. For Madison, individual freedom to worship as one pleased was the paramount concern. For Baptists, the primary aim was a life of holy fellowship. But both groups, as Isaac points out, "were united against the traditional conception of community as a hierarchy of head and members whose corporateness was symbolically expressed in an obligation to provide for public worship."[56]

The American Revolution also helped to subvert traditional authority relations. Although the patriot movement in Virginia was originally led by the gentry and "initially tended toward a revitalization of ancient forms of community [and] traditional authority," it soon became "a vehicle of popular assertion."[57] The war against the British offered the gentry a chance to assume visible roles of exemplary public leadership, but in encouraging the populace to reject British authority they inadvertently invited indiscriminate attacks on all traditional authority, particularly on those who had emulated the British in so many ways. The rejection of British finery in favor of homespun cloth dramatizes the manner in which a war against a distant hierarchy became subversive of domestic hierarchical relations.[58]

As substantial as these cultural changes were, particularly in the area of religion, one should not exaggerate the extent of the transformation. Even with these substantial jolts to hierarchy, as Isaac admits, "much remained the same—especially in fields and houses. Patriarchy, and its adaptation, paternalism, continued to be a powerful principle in a thoroughly agrarian society."[59] Politics, too, remained largely in the hands of an eastern slaveholding elite; indeed their grip on power remained strong up until the state constitutional convention of 1850. As the eighteenth century came to a close, hierarchy was bowed but not beaten. What emerged in the decades after the revolution was a more culturally pluralistic universe in which "the vivid culture of the gentry, with their love of magnificent display, had now to coexist with the austere culture of the evangelicals."[60]

Viewed against the backdrop of Virginia's seventeenth- and eighteenth-century past, the nineteenth-century paternal apologia for slavery that Hartz sees as a deceitful attack on the South's basically liberal tradition looks more like a consistent expression of the hierarchical culture that the Virginia gentry had long promoted. Hartz argues that proslavery theorists who rejected Jefferson in the name of Burkean traditionalism were caught in a self-defeating contradiction because Jeffersonianism was the South's tradition.[61] But Jefferson was far from being the whole or even the main of Virginia's cultural tradition. From the outset, especially for large plantation owners, patriarchy had played at least as large a role in everyday life as notions of individual contract.

When Hartz suggests that "in its Jeffersonian youth the South itself had considered slavery bad,"[62] he overlooks those early Virginians like William Byrd, whose hierarchical worldview readily incorporated the notion of slavery. For Byrd, explains Isaac, "slavery did not pose a problem [because] differences in kinds of social being and the state of total subjection itself were part of the nature of things."[63] There was no elaborate body of proslavery thought in the first century of Virginia's existence not because the gentry doubted that slavery was right but because few if any criticized slavery during this period. It is only with the coming of the American Revolution and the ascendancy of natural rights doctrines that Americans for the first time began to cast a more critical eye upon an institution that they had hitherto largely taken for granted.[64]

When slavery came under attack in the 1760s and 1770s, the institution's defenders, in the North as well as in the South, were able to draw upon their hierarchical heritage to defend slavery against the natural rights onslaught. "Absolute freedom," they argued, "is incompatible with civil establishments." "The nature of society . . . requires various degrees of authority and subordination." While it might be regrettable that all could not enjoy liberty to the same extent, yet life was "continually chequered with good and evil, happiness and misery." "Nature, governed by unerring laws which command the oak to be stronger than the willow, . . . has at the same time imposed on mankind certain restrictions, which can never be overcome."[65] Hierarchy was not, then, as Hartz would have it, a foreign doctrine injected into American life for the first time in the nineteenth century. Patriarchy, deference, noblesse oblige, and condescension were there from the outset. James Oakes is much closer to the mark when he describes one antebellum proslavery ideologue as "a nineteeth-century slaveholder with a seventeenth-century mind." "The legacy of the seventeenth century," Oakes explains, "was strong among those slaveholders who *continued* to view society as a complex organism in which reciprocal obligations defined the relations between the upper and lower classes. Antebellum conservatives struggled not to establish a new orthodoxy but to retrieve an old one."[66]

Scholars still differ about how representative this paternal worldview was in the antebellum years, and whether life on the plantation generated or undermined paternalism.[67] What is certain is that the South was not all of a piece. In frontier states like Alabama, where the political system was highly democratic, mobility was omnipresent, and settlers lacked hierarchical cultural traditions, individualism and entrepreneuralism predominated.[68] Hierarchy was more prevalent in those areas like eastern Virginia and lowland South Carolina, where plantations were larger and more stable, and where the political system was less democratic.[69]

That the hierarchical ethic expressed by many Southern planters in the antebellum era was more than a cynical effort to defend black servitude is suggested by their adherence to hierarchy in other areas of life. Among the most striking illustrations of this was the fervent Southern commitment to military professionalism. In a study of the American military tradition, Samuel Huntington finds that "the South gave military professionalism its only significant support in the pre–Civil War years."[70] In contrast to Jeffersonianism, which idealized a voluntary, citizen militia, and Jacksonianism, which attacked West Point and the officer corps as elitist, conservative Southern planters, particularly in Virginia, embraced the hierarchy, rank, and discipline that were required for a professional military.

The 1830s and 1840s witnessed an outpouring of military thought and writing which Huntington terms "the American Military Enlightenment." The intellectual sources for this creative outburst were, according to Huntington, "predominantly Southern." It was conservative Southerners who "dominated the serious thought and discusson of military affairs." Preeminent among the journals interested in promoting military professionalism was the Virginia-based *Southern Literary Messenger*, and the two outstanding military writers and thinkers of this era, Dennis Hart Mahan and Matthew Fontaine Murray, were both Virginians. So great was the South's interest in the military arts that by 1860 every Southern state, with the exceptions of Florida and Texas, had its own state-supported military academy patterned on the model of West Point. The first state to create such a local military academy was, not surprisingly, Virginia, which established the Virginia Military Institute in 1839.[71]

Southerners' greater interest in military affairs as well as their greater respect for professional military officers meant that talented Southerners were much more likely to choose the military as a career than were their counterparts in the North. The South, according to Huntington, "furnished a heavily disproportionate share of the cadets at West Point." This pattern was particularly striking for Virginia. In 1837, for instance, three of the four active generals, six of the thirteen colonels, and ten of the twenty-two highest ranking officers of the Army were from Virginia. Within the federal government, the military departments were dominated by Southerners up until the outbreak of the Civil War.[72]

The South consequently had a significant advantage over the North when civil war broke out in 1861. Fortunately for the North, somewhere between 40 and 50 percent of the Southern West Point graduates who were on active duty in 1860 remained loyal to the Union. Even so, the South initially had a decided advantage in its preparedness and the quality of its officer corps. The greater professional ethos of the South was evident throughout the war. Professional officers in the North, at least in the early stages of the war, were often passed over for political appointees. The South, in contrast, tended to promote from within the military hierarchy. According to Huntington's calculation, "sixty-four per cent of the Regular Army officers who went South became generals [and] less than 30 per cent of those who stayed with the Union achieved that rank."[73]

An important amendation to Hartz appears to be necessary. As evidence for America's "liberal absolutism," Hartz points to the "oblivion," the "vast and unbelieving neglect," that was the ultimate fate of "the feudal dreamers of the South" after the Civil War.[74] But this is one area in which Southern hierarchy most emphatically did not expire with the collapse of the Confederacy. For, as Huntington points out, it was Southerners' "gospel of professionalism" that "sparked the emergence of military professionalism as a concept and paved the way for the institutional reforms of the post–Civil War era."[75] The South, and particularly Virginia, has thus played a significant role in transmitting hierarchical values in American political culture. But the South's gentlemen-planters were hardly the only or perhaps even the most important carriers of a hierarchical tradition in the United States.

New England Federalists

The juxtaposition of the hierarchical or aristocratic culture of the Cavalier South and the entrepreneurial culture of the Yankee North has long been a staple of

American lore. This simple dichotomy is bound to fail, however, because both regions were culturally varied. An entrepreneurial ethos prevailed not just in the North but in much of the South, particularly in frontier areas like Alabama. And although hierarchy was strong in parts of the South, particularly in settled regions like tidewater Virginia and South Carolina, it was also vibrant in sections of the North, particularly in New England.

To view the slaveowners of Virginia and the Federalists of New England as being cut from the same cultural cloth may seem strange to those familiar with traditional stereotypes or impressed by regional differences in customs. Certainly this portrayal is at odds with Daniel Elazar's well-known scheme, which juxtaposes the "moralistic subculture" (defined roughly as a way of life that is participatory and communitarian) of New England with the "traditionalistic subculture" (a way of life that is deferential and communitarian) of the South. Elazar's framework does have the advantage of distinguishing the cultural patterns of New England from the "individualistic subculture" of much of the rest of the North. From the beginning, New England political culture had been, as Elazar posits, strongly collectivistic. But that collectivity, at least by the late eighteenth century, placed more emphasis on deference to authority than it did on popular participation. If distrust of widespread political participation and suspicion of an unregulated free market are the hallmarks of a traditionalistic political culture, as Elazar argues, then the Federalist party of New England was the embodiment of that traditionalistic, or, as I prefer to call it, hierarchical, culture.[76]

Although both Federalist New England and slaveholding Virginia were importantly hierarchical, there were significant differences between hierarchy in New England and hierarchy in Virginia. New England's Puritan heritage meant that New Englanders tended to be more conscious of the collectivity than were most Virginians. On the other hand, social relations in Virginia tended to be slightly more stratified and rank conscious than in New England. In terms of Douglas's typology, Federalist New England would be situated further to the right along the horizontal or "group" axis but at a slightly lower position along the vertical or "grid" axis than Virginia.

The shared commitment to hierarchical social relations that linked the conservative elites of New England and Virginia is highlighted by historian Larry Tise, who demonstrates the prominent role conservative New Englanders played in developing a defense of slavery as a positive good, a defense that has long been considered a distinctively Southern aberration from the liberal tradition. The earliest published defense of slavery in the colonies was written in 1701 by John Saffin, a justice on the Massachusetts provincial court. In response to an antislavery tract equating slavery with "manstealing," Saffin denied "that all men have equal right to Liberty." Such a doctrine was false because it denied "the Order that God hath set in the World, who hath Ordained different degrees and orders of men." Some, he insisted, were born to be "High and Honorable, some to be Low and Despicable; some to be Monarchs, Kings, Princes and Governors, Masters and Commanders, others to be Subjects, and to be Commanded; Servants of sundry sorts and degrees bound to obey; yea, some to be born Slaves, and so to remain during their lives."[77]

Three-quarters of a century later, Theodore Parsons, a Harvard senior and the son of a Congregational clergyman at Byfield, Massachusetts, wrote what Tise identifies as "the most astute proslavery refutation of Revolutionary ideology in the late eighteenth century." Parsons took aim at the "principle of natural equality, which is zealously contended for by the advocates of universal Liberty." "That Liberty to all is sweet I freely own," Parsons began; but he cautioned that "the nature of society . . . requires various degrees of authority and subordination; and while the universal rule of right, the happiness of the whole, allows greater degrees of Liberty to some, the same immutable law suffers it to be enjoyed only in less degrees by others." A "vast inequality," he continued, existed between "different individuals of the human species, in point of qualification for the proper direction of conduct." Human happiness is thus best secured not by "the enjoyment of equal liberty in each," but rather by vesting authority in those who are "found so far to excel others both in respect to wisdom and benevolence." Subordination and servitude were thus necessary parts of any just social order that recognized the principle of authority.[78]

A significant number of Southern slaveowners who articulated a paternalistic proslavery ideology in the antebellum era, Tise shows, were born or educated in the Northeast and were influenced by Federalism.[79] One such person was Henry Watson, Jr., who was born and raised in Connecticut but later moved to Alabama, apparently on account of poor health. Like most of those raised in a Federalist milieu, he spoke of society in terms of the "high and low" orders. He was appalled by the "thorough going democratic notions" espoused by Alabama's cotton farmers as well as the "vulgar style" of the evangelical Baptists and Methodists. Slavery, though, gave him few qualms. "Most people here," Watson explained to a northern friend,

> feel an attachment to their servants similar, in some respects, to that we feel for our children. We feed them, clothe them, nurse them when sick and in all things provide for them. How can we do this and not love them? They too feel an affection for their master, his wife and children, and they are proud of his and their successes. There seems to be a charm in the name of "master"—they look upon and to their master with the same feeling that a child looks to his father. It is a lovely trait in them. This being the case, how can we fear them?

Watson's "classically paternalist" worldview alienated him from many of his Alabama neighbors, who were wedded to an alternative "Herrenvolk" defense of slavery, but it connected him ideologically with the planters of Virginia's tidewater.[80]

The point is not that most or even many New England Federalists favored slavery, nor is it that an embrace of servitude followed inevitably from a commitment to hierarchy. Rather the point is that when New England Federalists defended slavery they did it in essentially the same language of paternalism and patriarchy used by Virginia's slaveholding elite. Committed to hierarchical stratification, neither group had a need to resort to the virulent racism that was required in order to reconcile equality and slavery: if all men were equal, slavery could be justified only by persuading oneself and others that blacks were not men. Even those Federalists hostile to slavery, who commonly saw it as a form of arbitrary despotism and not a paternal hierarchy of reciprocal obligations, still shared with conservative Virginians a belief in structured inequality and a deferential social order.

The preeminent place of hierarchy within the Federalist party in New England is brought out strongly in James Banner's rich study of Massachusetts Federalism, *To the Hartford Convention*.[81] "The basic motifs of Massachusetts Federalist thought," Banner finds, were "harmony, unity, order, solidarity." Federalists conceived of society "both as a structure of harmonious and mutually interdependent interests and as a collectivity in which individuals by occupying fixed places and performing specified tasks, contributed to the health and prosperity of the whole community."[82] A Federalist cleric explained,

> The social body is composed of various members, mutually connected and dependent. Though some may be deemed less honorable, they may not be less necessary than others. As the eye, the ear, the hand, the foot of the human body, cannot say to the other, I have no need of you, but all in their respective places have indispensable uses; so in the Commonwealth, each citizen has some gift or function, by which he may become a contributor to the support and pleasure of the whole body.[83]

Although each member of the collectivity was assigned a role to play in the whole, each was to keep to his proper sphere. Participation in the political sphere was ideally limited to the few who were virtuous, wealthy, and wise. "The Commonwealth's well-being," Federalists believed, "was directly proportionate to the participation of the better sort in public affairs." As the clergyman Jedidiah Morse taught, "distinctions of rank and condition in life are requisite to the perfection of the social state. There must be rulers and subjects, masters and servants, rich and poor. The human body is not perfect without all its members, some of which are more honorable than others; so it is with the body politic." It has pleased God, preached another, "to place mankind in different stations and to distinguish them from each other by a diversity of rank, power, and talent."[84]

Deference to one's betters, which was a defining feature of the Massachusetts Federalists' political culture, was instilled at an early age. It was common for children of Federalist families to bow before their parents and address them by a deferential "Honored Papa" and "Honored Mama." The "tightly regulated and patriarchal family," Banner shows, was seen as "the best place . . . to initiate this training in deference." Children were taught, Banner continues, "to honor and obey their elders, control their whims, and restrain their youthful energies in the general interest of the whole household. It was within the home that the growing child must first be exposed to the necessary distinctions among men and taught that only a few possess the privilege to rule."[85] Such "habits of subordination" were also to be inculcated in school. Schools were to teach people "to confide in and reverence their rulers." The whole purpose of education, according to another Federalist, was "to inculcate on . . . expanding minds the necessity of sub-ordination and obedience to their superiors."[86]

The Massachusetts Federalists cited previously closely resemble the pure type of hierarchy. Of course, not all of those who called themselves Federalists were this extreme (or articulate). Political parties are, after all, coalitions that house considerable ideological diversity. Other members of the party—John Marshall, for instance—tried to blend their preference for hierarchy with a preference for the individualist values of social mobility and competition. On balance, however, hierarchy remained the dominant cultural motif of the Federalist party elite, at least in New England.

Consistent with their hierarchical proclivities, most New England Federalists were deeply suspicious of competitive individualism. Federalists, Banner points out, "spiritedly rejected the argument that the general welfare of a republican society was enhanced by uninhibited competition and the pursuit of self-interest." The quest for individual gain, Federalists charged, diverted individuals' attention away from sacrificing for the collectivity. Their complaint against individualism was not (as egalitarians would contend) that it created inequalities but that it subverted the natural order of social rankings: "Rather than trying to better his condition, a man was to keep his place; rather than striving to transcend his station, he was to transfer to his betters the authority to decide at what moment he was to be permitted to rise."[87]

The Federalists' suspicion of the unfettered pursuit of private interests casts doubt upon the consensus thesis as formulated by Hartz. Federalists did, to be sure, believe that protection of the rights of property was, as Fisher Ames put it, "the essence, and almost the quintessence, of a good government."[88] But this only confirms how limited is the ideological constraint entailed by a belief in private property. "Forced to choose between the rights of property and the public interest," David Hackett Fischer points out, Federalists "chose the latter." But, as Fischer adds, "the choice rarely presented itself in these terms. The security of property was [in the Federalist view] the cement of society."[89] Federalists could thus agree on the need for securing the fruits of industry without embracing equal opportunity, unbridled competition, or the unregulated pursuit of individual gain. Agreement on property rights did not translate into agreement on ways of life.

The Federalists' suspicion of expanding political participation and their forthright defense of deference and authority belie Huntington's thesis that all Americans have adhered to anti-authority principles. There is no place in Huntington's account for the likes of Alexander Hamilton, George Washington, Gouverneur Morris, John Adams, Oliver Ellsworth, Rufus King, John Jay, Timothy Pickering, Theophilus Parsons, Fisher Ames, Jonathon Jackson, or George Cabot. Few if any of the important Federalists can be seen as anti-authority. Indeed they spent most of their waking hours trying to shore up authority.

It might be conceded that Federalists of the "old school" were hierarchical but countered that they quickly disappeared from the face of the American political landscape. America may not have been born modern but it quickly became so. The Federalists, in this view, represent a residual politics of the past that vanished virtually without trace soon after Jefferson's victory in 1800. There is some validity to this objection. Never again would hierarchy have the national political strength that it had during the Washington administration. And yet hierarchy did live on even after the demise of the Federalist party. At least in New England, as Daniel Walker Howe shows, hierarchy survived (in diluted form, to be sure) through the

Whig party as well as in nonpolitical arenas such as higher education, family, and religion.[90] Indeed, something of the Federalist's hierarchical vision survived in the ethos of the late-nineteenth-century Mugwumps.

Mugwumps, Bosses, and Capitalists in the Gilded Age

"Who is the boss?" asked Max Weber in his celebrated essay "Politics as a Vocation." Weber's answer: "He is a political capitalist entrepreneur who on his own account and at his own risk provides votes."[91] The entrepreneurial ethos of the urban boss is encapsulated in Tammany boss George Washington Plunkitt's infamous boast: "I seen my opportunities and I took 'em."[92] It is impossible to read Plunkitt, as Arthur Mann points out, "without sensing that politics was to him what buying and selling railroads was to Jim Fisk."[93] It is from these entrepreneurs, political and economic, that the post–Civil War era has taken its (largely unflattering) names: "the Gilded Age," the "Age of Excess" or, when viewed more positively, the "Age of Enterprise."[94]

During this period, perhaps the most vocal opposition to the excesses of the bosses and large industrialists came from those labeled by contemporaries as Mugwumps. The typical Mugwump tended to come from an old, established New England family; his parents and grandparents were likely to have been Whigs and/ or Federalists.[95] His wealth was more than likely inherited. Perhaps because of his secure economic and social position, he felt little need to "get ahead." He was already, in his own eyes at least, one of society's "best men."[96] He tended to be exceptionally well educated; more were from Harvard than from any other university.[97] Most were Anglophiles, and their favorite cause, civil service reform, owed a great deal to the British example.[98] Descended from respected merchants, clergymen, and public servants, the Mugwumps were mostly professionals. Those businessmen that did exist within the civil service reform movement tended to be, as Ari Hoogenboom points out, "latecomers rather than originators of the movement, and followers rather than leaders."[99]

The Mugwumps were their fathers' sons not only in their social and economic status but in their values also. Among the most important institutions for transmitting a hierarchical worldview was the university, particularly Harvard. Harvard, according to historian Geoffrey Blodgett, "was a seminal source for the whole cluster of ideas that gave Mugwumpery its drive." The faculty at Harvard instilled in their students a sense of intellectual and moral superiority as well as an "august sense of civic stewardship." As members of a natural aristocracy, students were taught that they had a solemn responsibility to provide disinterested leadership for the public good as well as to educate the lower classes in "their duties of citizenship." In their own mind, at least, they made up that small group of men that each generation of citizens produces "who are free from self-seeking and who, recognizing their obligation to the community, are prepared to give their work and their capacities for the service of their fellowmen."[100]

To men reared on hierarchical notions of obligation, public service, and duty, the politicos' "apparently mindless game of grab and barter" seemed a blatant breach of public trust.[101] The Republican party, lamented John F. Andrew, son of the famous Massachusetts war governor, had fallen into the hands of "mere trading politicians, whose only interest is to advance themselves."[102] Votes of New York legislators, echoed the *Nation*, are bought and sold like "meat in the market."[103] It was inconceivable to the Mugwumps that trades rooted in self-interest could sum to anything approximating the public interest.

The Mugwumps' distaste for the party boss and spoilsman was matched by their dislike for the successful capitalist entrepreneur.[104] In the Mugwump imagination, explains Richard Hofstadter, industrialists were "uneducated and uncultivated, irresponsible, rootless and corrupt, devoid of refinement or any sense of noblesse."[105] In his autobiography, Charles Francis Adams, Jr., recalled having "known, and known tolerably well, a good many 'successful' men—'big' financially—men famous during the last half-century; and a less interesting crowd I do not care to encounter. Not one that I have ever known would I care to meet again, either in this world or the next; nor is one of them associated in my mind with the idea of humor, thought or refinement. A set of mere money-getters and traders, they were essentially unattractive and uninteresting."[106] These men of new wealth, in E. L. Godkin's view, lacked the "restraints of culture, experience, the pride, or even the inherited caution of class or rank."[107] Their "insistent flaunting of money and showy wastefulness" were deplorable not only for vulgarity but for neglect of public responsibilities.[108] Charles Eliot Norton's revulsion against the cupidity of capitalists led him to question the justice of "systems of individualism and competition. We have erected selfishness into a rule of conduct," Norton complained, "and we applaud the man who 'gets on' no matter at what cost to other men." It was necessary, Norton believed, to make these men "more conscious of their duties to society."[109]

Mugwumps remained uncertain about how to deal with these powerful economic upstarts. Mostly they hoped to persuade the new rich to adopt a "wider conception of the responsibilities of property and the duties of trusteeship."[110] Downward revision of the tariff was also seized upon as a means to counteract the power of the large industrialist. Mugwumps had a much surer sense of how to combat the power of the boss: namely, replace the spoils system with a professional civil service. Throughout the 1870s and early 1880s, civil service reform was preeminent to the Mugwumps. Indeed it was the issue that virtually defined their identity.

The acknowledged leader of the civil service reform movement was George William Curtis, editor of *Harper's Weekly*, first chairman of the Civil Service Commission established under President Grant in 1870, and first president of the National Civil Service Reform League set up in 1881. Like many Mugwumps of New England descent, Curtis had a "sympathetic view of the national government as an agency for expressing and guiding the national will." Civil service reform, in Curtis's view, was a prerequisite to attracting the "best people" into government and thereby restoring public confidence in government. According to John Tomsich, Curtis "was flatly opposed to . . . the prevailing practice of arriving at the public interest through the conflict of various private interests." He distrusted the "hidden hand" not only in politics but also in the economic marketplace. In

setting wages, Curtis believed that employers should "mix morals with economics," and he refused to trust in "phrases about supply and demand" or "the laws of
political economy." "There must," he said, "be something wrong in laws which
leave the mass of men poor, ignorant, and dissatisfied."[111]

Another influential advocate of civil service reform was Curtis's close friend
Charles Eliot Norton. Descended from a long line of respected New England ministers, Norton helped found the *Nation* and later taught medieval literature and
architecture at Harvard. Like Curtis, Norton criticized individualism from the vantage point of hierarchy. Liberty, in Norton's view, was not freedom from government but freedom "from all restraints which may prevent the doing of what is right."
Norton did not, Tomsich tells us, "worry about government action to compel the
doing of what was right." The individual, Norton wrote in the best Burkean fashion, was "but a link . . . in a chain reaching back indefinitely into the past, reaching forward indefinitely into the future." In contrast to William Graham Sumner,
who concluded that talk of social classes owing each other was simply a ruse for
governmental coercion, Norton vigorously upheld the "duties of all classes toward
each other."[112] These duties, most Mugwumps would have agreed, included the
responsibility "of the gifted and educated classes" to look out for and educate "the
weak, the witless, and the ignorant,"[113] and the responsibility of the lower classes
to defer to the superior wisdom and moral virtue of the educated classes.

The Mugwumps' anti-individualist ethos should not be exaggerated. In contrast to their Federalist and Whig predecessors, Mugwumps opted for free trade
over protective tariffs. And Mugwumps generally envisioned a much more limited role for the central government than had their fathers and grandfathers. However, it would be a mistake to assume that Mugwumpery was simply another
expression of laissez-faire individualism. Two points are particularly important here.

First, important cultural variations existed among those who were called Mugwumps. There was certainly little of the hierarchical ethos in the self-made entrepreneur Edward Atkinson, for instance, who envisioned a future in which "the
functions of the officers of Government will become less important than they now
are, and it need not long be necessary for able men to make a great sacrifice in
order to take a share in the executive work."[114] Many businessmen were, like
Atkinson, attracted to civil service reform less out of genteel notions of public
service than out of a conviction that the public sector should be managed more
along the lines of the private sector.[115]

Second, the civil service reformers' increasingly negative attitude to governmental
power was significantly shaped by their distrust of governmental incumbents and
their resulting disillusionment with democracy. In the immediate aftermath of
the Civil War, most Mugwumps were advocates of a strong central government.
The *Nation*, for instance, which became a tireless advocate of laissez-faire in the
course of the late nineteenth century, began its life in 1865 firmly committed to
promoting the authority of the national government and counteracting localism. After
being spurned by Grant's administration, however, civil service reformers began
to lose their enthusiasm for positive government. They became, as one admitted,
"despondent . . . in view of the manner in which legislation is conducted."[116]

As they liked less and less the kind of men who entered government and the politics they pursued, they scaled back their notions of the appropriate functions of government.[117]

At bottom, the Mugwumps' prophylactic approach to power stemmed not from a doctrinaire belief in laissez-faire but rather from a growing skepticism about democracy. Like southern slaveholders, who feared what a democratic majority might do to the institution of slavery, Mugwumps worried that a strengthened democratic government would be overly responsive to the redistributive demands of immigrants and labor. The social vision of Charles Francis Adams, Jr., was every bit as hierarchical as that of his grandfather, John Quincy Adams, but the younger Adams's belief that "the mass—the mighty majority—of our fellow voters are ignorant and stupid and selfish and short-sighted" led him and his fellow Mugwumps to a more circumscribed view of federal authority.[118] This is the hierarchist's central dilemma in a democratic political system: how to support national authority without providing government the means to undermine a stratified society.

For all of their skepticism about democracy, there remained a readiness among the more genteel Mugwumps to look to government to secure societal ends. Charles Francis Adams, for instance, was intimately involved in railroad regulation.[119] Richard Watson Gilder's *Century* endorsed the newly founded American Economic Association's recommendation for an expansion of the functions of government, contingent upon civil service reform lessening the problems of patronage. Curtis continued to believe that government had a positive obligation to promote the general welfare, particularly where there was a clearly identifiable public interest, such as public safety. It was Alexander Hamilton, Curtis believed, who was "the greatest of our great men" and Jefferson who was "the least of them."[120]

Many Mugwumps felt frustrated at being men out of time. They looked back nostalgically to a hierarchical order in which statuses were better defined, and deference to men of education and breeding was more forthcoming.[121] Frederic J. Stimson, a Mugwump lawyer, idealized a pre–Civil War Massachusetts in which "though all were recognized to be of the same stock, *there were very nice grades of social distinction*."[122] Another civil service reformer hoped that a reformed civil service would help restore the practice of "the early days of the Republic" when executive officials were "generally selected from well-known families."[123]

But Mugwumps looked forward as well as backward. For if their social status and family heritage linked them with a patrician past, their "professional impulse" joined them with emerging bureaucratic structures and the Progressive "establishment reformers" of the early twentieth century.[124] The Mugwumps' professional impulse, historian Gerald McFarland explains, "sought to check the decline of hierarchy and specialization in such fields as the ministry, law, and medicine" that had resulted from the Jacksonian attack on restrictive rules for admission to and practice of these professions. In attempting to restore the high standards of professional training and practice advocated by their Federalist and Whig ancestors, Mugwumps were thus not merely reasserting traditional, hierarchical values but reshaping the hierarchical vision to make it viable in a technologically advanced world.[125]

Hierarchy in Modern America

Hierarchy may be significantly weaker in the twentieth century than in previous centuries but it would be incorrect to see hierarchy as just a trivial vestige of a bygone era. The modern army is perhaps the most obvious repository of hierarchical institutional structures and values.[126] Of course, as *The American Soldier* pointed out forty years ago, the American army's institutional characteristics contrast sharply with the civilian life around it. But as Samuel Stouffer and his colleagues also showed, there were important differences in how Americans reacted to the hierarchical life of the army. There was much greater acceptance of the military's hierarchical system among the army regulars (who had significantly lower education levels than the selectees) and the less well educated selectees than among the better educated selectees. The better educated draftees were much more likely to question the army's authority structure and status system. For many of the less educated, the hierarchical life of the army was evidently a less radical departure from their everyday life than the civilian–military distinction might suggest.[127]

The Catholic church is another institutional locus for hierarchical values in modern America. Admittedly, the Catholic church in the United States carries nothing like the political influence that it exercises in the predominantly Catholic countries of Europe (Poland, Ireland, Italy, France) and Latin America. And, too, the Catholic church's efforts to preach hierarchy within the church and economic equality outside the church often come into conflict. "It is not easy," as Aaron Wildavsky points out, "to say that all forms of inequality are bad but that popes and bishops are good."[128] Still, American Catholics do spend much of their time in a hierarchical institutional setting, and their political and social thought is often importantly hierarchical.

No Protestant religion (and America remains, of course, an overwhelmingly Protestant nation) boasts the elaborate hierarchy of the Catholic church, but hierarchy is not absent from Protestant fundamentalism. Frances Fitzgerald's journey to Jerry Falwell's Thomas Road Baptist Church in Lynchburg, Virginia, unearthed unmistakable signs of hierarchy. The Thomas Road pastors, Fitzgerald found,

> have a much more specific and detailed set of prescriptions for the conduct of everyday life than do most Protestant churchmen. They have—like many Southern Baptists—absolute prohibitions against drink, tobacco, drugs, cursing, dancing, rock and roll, and extramarital sex, but they also have special prescriptions for dress, child rearing, and the conduct of marriage. The Liberty Baptist College student handbook, for instance, decrees that men are to wear ties to all classes. "Hair should be cut in such a way that it does not come over the ear or collar. Beards or mustaches are not permitted. Sideburns should be no longer than the bottom of the ear." As for women, "dresses and skirts . . . shorter than two inches [below] the middle of the knee are unacceptable. Anything tight, scant, backless, and low in the neckline is unacceptable."

Authority is upheld in every social sphere. "In school, children should not challenge their teachers but should accept instruction and discipline. On the job, a man should work hard, show discipline, and accept the authority of his employer. He should accept the authority of the church and the civil government in the same way." *The Total Family*, a book by the family-guidance pastor of the church, counsels that "the Bible clearly states that the wife is to submit to her husband's leadership. . . . She is to submit to him just as she would submit to Christ as her lord." Parents are instructed to exercise absolute authority over their children. An organization chart, entitled "God's Chain of Command," depicts their hierarchical ideal of human relationships: "from 'God,' at the top, the lines of authority descend to 'local church,' on the one hand, and 'civil authority,' on the other; the lines then descend and converge upon 'total family'—father first, then mother, then children."[129]

Probably only a minority of American families today, even fundamentalist families, closely approximate this patriarchal model. Women are often as likely if not more likely to make important family decisions,[130] and most children have a greater role in family decision making than these hierarchical lines of authority would suggest.[131] Nevertheless, hierarchical notions of ascription and authority remain important organizing principles in many American families. The sharp division between male and female roles that the Lynds found in *Middletown* in the 1920s has been modified subsequently but continues to be a central aspect of family life both in Middletown and in the United States generally.[132] Similarly, although the emphasis upon "strict obedience" to parental authority that the Lynds found among Middletown parents is not as prevalent today as it once was, the idea that "you have to teach [children] to respect parental authority" is still prominent in many American families.[133]

Even in a vibrant capitalist economy, ascriptive status and hierarchical stratification persist, albeit in a diluted form. "The tendency [toward hereditary privilege and stratification]," as John Gardner points out, "is so deeply rooted in human interaction that if one could eliminate every trace of it today, it would begin to creep back tomorrow." Alternative principles of organization like individual merit or equality are difficult to abide by on a regular basis. "People find that there are many times in life when they are delighted to accept some good thing that comes to them not because they earned it, and not because it is their fair share of something everyone else received too, but simply because they stand in a family relationship to the donor, or occupy a certain position in the community, or are members of the same lodge as the donor." What Gardner refers to as "aristocracies of profession" provide particularly glaring evidence of the persistence of hierarchy. "By even the second generation of an academic, military, or banking family, a private world of presuppositions, exclusiveness, endogamy, and superiority flourishes."[134]

Moreover, the spread of a market economy has coincided with a growth in hierarchically structured business and governmental organizations.[135] Within these organizations, authority inheres in office. Those in positions of authority often command by virtue of the rank they hold rather than any specialized ability or knowledge.[136] The hierarchy of modern bureaucracy is, to be sure, different from the all-encompassing hierarchy of "traditional" societies described by scholars like Louis Dumont.[137] In the modern world, one may work in a hierarchical organization from nine to five but come home to a family organized along egalitarian lines, participate in a competitive political system, and purchase goods in a competitive market. The separation between spheres of life—between the family, religion, the polity, and the economy—that seems to distinguish "modern" societies perhaps militates against the pervasive dominance of a single cultural bias found in some traditional societies.

The aim of this concluding section is not to lose sight of Hartz's important comparative point (indeed I shall end by affirming it)—-compared to most of the rest of the world, the United States is distinguished by the relative weakness of hierarchical institutions and values—but rather to stress the ineradicable nature of the hierarchical mode of organizing. Some of these hierarchical institutions, like the Catholic church and the patriarchal family, might be identified as residues of a hierarchical past but others, like large-scale bureaucracy, have grown out of (or in reaction against) competitive capitalism. Hierarchy exists, moreover, not just as a form of organization but in the values and beliefs of Americans.

Public opinion surveys consistently show the existence of a small but significant minority of Americans willing to support hierarchical or pro-authority ideas. In *The American Ethos*, Herbert McClosky and John Zaller found that 29 percent of the general public (and 25 percent of opinion leaders) agreed that "a person who holds a position of great responsibility, such as a doctor, judge, or elected official, is entitled to be treated with special respect." This was less than half of the 64 percent (63 percent among opinion leaders) who believed such a person "should be treated the same as anyone else" but represents nonetheless a significant chunk of the American population.[138] These results parallel those uncovered thirty years earlier in a survey carried out immediately after World War II by Samuel Stouffer and his colleagues. The poll found that 23 percent of enlisted men agreed that "officers deserve extra rights and privileges because they have more responsibility than enlisted men."[139] In most other countries, the proportion assenting to such propositions would undoubtedly be significantly higher. But the validity of this comparative point should not lead one to ignore altogether the existence of such sentiments in the United States.

Those who would portray America as a land of restless, bustling entrepreneurs and rugged individualists should be given pause by the results of a survey taken in the late 1950s that found 70 percent of the general public assenting to the proposition that "security is more important to me than advancement," 53 percent disagreeing with the statement "I have no use for a man who is satisfied to stay where he is all his life," and 61 percent agreeing that "it is better just to be 'one of the group' than to be singled out for special attention."[140] The man on the make is part of the American story, but he is far from the whole story.

Studies of American values employing in-depth interview techniques confirm the existence of the hierarchical strains of thought suggested by these survey results. Jennifer Hochschild's *What's Fair?* uncovers an unmistakable hierarchical voice in the person of Eleanor Fox, a seventy-year-old widow descended from a long line of aristocratic Yankees. Fox favors bolstering the authority of parents over children and teachers over students, trusts politicians to "do their very best," and advocates clearly differentiated male and female spheres. "I don't see any sense in woman equality," she explains. "A woman was born to raise her children. . . . I wouldn't dream of having my husband scrub the floor or clean the stove. I'd do that myself. And why shouldn't a woman? If she's a housewife, that's her job."[141] Although she feels "sorry for the people who are not getting everything you're getting," she insists that "you just can't have a country without labor, without different classes in it. You've got to have a laboring class with less pay." Each role, no matter how menial, is essential to the whole. After all, a nation "cannot live without people who are willing to collect the garbage."[142]

Of the twenty-eight Americans Hochschild interviewed, none matched Eleanor Fox's unabashed advocacy of class and sexual distinctions, but several did express a more ambivalent attachment to hierarchical values. Wendy Tonnina, a young Italian-American saleswoman in a clothing store, dislikes huge corporations but still believes that "hierarchy within a firm is essential." As for government, she believes that it "needs to be powerful because you have to have something as the head of the society." And Sally White, an unemployed clerical worker, retains a strong faith in those in positions of authority: "They have all those smart people up there that are not just going to throw money away." Pamela McLean, a part-time secretary and Irish Catholic, believes the rich "have a certain responsibility . . . to be good-citizen-type people . . . to share their talent, to use their knowledge and talents to solve problems in the world." She combines what Hochschild labels her "ascriptive awe" of the wealthy with a feeling that society has an obligation to help "poverty people."[143]

Studs Terkel's oral history *The Great Divide* introduces a number of recognizably hierarchical voices in America of the 1980s. Dennis McGrath, a forty-year-old radio engineer who attended Catholic parochial schools and is now active in a fundamentalist church, flatly sides with authority: "I tend to believe we should obey our government. Scriptures say obey our leaders." And a teacher at the University of Arkansas–Little Rock, Robert Franke, explains that for his students, almost all of whom come from fundamentalist families, "Church and family is all they have. It's out of this milieu came their values of what is right and wrong. They believe in Authority. Authority is knowledge. Authority is government. Authority is parent. Authority is church. You do not question Authority." Lawry Price, who runs the neighborhood tavern that his parents opened in 1938 and whose kids attend the same local parochial school that he did, defends clearly demarcated male and female roles: "These women are doing everything guys are. They're out riding motorcycles and parachuting and everything. I'm from the old school. . . . I'd prefer to have it the old way because it was steady every day. You knew what was going to happen."[144]

Interviews with "prolife" activists have also uncovered telltale signs of hierarchical proclivities. Competing positions on the abortion issue, as Kristin Luker shows, are grounded in rival worldviews. "Each side of the abortion debate," Luker finds, "had an internally coherent and mutually shared view of the world that is tacit, never fully articulated, and, most importantly, completely at odds with the world view held by their opponents."[145] Prolife activists, 80 percent of whom were Catholics,[146] emphasized the intrinsic differences between men and women. Theirs was a world in which, as Archie Bunker had it, girls were girls and men were men. They deplored the feminist movement's "restless agitation against a natural order," and believed that a division of labor in which women stay home and men go to work is "appropriate and natural." "I think they've made women into something like the same as men, and we're not," complained one. Moreover, from their point of view, as Luker points out, "any policy that seeks to address the members of a family as separate entities, rather than as an organic whole, is a priori harmful."[147]

The distinction between hierarchy, often termed social conservatism, and competitive individualism, or economic conservatism, exists not just in the mind of the theorist but in the minds of actual American citizens, as several recent studies attest. One survey of Republican party national convention delegates, for instance, found that "those delegates who were clearly 'conservative' on economic issues were not reliably 'conservative' on [social] issues, while, conversely, those who were clearly 'conservative' on [social] issues were not reliably 'conservative' on economics."[148] In *Women of the New Right*, Rebecca Klatch finds a "fundamental division" between women she labels "social conservatives," who believe family to be "the sacred unit of society" and "envision gender as a divinely ordained hierarchical ordering in which men have natural authority over women," and those she identifies as "laissez faire conservatives," who reject ascriptive differences between the sexes and wish to make it possible for both men and women, as rational, self-interested, and autonomous actors, to "climb to the height of their talents in a marketplace free from outside interference."[149] Similarly, Fred Kerlinger documents the clustering of questions tapping themes of discipline, duty, and authority on a factor he labels "traditional conservatism" and the loading of items relating to profit, money, business, and capitalism on a separate factor, which he labels "economic conservatism."[150]

Although analytically and actually distinct, there nonetheless exists a strong affinity between hierarchy and capitalism in modern America. As McClosky and Zaller explain:

> Capitalism appealed to conservative [hierarchical] sensibilities in several ways. For example, the system of social stratification and differential rewards under modern industrial capitalism is consistent with conservative notions of how a just society ought to be organized. In creating and maintaining a powerful financial and industrial class, capitalism fulfills what conservatives regard as society's essential need for strong and meritorious leaders. Capitalism also emphasizes the need for everyone either to work hard or to suffer material loss—thus providing the strong incentives that conservatives believe are necessary for developing self-discipline and the restraint of appetites. Then, too, capitalist institutions, such as private property, have become vital features of social stability and the established order in America. Since conservatives especially fear instability and are strongly averse to schemes and theories for reconstituting society, . . . capitalism seems to them far preferable to the radically egalitarian, collectivist, and still experimental systems that represent themselves as the principal alternatives to private enterprise.[151]

Supporters of hierarchy in this country would evidently rather bear those ills they have than fly to others that they know not of.

Hierarchy in America is distinguished not only by its often relatively uncritical embrace of competitive capitalism but also by its suspicion of governmental authority. Just as the New England Whigs supported authority in all walks of life yet distrusted executive power[152] and the Virginian planters defended deference and authority while denying the authority of the federal government, so contemporary fundamentalists express respect for authority in the family, school, church, and country, yet often evidence little trust in government and openly attack institutions such as the Supreme Court.[153] Hierarchy often looks different in America than it does in Europe (deference to authority, for instance, is only infrequently reciprocated by an insistence on noblesse oblige toward those below)[154] because of its position on the periphery of American political culture. Because contemporary hierarchical advocates are an estranged minority, to support national authority would be to further values antithetical to their own. Unable to enlist government on their side, it is little wonder many reject it. This is the central, tragic dilemma of hierarchy in the United States.

7

Fatalism in America: The Case
of Slavery

Studies of political ideas tend to be biased toward the articulate, toward those who leave records and write books, letters, or memoirs. Activity is unavoidably privileged over passivity. Fatalism as a cultural orientation is thus consistently underreported in most history.[1] Hierarchy, individualism, egalitarianism, even hermitude —all have eloquent exponents of their ideas. Fatalists, however, remain peculiarly silent. What true fatalist would take the time to commit thoughts to paper? What would be the use? Indeed the very act of speaking out would count as evidence of a shift away from the fatalistic bias.

The difficulties of reconstructing a fatalistic way of life have long presented a formidable obstacle to the study of slave culture in the South. While slaveowners left a wealth of diaries, letters, books, and essays to help scholars reconstruct their worldview, slaves left relatively little in the way of such records. Some scholars chose to resolve the paucity of direct evidence about what went on in the heads of slaves by relying on slaveholder accounts. But, as one former slave aptly put it, "Tisn't he who has stood and looked on, that can tell you what slavery is,—'tis he who has endured."[2]

Because slaves left relatively few written records, there is a great deal of scope to interpret slave culture in such a way as to accord with one's own cultural bias. The social construction of slave life reached a fever peak in the decades prior to the Civil War as proslavery ideologists insisted slaves were content members of a caring hierarchy and abolitionists responded that slaves were an oppressed people with special virtues. So heavily did abolitionists edit slave narratives that, as Gilbert Osofsky observes, "scholars have rightly wondered where the slave's experience began and that of the antislavery recorder left off."[3] But the social construction of slave culture is not a peculiar feature of a bygone era. It has continued down to the present day in the academic debate about how best to characterize slave culture in the South. Four different interpretive themes emerge: slaves as loyal members of a benevolent hierarchy; slaves as resigned victims of an arbitrary and despotic system; slaves as success-oriented, self-reliant individualists; and slaves as members of a tightly bounded, noncoercive "countercultural" community that attempted to subvert the dominant master class.

Slavery as Hierarchy

The view that slaves were an integral part of a caring hierarchy was first advanced by slaveholders. Apologists for the peculiar institution portrayed slaves as children: irresponsible, loyal, affectionate, careless. Like children, slaves needed the guidance, discipline, and direction that only their superiors could provide. The authority of the master was thus rooted in the same hierarchical principles as the authority of the father. As Christopher Memminger, a Charleston layman, explained to a gathering of young men of Augusta, Georgia: "Each planter in fact is a Patriarch—his position compels him to be ruler in his household," guiding children and slaves alike with a steady hand and loving voice. Slaveholders would concede that some masters treated their slaves as mere "chattel for profit," but they insisted the cultural norm was paternalism. Most slaves were not oppressed victims, as abolitionists charged, but rather were part of a complex web of reciprocal duties and obligations in which the kindness and humanity of masters were matched by the gratitude and loyalty of slaves.[4]

Contrary to Hartz, this view of slavery as the benevolent institutionalization of inequality did not disappear with the collapse of the Confederacy. Rather the influential writings of Ulrich Bonnell Phillips helped make it the standard interpretation of southern slavery by the opening decades of the twentieth century. At the time of his death in 1936, Phillips was acknowledged within the historical profession as the foremost authority on plantation life. Although aspects of Phillips's interpretation were questioned during the late 1930s and 1940s, his *American Negro Slavery*, published in 1918, remained the only comprehensive scholarly work on the subject until the appearance of Kenneth Stampp's *The Peculiar Institution* in 1956.[5]

Phillips maintained that plantation life benefited both black and white races. Because "the average negro has many of the characteristics of a child," Phillips believed that "the presence of the planter . . . is required for example and precept."[6] By the same token, the responsibility of living among and managing Negro slaves "promoted, and wellnigh necessitated [among whites], a blending of foresight and firmness with kindliness and patience." "The standard [plantation] community," Phillips noted approvingly, "comprised a white household in the midst of several or many negro families. The one was master, the many were slaves; the one was head, the many were members; the one was teacher, the many were pupils."[7] The analogies Phillips selects to make sense of plantation social relations—the family, the school, the human body—all suggest a hierarchical relationship between master and slave.

Like the proslavery ideologues, Phillips accents slave loyalty. The "predominant plantation type," according to Phillips, displayed "a courteous acceptance of subordination [and] a readiness for loyalty of a feudal sort." The modal slave, in Phillips's interpretation, was devoted to his master. Upon returning after an extended absence ("as they might well be in the public service"), masters were likely to be greeted by slaves much as an absent parent would be welcomed by a child. The experience of Henry Laurens is held out as not untypical: "My knees were clasped, my hands kissed, my very feet embraced. . . . They held my hands, hung upon me; I could scarcely get from them."[8]

Phillips stresses the reciprocal obligations and duties that define an inclusive hierarchy. Plantation slavery was "shaped by mutual requirements, concessions and understandings, producing reciprocal codes of conventional morality." Masters, although possessed of "by far the major power of control," did not simply impose their will upon slaves. Rather "the adjustments and readjustments were mutually made." In their lives together, typically "there was occasion for terrorism on neither side. The master was ruled by a sense of dignity, duty and moderation, and the slaves by a moral code of their own."[9]

The hierarchical interpretation of slavery articulated by Phillips and proslavery ideologues is simultaneously a claim about slaveholder culture and slave culture. It is at once a denial that slaveholders were heartless, impersonal, profit-maximizing capitalists and a refutation of the view that slaves were oppressed and brutalized fatalists. The cultures of the two groups are portrayed as tied together by an organic relationship in which the interests of the hierarchs are inseparable from the welfare of the lowerarchs. Just as the master is concerned with the well-being of his slaves, so "very many [slaves] made the master's interests thoroughly their own."[10]

Although Phillips claims to describe the culture of both slave and slaveowner, he presents a more reliable and informed guide to the culture of slaveholders than he does to the culture of slaves. This is hardly surprising in view of his overwhelming reliance on the testimony of slaveholders for evidence. When Phillips relates the story of a South Carolina planter who tells an English visitor, "I respect [the slaves] as my children, and they look on me as their friend and father,"[11] it tells us much more about the paternal self-conception of masters than it does about how slaves viewed their masters. An accurate account of slave culture, as Richard Hofstadter recognized almost a half-century ago, "must be written in large part from the standpoint of the slave."[12] Relying on what slaveholders thought slaves thought, as Phillips did, is inadequate.

Hofstadter's advice went unheeded for so long in part because slaveholders left more plentiful records than did slaves. In the 1960s and 1970, scholars would begin to tap underutilized or unconventional sources in an attempt to sympathetically listen to the voices of the hitherto "inarticulate." But in the meantime, the relative absence of records left by slaves was construed by many scholars as evidence of the tyrannical and debilitating character of slavery.

Slaves as Fatalists

Antebellum abolitionists had countered slaveholders' claims by alleging that slaves were in fact defenseless victims of tyranny and not grateful recipients of patriarchal beneficence. The relationship of master and slave, according to slavery's contemporary critics, was characterized not by the reciprocal duties and obligations of hierarchy but rather by the arbitrary despotism of fatalism. "Between the slave and the slaveowner," observed Frederick Douglass, there were no institutions or laws "to restrain the power of the one, and protect the weakness of the other."[13] Antislavery advocates would have agreed with the North Carolina planter who sadly conceded that "Slavery and Tyranny must go together. . . . There is no such thing as having an obedient and useful Slave without the painful exercise of undue and tyrannical authority."[14]

The abolitionists' view of slavery as a degrading despotism received its most significant scholarly support in Kenneth Stampp's *The Peculiar Institution* (1956) and Stanley Elkins's *Slavery* (1959). Although these books differed on several important questions, both took issue with the view of slaves as part of a caring hierarchy and placed emphasis instead on the cruelty and arbitrariness of the plantation regime. Not loyalty and affection but fear and coercion were the primary social cement holding master and slave together. Slavery's arbitrary and exploitative character was directly antithetical to the mutual obligations of hierarchy. Stampp, for instance, pointed out that while "white marriages were recognized as civil contracts which imposed obligations on both parties . . . slave marriages had no such recognition in the state codes; instead, they were regulated by whatever laws the owners saw fit to enforce." In a few cases, masters even "arbitrarily assigned husbands to women who had reached the 'breeding age.'"[15]

Elkins pushed the thesis that southern slaves inhabited an arbitrary regime still further. In contrast to Stampp, who allowed for the "day to day resistance" of slaves, Elkins argued that the absolute power of the slave system destroyed the slave's personality, reducing the black man to a docile Sambo. Comparing slavery in the United States with slavery in other New World countries, Elkins argued that what distinguished the United States was precisely the *absence* of those hierarchical institutions and traditions that could protect the weaker elements of society. Whereas slavery in Latin America emerged in the context of the paternalistic, feudal, Catholic culture of Spain and Portugal, slavery in the United States established itself in an environment of what Elkins called "unopposed capitalism."[16]

In contrast to Phillips and proslavery apologists who saw slavery as bringing civilization to the backward African, Elkins viewed slavery as destructive of pre-existing, sophisticated African cultures. Working against the background of anthropological accounts of "the energy, vitality, and complex organization of West African tribal life," Elkins felt it necessary to reverse the question altogether by asking "how it was ever possible that all this native resourcefulness and vitality could have been brought to such a point of utter stultification in America."[17] Slaveholders enforced this transformation, Elkins argued, by severing personal ties and annihilating the past.[18]

Elkins maintained that hierarchical analogies such as the family, school, or human body were inadequate to capture the social reality of the master–slave relation. A better comparative reference is a despotic social system in which power is absolute. Elkins selected the Nazi concentration camp as the most illuminating analogy. Others have suggested that a prison or mental hospital would be a more apt analogy.[19] The validity of these analogies, particularly the concentration camp analogy, has been hotly challenged by critics who contend that they exaggerate the psychological damage and underestimate the creative space available to most slaves. But if the concentration camp analogy was flawed, the questions Elkins posed about the relationship between social structure and cosmology remain important. Elkins's enduring contribution was to focus theoretical attention on the way in which "the social system represented by American plantation slavery might have developed a sociology and social psychology of its own."[20] Where Phillips attributed slave behavior to innate racial features, Elkins highlighted the structural elements of the plantation system "that could sustain infantilism as a normal feature of behavior."[21] Recent research on slave culture has been so intent on affirming slave autonomy and community that it has sometimes slighted the important problem of specifying the effect that slavery actually had on slaves' beliefs and behavior.

Slaves as Individualists

A dramatically different portrait of slave culture is presented by Robert Fogel and Stanley Engerman in *Time on the Cross* (1974). If Phillips stresses loyalty as the motivation for slave behavior, and Stampp and Elkins emphasize fear, Fogel and Engerman focus on incentives. According to Fogel and Engerman, slaves partook of an entrepreneurial culture that was little different from that inhabited by slaveholders of the South or laborers of the North. Slaves were "diligent workers, imbued like their masters with a Protestant ethic." And like whites, slaves struggled "to develop and improve themselves in the only way that was open to them."[22] Slaves, like wage laborers, were offered opportunities for advancement. A "flexible and exceedingly effective incentive system" rewarded hard work and allowed for significant occupational mobility. "Climbing the economic ladder brought not only social status, and sometimes more freedom; it also had significant payoffs in better housing, better clothing, and cash bonuses." As in free society, those who worked diligently were rewarded with economic and social benefits and those who loafed were punished.[23]

Fogel and Engerman lead the reader astray when they declare their aim is "to strike down the view that black Americans were without culture."[24] This formulation misleads for what is in question is not whether slaves had a culture but the *content* of that culture. To live without culture would be to live without bias, seeing all sides of every issue at once. As philosophers of science remind us, mere mortals can never attain such an Archimedean position of objectivity. In this respect, slaves are no different from the rest of us. Their social environment inevitably biases the way they see the world. Elkins hypothesizes that slavery generates a slave culture of resignation and submission; Fogel and Engerman posit that slavery produces a success-oriented, "achieving" slave culture.

Fogel and Engerman's thesis has drawn heavy fire from many quarters. Among the most telling criticisms is that their portrayal of slave culture as individualistic and entrepreneurial "rests on little direct evidence."[25] Instead inferences about slave attitudes are drawn from demographic data such as aggregate fertility and mortality. Little effort is made to get at values and beliefs directly. Slave narratives as well as slaveholder reports are regarded as too impressionistic to form the basis for a scientific portrait of slave culture. Unfortunately, statistical analysis of aggregate demographic data, although powerful for certain purposes, is inadequate for the purpose of re-creating the world as slaves saw it.

Knowing the incidence of whipping on a plantation, for instance, reveals little about how slaves perceived these whippings. Was an average of "0.7 whippings per hand per year" a lot or a little? Did slaves experience punishment as an ever-present threat that profoundly shaped their behavior (something that happened to a slave on the plantation once every four days) or was it at the periphery of their attention (something that happened to any given slave only once every year or so)?[26] Herbert Gutman and Richard Sutch point out another key unasked question: "Were the punishments clearly related to incidents of individual malfeasance or were they indiscriminately applied?"[27] That is, the average itself cannot tell us whether slaves experienced punishment as an incentive system, carefully calibrated to fit the crime, or as a relatively random use of violence. Much slave testimony suggests that slaves frequently experienced violence as an arbitrary striking out by those in positions of authority. Frederick Douglass, for instance, recalls the necessity of keeping his distance when his master was in a violent mood for "the victim had only to be near him to catch the punishment, deserved or undeserved."[28]

Fogel and Engerman's conclusion that whipping was part of an incentive system used "to achieve the largest product at the lowest cost"[29] is less an empirical finding than a deduction from the unproven premise that maximization of profits was the planters' guiding aim. Given this goal, acts of cruel or spiteful brutality that might damage a planter's work force are irrational. Granted many planters took measures to ensure a productive labor force, but they were at least as concerned to maintain a quiescent labor force. It was for this reason that an Alabama planter insisted that slaves had to be flogged until they showed "submission and penitence."[30] Although instilling an individualist ethic might improve slaves' capacity to work without supervision, it also risked an increase in independence and rebelliousness. Thus even to the extent that punishment was instrumental, it was often geared less at improving work performance than at instilling fear and breaking the spirit of slaves. Frederick Douglass vividly recalls the experience of being "broken":

> I was somewhat unmanageable when I first went there; but a few months of discipline tamed me. Mr Covey succeeded in breaking me. I was broken in body soul and spirit. My natural elasticity was crushed; my intellect languished; the disposition to read departed; the cheerful spark that lingered about my eye died; the dark night of slavery closed in upon me; and behold a man transformed into a brute![31]

Violence such as that experienced by Douglass is more likely to illicit submission and resignation than encourage bourgeois notions of individual initiative and responsibility.

What is presented as a scientific study of slave culture turns out to bypass the crucial questions that any student of slave culture must confront: What were slaves' attitudes to the future, to the past, to time, to the body, to risk, to innovation, to cooperation, to death, to authority, to equality? The analysis is not cultural but rather is based on a universalistic rational actor model. Far from endowing blacks with a culture, Fogel and Engerman have denied the utility of the concept of culture altogether. They deny the possibility that culture might bias how the same objective situation is experienced or that divergent systems of social relations might generate different ways of perceiving the world.

To say that Fogel and Engerman provide little evidence for an entrepreneurial slave culture is not to deny that some slaves may have fit this mold. Some slaves did inhabit an environment where there was a rational, predictable relation between work and advancement. Hard work for these slaves did have its rewards. Moreover, some masters opened up delimited areas of individual initiative for slaves by permitting them to cultivate their own crops during spare hours. A few even allowed slaves to sell their goods in town.[32] But this was not the modal slave experience.

Slaves as Communitarians

Fogel and Engerman's portrayal of slaves as aspiring individualists has met with much controversy but little acceptance. Of much more lasting impact has been an interpretation stressing slave "community." This interpretive category includes much of the work on slavery published in the 1970s: George P. Rawick's *From Sundown to Sunup: The Making of the Black Community* (1971), John Blassingame's *The Slave Community: Plantation Life in the Antebellum South* (1972), Eugene Genovese's *Roll, Jordan, Roll: The World the Slaves Made* (1974), Herbert G. Gutman's, *The Black Family in Slavery and Freedom, 1750–1925* (1976), Lawrence W. Levine's *Black Culture and Black Consciousness: Afro-America Folk Thought from Slavery to Freedom* (1977), and Thomas L. Webber's *Deep Like the Rivers: Education in the Slave Quarter Community* (1978), for example. Much of this work is among the best scholarship produced on slave culture. Where Fogel and Engerman largely inferred slave culture from aggregate demographic data and Phillips built his portrayal upon the testimony of slaveholders, these scholars turned to the testimony of slaves. They mined folktales, songs, autobiographies of runaway slaves, and interviews of former slaves conducted in the 1920s and 1930s in order to reconstruct slave culture.

Although the books just listed differ in many important respects, they are united in highlighting slave community and resistance rather than the damage and resignation stressed by Elkins. The paradox of a cohesive and resourceful community existing in the midst of a brutally exploitative labor system is resolved by distinguishing between the work environment and life in the "slave quarters." Whatever the oppression slaves experienced in the fields, in the living quarters slaves found themselves amid a noncoercive and supportive community. "While from sunup to sundown, the American slave worked for another and was harshly exploited," Rawick explains, "from sundown to sunup he lived for himself and created the behavioral and institutional basis which prevented him from becoming the absolute victim."[33] Blassingame goes still further, arguing that "the social organization of the quarters was the slave's primary environment which gave him his ethical rules and fostered cooperation, mutual assistance, and black solidarity." Within the slave quarters, "recreational activities led to cooperation, social cohesion, tighter communal bonds, and brought all classes of slaves together in common pursuits."[34]

In the face of adversity, slaves thus created their own autonomous life of community, solidarity, cooperation, and equality.

The thesis that the slave quarters represented an autonomous, noncoercive community of equals is most fully elaborated in Webber's *Deep Like the Rivers*. Within these quarters, Webber argues, slaves carved out "a society within a society," which gave slave culture "a community structure [that] set blacks apart from whites and enabled them to form and control a world of their own values and definitions." A slave's identification with the slave community was furthered in a number of ways: clandestine religious meetings gave blacks a feeling of cohesive fellowship and allowed slaves to "experience moments of peace, ecstasy, fellowship, and Freedom"; and storytelling "frequently took the form of a community happening" that gave slaves "a sense of communal spirit and camaraderie with their fellows" as well as "a feeling of identification with the larger community of American slaves." No matter how far slaves might travel "from their quarter home, these feelings of communality and kinship would always be rekindled with the soulful singing of a quarter song or the . . . performance of a quarter story." Webber is particularly impressed by the egalitarian spirit within the slave quarters. "There is little in the slave narratives," Webber finds, "to suggest that the quarter community made a strong delineation among people, either with respect to innate ability or role, according to their sex. One is struck by the absence of the familiar theme of male superiority and by the lack of evidence to support the view that the quarters was a female-dominated society. The black source material suggests, rather, a general equality between females and males."[35]

In *Roll, Jordan, Roll*, Genovese, too, accents slave community and creative resistance. Slaves, according to Genovese, "developed their own values as a force for community cohesion and survival." At another point he suggests that slaves developed a "protonational black language" that "created a special bond among all blacks." This slave community was not only tightly bounded but noncoercive. Genovese marvels at the way slaves "transformed the ghastly conditions under which they labored into living space within which they could love each other." Much like the abolitionists who, as Stampp notes, frequently portrayed blacks as "innately gentle and Christian people," Genovese suggests that slave family members treated one another with "tenderness, gentleness, charm, and modesty." Their distinctive religion, moreover, not only possessed a "humanism that affirmed joy in life in the face of every trial" but constituted "a critical world-view in the process of becoming."[36]

The primary intellectual foil for the communitarian interpretation is Elkins's *Slavery*. It is primarily Elkins who Rawick has in mind when he complains that historians "have presented the black slave as dehumanized victims, without culture, history, [or] community." And it is at Elkins that Levine takes aim when he criticizes historians for treating black history "not as cultural forms but as disorganization and pathology."[37] This dichotomy between culture and pathology, however, is false and misleading. *Every* cultural form has its pathologies. The question should not be "Did slavery produce pathology or culture?" but rather "What cultural forms with what associated pathologies did slavery engender?"

Advocates of the communitarian thesis are often torn between a desire to demonstrate the positive aspects of slave culture and a need not to lose sight of the brutal oppression of slavery. This is often a difficult balancing act to perform. For as the slave's power and creative capacities are made more prominent, so the menacing shadow of the slaveholder and overseer recedes into the background until it sometimes becomes easy to forget that these men and women were after all slaves.[38]

Communitarian interpretations of slave life often manifest the egalitarian proclivity to identify with and idealize the culture of the oppressed. Just as Russian populists looked upon serfs not only as damning evidence of the brutality of the system but "as embodiments of simple uncorrupted virtue, whose social organization . . . was the natural foundation on which the future of Russian society must be rebuilt,"[39] so many contemporary historians of slavery have presented a view of slave culture that often says as much about their own cultural biases as it does about the biases of the slaves.

Much of the scholarship of the 1970s, as Paul David and Peter Temin recognize, was part of an effort "to retrieve for American blacks some 'usable past,' some reconstruction of history that would leave them more than victims of a social tragedy, brutally infantilized, cut off from their cultural roots, dependent upon white society."[40] That historians sought a usable past is hardly surprising. What is of interest from the standpoint of the sociology of knowledge is the form that this usable past took. Fogel and Engerman's findings that slaves were bourgeois individualists was evidently not a suitable "usable past." What was usable for many historians of this period was not individual self-improvement but communal solidarity, not acceptance but rejection of bourgeois, competitive individualist values.

To draw attention to the societal conditions that made the communitarian interpretation attractive to historians is not to suggest that this description of slave culture is without validity. The slave quarters, at least on large plantations where the slave quarters were clearly set off from the master's residence,[41] did enable slaves to carve out pockets of autonomy separate from the influence of their masters. This emphasis is a useful corrective to Elkins's concentration camp analogy, which greatly underrepresents the extent of psychological and cultural breathing space allowed slaves. But there is a danger, too, of exaggerating the extent of autonomy and community. A comparative perspective can serve as a useful check on some of the excesses of the communitarian interpretation of slave culture. Particularly valuable is Peter Kolchin's comparative study of Russian serfdom and American slavery, *Unfree Labor*. By comparing the experience of American slaves with that of Russian serfs, Kolchin shows how relatively circumscribed was the autonomy of slaves and how relatively undeveloped was slave community.

Compared to Russian serfs, Kolchin finds that American slaves "found it more difficult to create their own collective forms and norms, and their communal life was more attenuated." Although conceding that the slave quarters often functioned as "a refuge from white control," Kolchin also points out that compared to the peasant commune, "institutionally the slave community remained undeveloped, never assuming the concrete forms and functions that would enable it to serve as a basis around which the slaves could fully organize their lives." When serfs wished to redress a grievance, "they turned instinctively to the commune, collectively petitioning their owners, local government officials, or the tsar." Slaves, in contrast, sought redress primarily as isolated individuals. "Seen in the light of the peasant [commune]," Kolchin continues, it is "the absence of any communal organization among American slaves [that] is striking." Slave drivers, for instance, were chosen not by other slaves but by the master. In sum, the slave community "lacked the formal institutional basis of the peasant obshchina." This absence, Kolchin explains, did not "preclude the existence of communal sentiment and behavior among the slaves." But it did "severely restrict the ability of slaves to express their communal feelings, in the process limiting the collective nature of their life and culture."[42]

Not only was group life less cohesive and less developed among American slaves than among Russian serfs; the slave community was also far less autonomous vis-à-vis dominant groups in society. The slave community was always precarious and often ill-equipped to resist interference by slaveholders. It is true that absolute regulation of the slaves' lives was impossible and that pockets of autonomy inevitably opened up. But, as Kolchin reminds us, slaves were significantly worse off on this score than Russian serfs, who "were usually subject to less regulation and were therefore freer to lead their lives as they wished."[43]

The Social Relations of Slavery

None of the four interpretations just reviewed above are without some basis in fact. The slave experience was extremely varied; there was no single "slave culture." Slaves on small agricultural units sometimes worked side by side in the field with their masters; those on larger estates might only infrequently get a glimpse of their master. Some slaves worked under the close supervision of watchful overseers; others were allowed to hire out their own time. Some owners intervened extensively in the personal lives of their slaves; others preferred to let the slaves manage their own domestic affairs. Some lived under the master's roof; others lived in separate slave quarters at a considerable distance from the Great House. Some slaves lived out their entire lives on a single estate in relatively stable families; others were repeatedly torn from family and friends. Some were brutally whipped into submission; others avoided punishment altogether. Some labored in the city where opportunities for contact with outsiders were relatively plentiful; others lived on isolated, self-contained plantations cut off from the outside world. The degree of individual autonomy and group life that a slave experienced thus varied immensely.

So great is the variety of slave experience that some historians have suggested that any attempt to isolate a dominant behavioral or personality pattern is bound to fail.[44] It is no doubt true that no model can capture all of the complex reality of slavery. "Neither slavery nor slaves," Genovese wisely cautions, "can be treated as pure categories, free of the contradictions, tensions, and potentialities that characterize all human experience."[45] But this does not mean that we must accept Blassingame's conclusions that "the slave was no different in most ways from most men" and that "the same range of personality types existed in the quarters as in the mansion."[46] Can it be that slavery matters so little? How can there be so little connection between the way people live and the way people think? The impressive diversity of the slave experience should make us wary of sweeping generalizations about slave culture, but we ought not to shrink from the important task of specifying the effects that slavery as a system of social relations has on values and beliefs, and how these values and beliefs in turn sustain slavery as a mode of social organization.

The life of a southern slave tended to be tightly prescribed. A slave generally could not leave the estate without a pass stating the destination and time of return. No slave was to be out of his cabin after "hornblow," and it was considered prudent for the overseer or master to inspect cabins at night to see that none were missing. Overseers were required to stay in the fields as long as the hands were at work. A slave was not to sell or trade anything without a permit. Gambling with whites or other slaves was illegal. Some slave codes made it illegal for slaves to raise cotton, swine, horses, mules, or cattle. Many masters forbade slaves to have whiskey in the cabin, quarrel or fight, or use abusive language.[47]

Slaves' interaction with those of a different status was closely regulated by slaveowners. Free blacks and whites were often forbidden to work or even talk with slaves. One planter, for instance, fired a white mechanic for "talking with the negroes." To insulate slaves, some planters bought up the lands of lower-class whites living on the outskirts of the plantation. Masters rarely allowed slaves to marry free blacks; most were reluctant even to allow slaves to marry slaves living on nearby plantations.[48]

To be sure, these laws and rules were not always adhered to in practice. For instance, although slave codes made it illegal for any person, even the master, to teach a slave to read or write, some masters did so anyway. Other slaves learned to read at the risk of being severely punished. Slaves in urban environments tended to find their opportunities for interaction with others of different circumstance and station to be much more plentiful than did slaves living on large, isolated plantations in the Deep South. There were more areas for initiative and autonomy in actual behavior than is suggested by looking only at formal legal codes and instructional manuals.

Nevertheless, if these ideal prescriptions and formal codes were not always followed to the letter, it remains true that the lives of most slaves were closely regulated. "It was the lot of the ordinary bondsman," concludes Stampp, "to work under the close supervision of his master or of some employer who hired his services." Comparing them to Russian serfs or even Jamaican slaves, who often supported themselves on private plots of land, Kolchin finds that American slaves "had little occasion to engage in their own economic activity and found it correspondingly difficult to accumulate property." Instead "most slaves had resident masters who constantly meddled in their lives and strove to keep them in total dependence."[49]

Not only were slaves' lives closely regulated and tightly circumscribed (high grid), but slaves were excluded from the group making the rules that bound them (low group). The slaves' exclusion from the decision-making group belies slaveholder claims to an inclusive hierarchy in which there was a place for everyone and everyone in his place. In contrast to Russia, where the serf's "commune represented the lowest level of authority in a chain of command that linked the peasant to the tsar,"[50] American slaves faced a rigid wall of exclusion. Rather than the finely graded distinctions among people according to station that are characteristic of hierarchy, southern law tended to define slaves as chattel rather than people.

There is merit in recent research that has shown how slaves carved out something of a communal life in the slave quarters or gained a sense of community in their spirituals. As we have already seen, however, a comparative perspective reveals the relative weakness of these slave communities compared to the high group solidarity (as well as local chauvinism and hostility to outsiders) common among peasants.[51] That slave life was distinguished more by individual atomization than by group solidarity can be seen even in the patterns of resistance among slaves, which were overwhelmingly individualistic. Collective action against slavery was extremely rare.

A common experience of oppression is frequently assumed to naturally mold an oppressed group into a cohesive unit. But the evidence collected by scholars examining social interaction in such total institutions as prisons and mental hospitals strongly suggests that atomization is the more likely outcome. Erving Goffman, for instance, found that "although there are solidarizing tendencies" in total institutions, they tend to be limited. "Constraints which place inmates in a position to sympathize and communicate with each other do not necessarily lead to high group morale and solidarity. [Often] the inmate cannot rely on his fellows, who may steal from him, assault him, and squeal on him." "In mental hospitals, dyads and triads may keep secrets from the authorities, but anything known to a whole ward of patients is likely to get to the ear of the attendant."[52] Much the same obstacles to group solidarity are evident in the lives of southern slaves.

Individual flight was by far the most common form of slave resistance. The decision to flee was usually reached individually rather than collectively for the very good reason that cooperation involved the risk of betrayal. Frederick Douglass was one of many slaves to discover the perils of joint action when one of his "band of brothers" informed the authorities of their plan of escape. Douglass learned his lesson: next time he kept the plan to himself and escaped on his own.[53] "The difficulty of movement for southern blacks," Kolchin observes, only "served to reinforce their noncollective ethos." Closely controlled and insulated from free social intercourse, "most fugitives escaped on foot, traveled by night and slept in the woods by day, and were properly leery of both whites and blacks."[54] This individualistic pattern of resistance differs markedly not only from the Russian experience, where it was not uncommon for all the serfs on an estate to flee together, but also from the experience of American slaves born in Africa, who not infrequently "ran off in groups or attempted to establish villages of runaways on the frontier."[55] The distinctive pattern of resistance offered by American-born slaves, Kolchin concludes, indicates both "the difficult conditions faced by slaves seeking to engage in communal endeavors and of the generally noncollective nature in which they responded to their bondage."[56]

Br'er Rabbit and Amoral Individualism

Nowhere is the atomized aspect of slave culture more clearly revealed than in the Uncle Remus stories. Those who have given an egalitarian or communitarian con-

132 *American Political Cultures*

struction to slave culture have interpreted these as tales of resistance, a form of wish fulfillment in which weakness overcomes strength, slave defeats master.[57] That this explains something of the slaves' fascination with the Br'er Rabbit stories seems plausible. But to leave the analysis at this point is woefully inadequate for it misses the amoral, atomized world depicted in these folktales.

The recurring motif in these stories, as Michael Flusche documents, is less that the weak triumph over the strong than that survival in this world necessitates distrust and deceit. Cunning and deception triumph not over strength and oppression but over gullibility and trust. It is not "the theme of the ultimate victory, of the Lord delivering Daniel" that one finds in these tales. Rather the tales are "a humorous statement of the treacherous ways of society. They point to the fate of the man who trusts his neighbor exceedingly." "Most of the stories," Flusche explains, "took for granted that deep hostility existed between the characters, however they might disguise it." The relationship between the animals in the stories is invariably portrayed as adversarial. There is no friendship, no cooperation, no pooling of resources, and no collective action. "The perpetual struggle constantly prevented the animals from working together, in spite of the supposed geniality. Attempts at co-operation for parties, for farming, or for storing food, constantly fell apart because someone sabotaged the efforts for his own advantage—typically to eat the butter." Br'er Rabbit's world was one in which "alliances were illusory and each man had to fight his battles alone; every other man was a potential enemy or rival in spite of his smiles and grins, and only the fool was unwary."[58]

Those animals who place their trust in the word of another animal invariably end up meeting a sticky end. Br'er Rabbit usually comes out ahead because he discerns and exploits the other animals' weaknesses and presumes that no one means him any good. "His completely cynical view of manners and social relations," Flusche explains, "enabled him always to get the better of his opponents; repeatedly they presumed good will on his part when he provided a facade of consideration and altruism." When, for instance, Br'er Wolf begs Br'er Rabbit to shelter him from pursuing dogs, Br'er Rabbit offers Wolf a chest in which to hide. Locking Wolf in the chest, Br'er Rabbit then scalds his victim to death by pouring boiling water through a hole in the top, all the while assuring Br'er Wolf that the pain is only fleas biting him.[59]

When Br'er Rabbit temporarily drops his guard and trusts another to play by the rules he, too, ends up a loser. Br'er Tarrypin bests Br'er Rabbit in a race by positioning members of his family at regular intervals along the course to be there when the rabbit passes. Remaining near the finish line throughout the race, Br'er Tarrypin steps over just as Br'er Rabbit comes into sight.[60] Unlike the Aesop tortoise and the hare fable, which teaches the virtue of perseverance ("slow but sure"), the Br'er Rabbit variant lauds treachery. Those gullible few who try to run a fair race invite defeat and humiliation. Competition, like cooperation, is a fraud in which victory goes not to the swift or to the strong or to the persevering but to the deceitful.

The social environment depicted in these tales is violent, cruel, and remorseless. By Flusche's count, "in about half the tales, by sudden inspiration or deliberate forethought, one animal burned, scalded, or hacked another."[61] Violent acts are not justified as the just comeuppance due to an oppressor or bully or as the righteous wreaking vengeance on the wicked. Instead the violence is spiteful, vindictive, even gratuitous and purposeless. Br'er Rabbit not only persuades other animals to help him out of traps but tricks them into taking his place. A lady who resists his wooing he kills, skins, and smokes over hickory chips. After persuading his neighbors to help him build a spring house, Br'er Rabbit has them all drowned. And on and on.[62]

Trickery may enable a slower or weaker creature to outsmart a faster or stronger one, but the cunning is corrosive of all social relationships. There is nothing Br'er Rabbit will not do to his fellow animals in order to survive. Loyalty and self-sacrifice, even pity and regret, are emotions totally foreign to Br'er Rabbit's amoral world. Even his own family members are not exempt from his self-interested scheming. In one story, Br'er Wolf seeks revenge upon Br'er Rabbit for having boiled Wolf's grandmother and tricked Wolf into eating her. To save himself, Br'er Rabbit sacrifices his own wife and children.[63]

Why did slaves find these stories so appealing? Why did they relish telling their young tales with such overtly amoral even antisocial messages? Interviews with former slaves suggest that many believed the stories accurately portrayed a hostile, compassionless world in which cooperation was futile and trust dangerous. Br'er Rabbit was admired for his ability to survive and even advance his material fortunes in the face of a harsh, unforgiving social environment in which one animal's gain was invariably another's loss. Asked to explain what the Br'er Rabbit stories meant to him, one former slave explained: "I is small man myself; but I ain't nevver 'low no one for to git 'head of me. I allers use my sense for help me 'long, just like Br'er Rabbit." That his master had made him a driver on the plantation so he did not have to work as hard and got more rations of food and clothes than other slaves, he attributed to "usin' my sense." Another former slave believed that the moral of these stories was: "learn not to trust . . . people whom you do not know." And another reported the moral as: "when two fellows are in your way, you must make them fight, then you will always save your skin."[64] This testimony suggests that many slaves understood the Br'er Rabbit stories as an allegory for all social relationships, not just the relationship between master and slave.

A passage from the slave narrative of William Wells Brown shows how the amoral individualism[65] of Br'er Rabbit was sometimes acted out in real life by slaves. Brown was at the time hired out to a slave trader by the name of Walker. On this particular occasion, Brown was called upon to serve wine to several men negotiating with Walker about the purchase of some slaves. When Brown filled the glasses perilously high, the visitors spilled wine on their clothes. Walker deemed it necessary to punish the slave's act of carelessness, and so the next morning he sent Brown to deliver a note and a dollar to the jailer. Suspecting "that all was not right," Brown showed the message to a sailor who informed the slave that the note instructed the jailor to whip him. The slave's response to this predicament was straight out of the pages of Br'er Rabbit:

> While I was meditating on the subject I saw a colored man about my size walk up, and the thought struck me in a moment to send him with my note. I walked

up to him, and asked him who he belonged to. He said he was a free man, and had been in the city but a short time. I told him I had a note to go into the jail, and get a trunk to carry to one of the steamboats; but was so busily engaged that I could not do it, although I had a dollar to pay for it. He asked me if I would not give him the job. I handed him the note and the dollar, and off he started for the jail.

The result was that the unsuspecting black man got a severe flogging and the cunning Brown escaped without physical harm. Unlike Br'er Rabbit, however, Brown later felt deep remorse at the deception he had practiced on his fellow man, citing this incident as an example of how "slavery makes its victims lying and mean."[66]

Even those slave stories that unambiguously referred to the master–slave relation share the atomized structure of the Br'er Rabbit stories. In the Old Master and John tales, for instance, John the slave tries to trick his master for his own advantage, not for the benefit of the slave community as a whole.[67] Notably absent from slave folklore, observes Kolchin, were "mythological stories [that] emphasized classical elements of heroism, self-sacrifice, and group solidarity. . . . Slave tales contain few depictions of courageous or noble behavior; there are no dragon slayers, giant killers, or defenders of the people." Even the animal trickster tales told by Russian serfs "often expressed a communal solidarity lacking in the slave stories. Animal tales that reflected a devil-take-the-hindmost amorality constituted a much smaller fraction in Russia . . . than in America." In sum, Kolchin concludes, "peasant folklore, unlike that of American blacks, contained substantial elements of social consciousness expressing lofty ideals of heroism, generosity, and struggle for the common good."[68]

The folklore of American slaves is not about affirming collective identity or about arousing collective resistance to slavery. Rather it speaks to individual survival in a hostile environment of atomized subordination. The recurring themes of these stories, as Flusche argues, "suggest that slavery tended to engender an atomistic, individualistic world view among the slaves."[69] But this is not the individualism of the aspiring entrepreneur, who sees the world as his oyster; it is instead the individualism of the ineffectual, who finds himself prevented from joining with others by virtue of a tightly regulated environment. Neither the entrepreneur nor the ineffectual puts much store in group solidarity. But the entrepreneur has the social freedom to negotiate for himself, to join with others to advance his fortunes, to build networks, or even to drop out. The ineffectual has no such freedom, for his life is prescribed by others.

Fatalism as a Rational Response to a Capricious Environment

Contemporary observers of slavery often described American slaves as fatalistic, apathetic, indifferent, resigned. Some analysts have dismissed this testimony as nothing more than slaveholder propaganda designed to justify the enslavement of blacks. It is true that those efforts by members of the southern medical profession to explain slave passivity in terms of innate racial makeup were unambiguously racist. But it is also true, as Genovese points out, that "too many contemporaries, black and white, described this fatalism to permit its dismissal as a figment of white racist imagination."[70] It was Frederick Douglass, not a proslavery apologist, who recalled living "among slaves, who seemed to ask, 'Oh, what's the use?' every time they lifted a hoe."[71]

Fatalism is a learned response to a social environment in which there is only a tenuous connection between preferences and outcomes. If slaves felt that there was little they could do to significantly alter their circumstances, it was because for most of them this was the hard reality. They felt that life was like a lottery because it often was. Living in a capricious world in which one could be punished on a whim or separated from family members without consent made it difficult for slaves to develop a self-reliant cosmology. Some unusual individuals in unusual circumstances did; most did not.

Fatalism is not, of course, an attitude limited to slaves. Misfortune, disaster, sickness, and death strike all human lives. Our control over the physical environment is often precarious, a truth which earthquakes, floods, and other natural disasters periodically bring home to even the most self-reliant individualists. Lack of control over nature's course is part of the human condition and thus creates a degree of fatalism in all of us. Indeed an understanding that there are things beyond an individual's control is perhaps the beginning of wisdom. But if no individual has total control over his physical environment, some individuals have greater control than others. Poor people usually live with much greater uncertainty, many fewer resources to resist nature's vagaries, and a much higher incidence of death and sickness. As one moves in the direction of these more vulnerable social worlds, a fatalistic orientation becomes increasingly rational.

The death of a master was one of many events in a slave's life that revealed the capriciousness and vulnerability of a slave's existence. In his autobiography, Frederick Douglass vividly recalls the terrifying sense of uncertainty and lack of control that he felt upon his master's death. The master's estate was to be divided among two children, Andrew and Lucretia. The slaves knew Master Andrew to be a cruel and profligate drunkard; Mrs. Lucretia, on the other hand, was widely acknowledged to be kind and considerate. To fall into the hands of Master Andrew meant not only brutal treatment but also the strong possibility of being sold away to the far South if he fell in debt. The prospects for those slaves who went to Mrs. Lucretia were inestimably brighter. But the slaves' preferences counted for nothing. The master's property—slaves together with cattle, pigs, and horses—was valued and divided without consulting the slaves in any way. "We had no more voice in the decision of the question," Douglass remembers, "than the oxen and cows that stood chewing at the haymow. One word from the appraisers, against all preferences or prayers, was enough to sunder all the ties of friendship and affection. . . . Thus early, I got a foretaste of that painful uncertainty which slavery brings."[72]

It is not that slaves lack preferences. Rather it is, as Douglass saw, that "neither their aversions nor their preferences avail them anything." Being a slave, Douglass lamented, "my wishes were nothing."[73] It is difficult to sustain preferences where they have little or no effect upon outcomes. Over a lifetime, many will learn that to formulate a preference serves no positive purpose.[74] Resignation to one's lot becomes a rational adaptation, reducing cognitive dissonance by teaching the slave not to want what he can't have.[75]

Chance figured prominently in the lives of most slaves. It was a world, Kenneth Stampp observes, "full of forces which [they] could not control."[76] This capricious environment generated a cosmology that attributed success less to individual effort than to good luck. Douglass recalls, for instance, how his grandmother's success in preserving seedling sweet potatoes was widely attributed by her neighbors to good fortune: "It happened to her—as it will happen to any careful and thrifty person residing in an ignorant and improvident community—to enjoy the reputation of having been born to 'good luck.'"[77] The slaves' arbitrary social world gave scant encouragement to the belief that there was a just or predictable relation between individual effort and success.

Although it is true, as Fogel and Engerman stress, that slaves who were especially diligent were sometimes promoted from field work to managerial or artisan positions, many other hardworking slaves were abruptly and arbitrarily shifted or sold with little or no relation to their work performance. Slave autobiographies are filled with such stories. One slave, who was the son of a groom's assistant, started out his youth taking care of horses. Upon the owner's death, however, the widow abruptly sold the horses and assigned those responsible for tending them to field work. Another slave from Kentucky began life as the pet of his white father owner only to be suddenly sold away to Alabama to work in the fields.[78]

The narrative of William Wells Brown illustrates the random character of slave mobility in a particularly dramatic fashion. Born in Kentucky, Brown was hired out by his master at a young age. First he was hired out as a house servant to a Major Freeland, a drunkard and a gambler who cruelly abused his slaves. After a short time, Major Freeland failed in business and Brown was then employed on a steamboat, where he spent what he described as "the most pleasant time . . . I had ever experienced." At the end of the sailing season, he was hired out to a hotelkeeper, who was perhaps the most "inveterate hater of the negro [who] ever walked on God's green earth." After a brief, unhappy time at the Missouri Hotel, Brown was hired to Elijah P. Lovejoy, a well-known abolitionist publisher and editor. In Lovejoy's employment, Brown was well treated and even obtained a little learning. After only a short time with Lovejoy, however, Brown was severely beaten in the street and forced to return to his master. After recovering from the beating, he was hired out as a steward to a steamboat captain, a situation Brown again found pleasant. But this, too, was short-lived, for the captain left his boat at the end of the summer, and Brown was returned home to work in the fields on his master's farm. Shortly thereafter he was moved from the field to the house to work as a waiter. Some time later, he was hired out to work for a slavetrader, a job Brown found odious in the extreme. After serving out the year with the slavetrader ("the longest year I ever lived"), Brown returned home and attempted to escape. The effort failed and he was ordered to work in the field but soon thereafter was sold as a house servant to a merchant tailor (from whom he soon escaped).[79] Of these many changes in employment, none is traceable to personal effort or skill. There is no clear progression of upward or downward mobility. Working conditions seem to vary randomly: sometimes better, sometimes worse. Decisions about careers are imposed from without, and to the slave at least there seems no rhyme or reason behind them. In such an uncertain and random environment an ideology of resignation and chance appears more rational than an ideology of self-improvement.

Slaveowners often complained of their slaves' passivity and indifference toward the future, but at the same time they promoted the very fatalistic characteristics they deplored. The editor of the *Southern Planter*, for instance, advised his readers that "no laborer should have his attention distracted or his time occupied in thinking about what he should do next: the process of thought and arrangement should devolve wholly upon the superintendent."[80] Another defended restricting slave movement on the grounds that such a policy ensured that "they do not know what is going on beyond the limits of the plantation, and [thus] feel satisfied that they could not . . . accomplish anything that would change their condition."[81] Initiative and foresight were double-edged swords: they were qualities that could make for a more responsible and productive work force as well as a more rebellious and unmanageable one.

If fatalism was pervasive among southern slaves it was not uniformly so. Large plantations with absentee owners in newly settled areas of the Southwest were particularly prone to produce an exaggerated fatalistic bias. Slaves on these plantations were often ill-treated and "profoundly apathetic, full of depression and gloom, and seemingly less hostile than indifferent toward the white man who controlled him."[82] Where a modicum of social space opened up, as occurred when slaves were permitted to hire themselves out, nonfatalistic personalities were capable of developing.[83] The greater degree of social space afforded the skilled slave (coopers, blacksmiths, bricklayers, and carpenters) explains their greater spirit of independence and initiative when compared with the closely supervised field hands.[84] Variation in cultural bias corresponded to variation in social experiences.

Frederick Douglass is a prominent example of a slave who was certainly no fatalist. But his social experiences were also exceptional. At the age of ten, Douglass was removed from a large Maryland plantation and sent to live with a Baltimore family. Here he found that "the crouching servility of a slave, usually so acceptable a quality to the haughty slaveowner, was not understood nor desired." His new mistress even began to teach Douglass to read before her husband put a stop to it. The relative freedom of Baltimore ("a city slave," Douglass wrote, "is almost a free citizen . . . compared with a slave on [a] plantation") afforded him the opportunity to work and associate with free men and to read speeches "filled with the principles of liberty." Douglass himself acknowledged that "but for the mere circumstance of being thus removed before the rigors of slavery had fastened upon me; before my young spirit had been crushed under the iron control of the slave-driver, instead of being, today, a freeman, I might [still] have been wearing the galling chains of slavery." Douglass's autobiography, in short, is a fascinating and self-conscious case study in the social conditions and qualities of mind necessary to escape what Douglass described as the "spirit-devouring thralldom" of bondage.[85]

It was the exceptional slave who, like Douglass, had experienced a substantial degree of social autonomy who almost invariably led slave uprisings or engineered escapes. Douglass testified that "perhaps not one of [my co-conspirators], left to himself, would have dreamed of escape as a possible thing. Not one of them was self-moved in the matter. They all wanted to be free; but the serious thought of running away, had not entered into their minds, until I won them to the undertaking."[86] The few slave revolts that did occur, as Elkins points out,

> were in no instance planned by plantation laborers but rather by Negroes whose qualities of leadership were developed well outside the full coercions of the plantation-authority system. Gabriel, who led the revolt of 1800, was a blacksmith who lived a few miles outside Richmond. Denmark Vesey, leading spirit of the 1822 plot at Charleston, was a freed Negro artisan who had been born in Africa and served several years aboard a slave trading vessel; and Nat Turner, the Virginia slave who fomented the massacre of 1831, was a literate preacher.[87]

Rebellions as well as runaways were disproportionately concentrated among urban, skilled slaves who had greater social space and had experienced more lenient treatment.[88] From a cultural point of view, the social source of resistance is as significant as the relative absence of slave rebellions. Both testify to the reciprocal relationship between a fatalistic bias and the social relations of slavery.[89]

Fatalism and Freedom

That antebellum slaves were fatalistic does not mean they did not desire freedom. Nor does it mean that they did not care whether they were slave or free. There is no reason to doubt the word of a former slave who told an interviewer that slaves used to "pray constantly for the 'day of their deliverance.'"[90] But if most slaves dreamed and even prayed for freedom, few had much reason for thinking such an objective attainable through their own efforts.[91] They could pray, wait patiently, even hope but they could not act on this dream. Deliverance could come from without or above but not from within. The great mass of slaves, reported former slave Henry Bibb, "know that they are destined to die in that wretched condition, unless they are delivered by the arm of Omnipotence."[92] Raising one's own arm in resistance was insufficient without outside assistance and unnecessary with it.

Edward Banfield uncovered a parallel pattern of behavior among peasants of southern Italy. Although "getting ahead" is a recurring theme of peasant existence, Banfield found that the peasant "sees that no matter how hard he works he can never get ahead.... He knows ... that in the end he will be no better off than before." The peasants studied by Banfield wait for fortune to smile upon them. They wait, for instance, for the "call" that will enable them to take their family to the more prosperous regions of the north or, even better, for the "call" from a relative in America that will enable them to migrate to the United States. They wait and they hope and they pray, but there is little they can do, or so they believe, to realize the desired outcome. "The idea that one's welfare depends crucially upon conditions beyond one's control—upon luck or the caprice of a saint"—acts as a tremendous check upon individual initiative.[93] What distinguishes fatalists from adherents of other ways of life then is not the desire for a better life, but the feeling that fate and society conspire to prevent them from improving their condition.

Much the same is true of today's so-called underclass. No doubt the urban poor wish to be rich. No doubt they hope to get ahead and escape poverty. What is often missing, however, is the belief that getting ahead and escaping the coils of poverty is something that is attainable through sustained individual or collective effort. The belief that individual initiative and collective action are inefficacious becomes self-fulfilling: believing that there is little they can do to significantly alter their lives, their lives in fact remain unaltered.

This is not to suggest fatalism is all in the head. Fatalism as a cultural bias can flourish only where social institutions sustain that bias as an adaptive and rational posture. In a world in which there are no escapes and few rewards, passivity or resignation is more rational and adaptive than the individualist's incurable entrepreneurial optimism. As the young Indian boy living on the Bombay streets in the film *Salaam Bombay* eventually discovers, certain social environments reward resignation and withdrawal over initiative and cooperation.

The fatalistic cultural bias of southern slaves was sustained by a hostile, random, and tightly prescribed social environment. Although slaves sometimes benefited from the pockets of social space that opened up in urban areas, in the slave quarters, or in the practice of hiring out, most slaves most of the time had little choice over the type of work to be done or how it was to be performed. Excluded from the group that made the key decisions affecting their lives—where to live, when to move, where to work—fatalistic resignation was an adaptive and rational cultural response for most slaves.

8

A Life of Hermitude: Thoreau at Walden Pond

Few Americans have proven as difficult to categorize as Henry David Thoreau. He has become, in the words of John Diggins, "a man for all persuasions."[1] For some, Thoreau is the bourgeois individualist par excellence. This was the view of such early twentieth-century Marxists as V. F. Calverton, who saw Thoreau as "the best individual product of the petty bourgeois ideology" of his period.[2] Much the same view is adopted by Sacvan Bercovitch, who argues that "*Walden* embodies the myth of American laissez-faire individualism." It is a work "intended not to change the profit system but to cure its diseases," "to wake his countrymen up to the fact that they were desecrating their own beliefs." Thoreau and Aristablus Bragg, James Fenimore Cooper's personification of the restless, acquisitive middle class, may "march to different drummers" but, Bercovitch insists, they march "along the same free and enterprising American way."[3]

But if the withdrawn Thoreau and the entrepreneurial Bragg are marching down the same individualistic path, then we need to ask whether the category of individualism lacks the capacity to discriminate between social phenomena. Is it the enterprising American way to excoriate trade for cursing everything it touches, or to throw scorn upon the accumulation of possessions? What sort of enterprising nation would this be if its inhabitants followed Thoreau's example of "studying [not] how to make it worth men's while to buy my baskets, [but] rather how to avoid the necessity of selling them," or heeded his counsel to "cultivate poverty like a garden herb, like sage" and "not [to] trouble yourself to get new things, whether clothes or friends"?[4] Thoreau is clearly more than just a quirky capitalist.

Many on the left have attempted to claim Thoreau as one of their own. In its early years, the British Labor party used *Walden* as "a pocket-piece and traveling bible of their Faith."[5] More recently, Staughton Lynd has stressed the similarities in the analyses of Thoreau and Marx. Lynd detects parallels not only in the two men's diagnoses of the alienating effects of capitalism but also in their "visions of an alternative" society without a division of labor.[6] "Here obviously were the spokesmen not of two utterly alien traditions with nothing to say to one another, but of two variants of one tradition springing from Rousseau's insight that (as

140

Thoreau expressed it) 'just in proportion as some have been placed in outward circumstances above the savage, others have been degraded below him.'"[7]

But what sort of a Marxist is it who rejects all forms of collective action, indeed refuses to join any group? Although Marx and Thoreau were agreed on what they didn't like—competitive capitalism—the cultural biases from which they launched their critiques were entirely different. For Marx, a primary defect of competitive individualism is its asocial nature; men are separated from each other and from the community.[8] For Thoreau, in contrast, capitalism is vile because its expansive materialism distracts from man's ability to attend to his inner, spiritual life. For Marx, capitalism's malady is the isolation of man in an "egoistic life"; for Thoreau, greater isolation is a solution to capitalism's ills. Socialism? His affections do not that way tend.

Others have tried to enlist Thoreau as an exponent of classical republicanism. Leonard Neufeldt, for instance, argues that Thoreau "sought to purify republicanism by recovering what he considered to be its true meaning and imperatives."[9] Thoreau does share classical republicanism's suspicion of commerce, but his critique of the marketplace begins from a radically different premise. Classical republicanism criticized commerce in the name of civic virtue, that is, the disinterested pursuit of the public good; Thoreau attacks commerce in the name of private well-being. A vision of life in which individual interests are subordinated to the good of the group is radically different from a vision in which individual conscience and identity are paramount. Classical republicanism taught self-sacrifice; Thoreau taught self-absorption.

Thoreau at Walden Pond, in sum, is neither classical republican nor proto-Marxist nor bourgeois individualist; he belongs neither to the right nor to the left; he is neither conservative nor liberal nor radical nor reactionary. Thoreau's experience does indeed, as one critic has suggested, "explode all our conventional political categories."[10] Some have tried to sidestep Thoreau as an unimportant anomaly in American political culture; thus Thoreau goes unmentioned in the important works of Louis Hartz, Seymour Martin Lipset, and Samuel Huntington.[11] But this seems unsatisfactory, for Thoreau expresses a worldview that recurs throughout American history.

Making sense of Thoreau requires us to rethink the categories that are conventionally used to analyze American political culture. We may better understand Thoreau if we see him as an instance of what Mary Douglas identifies as the voluntary recluse or hermit. The hermit resembles the competitive individualist in that his social experience is not constrained by either group boundaries or ascribed status. But while both the hermit and the competitive individualist are free to transact for themselves, the hermit repudiates the rewards of the competitive system. Whereas the individualist tries to make himself the center of a network, the hermit consciously withdraws from relationships of control and manipulation.[12]

In his withdrawal from the world, the hermit resembles in some ways the egalitarian community that walls itself off from the larger society. This was what biographer Joseph Krutch had in mind when he said that Thoreau "set up a kind of one-man Brook Farm."[13] Although both the hermit and the egalitarian share a critical view of the division of labor and competition that accompanies the individualist way of life, the hermit differs fundamentally from the egalitarian in preferring a solitary existence to social solidarity.

Thoreau's voice (at least the voice that emerges from the pages of *Walden*) is the voice of the hermit, the autonomous individual who withdraws from coercive social involvement in order to live a life of austere self-sufficiency. My aim in this chapter is to explicate the hermit cosmology of Thoreau and to show how it differs from the individualist and egalitarian currents prevalent in antebellum America.

"A little world all to myself"

Adherents of the autonomous way of life seek physical removal from the temptations and manipulation of society. This separation may take the form of isolated mountain retreats or overlooked garrets closer at hand. Thoreau's mountain fastness was a ten-foot by fifteen-foot one-room cabin on the shore of Walden Pond, nestled in the woods "a mile from any neighbor."[14]

The New England that Thoreau sought to remove himself from, Ralph Gabriel has written, was "bustling with industrial enterprise. . . . Almost every New England stream was turning the wheels of some mill or factory. Boston was the rendezvous of entrepreneurs great and small, men skilled in the . . . art of pecuniary competition."[15] This competitive way of life, Thoreau believed, robbed individuals of true autonomy. Interdependency replaced self-sufficiency. It turned even the most prosperous among them into little better than slaves: "It is hard to have a southern overseer; it is worse to have a northern one; but worst of all when you are the slave-driver of yourself."[16]

If Thoreau rejected the self-seeking, competitive world he saw around him in Concord, he was equally adamant in opposing communitarian alternatives. Thoreau refused to join any collective movement, even when (as with the abolitionists) he fully sided with their cause. While other New England Transcendentalists flirted with utopian communes such as George Ripley's Brook Farm (1841–47) and Amos Bronson's Fruitlands (1843–44), Thoreau affirmed his preference for "the solitary dwelling." Group obligations impeded individual autonomy. "The man who goes alone," Thoreau explains, "can start to-day; but he who travels with another must wait till that other is ready." Joining a group inevitably diminishes the man: "Wherever a man goes, men will pursue and paw him with their dirty institutions, and, if they can, constrain him to belong to their desperate odd-fellow society."[17] Cooperation no less than competition, Thoreau believed, would subject the individual to coercion and manipulation by others.

The difference between the group-oriented ideal of abolitionists and the autonomous vision of Thoreau is nicely captured in the contrast between Charles Follen's description of the abolitionist movement as "a world in ourselves and in each other" and Thoreau's characterization of his "little world all to myself."[18] Abolitionists believed they must shun those compromised by the sin of slavery, but they also felt a simultaneous urge to join in "pious fellowship" with those who were free of moral taint. Thoreau, in contrast, professed to have "never found the companion that was so companionable as solitude."[19] Whereas the abolitionist Wendell Phillips fondly recalled "the large and loving group that lived and worked together [and had been] all the world to each other," Thoreau glorified the solitary individual, beholden to no one. Instead of the tightly bound, cohesive fellowship of abolitionism, Thoreau preferred life in "the great ocean of solitude."[20]

The hermit does not necessarily spurn human contact entirely. Indeed he may have numerous friends but, as Mary Douglas explains, "the voluntary isolate expects no more from friends than friendly unannounced appearances, like the sparrows on the sill."[21] This precisely describes Thoreau's relationship with his friends. Thoreau enjoyed the occasional visitor, but he wished to keep these relationships unpredictable and spontaneous, thereby avoiding obligations and those deep "ruts of tradition and conformity." Particularly exemplary in his view was a poet friend (Ellery Channing) who would visit him at the least likely times, during the "deepest snows and most dismal tempests." "Who," he asked approvingly, "can predict [the poet's] comings and goings?" Thoreau seemed to take as much if not more pleasure from his unpredictable animal visitors: from the birds, for instance, that "sang around or flitted noiseless through the house," or from the red squirrels that would chuckle and chirrup beneath his feet. Such animals made ideal friends for Thoreau because they imposed no duties and made no demands, thus leaving him beholden to no one.[22]

"My greatest skill has been to want but little"

For the individualist, needs are limitless. Yesterday's luxuries become today's necessities. What the individualist buys is limited only by what he can afford. Increase his resources and he will increase his demands. Economizing is thus brought about only by the finiteness of resources.[23]

In the opening chapter of *Walden*, entitled "Economy," we are immediately confronted with the completely different voice of the hermit. Thoreau sets himself the task of distinguishing the "necessaries of life" from what are "luxuries merely." Having distinguished between those things that are truly necessary from those that we desire only out of "a regard for the opinions of men," Thoreau proceeds to figure out what resources are necessary to meet these irreducible needs. Where the expansive individualist increases his needs to meet his growing resources, Thoreau decreases his resources and his needs so that the latter will nestle comfortably within the former. "He chose to be rich," Emerson explained, "by making his wants few."[24]

"Simplify, simplify" was Thoreau's mantra. "Instead of three meals a day, if it be necessary eat but one; instead of a hundred dishes, five." Thoreau asks his reader also to "consider . . . how slight a shelter is absolutely necessary" to protect oneself from the elements.[25] It is the same cultural bias we hear when we listen to Po Chü-i, a ninth-century Chinese ascetic:

> What I shall need are very few things.
> A single rug to warm me through the winter;
> One meal to last me the whole day.
> It does not matter that my house is rather small;
> One cannot sleep in more than one room![26]

For Thoreau, like Po Chü-i, worldly possessions are not a sign of success (as they are for the individualist), but rather an affliction that weighs down the human spirit. "I see young men, my townsmen," laments Thoreau, "whose misfortune it is to have inherited farms, houses, barns, cattle, and farming tools. . . . How many a poor immortal soul have I met well nigh crushed and smothered under its load, creeping down the road of life, pushing before it a barn seventy-five feet by forty." Possessions are a burden to be shed, not accumulated. Exemplary in Thoreau's view is the practice of the Mucclasse Indians, who periodically burned all of their belongings—clothes, utensils, furniture, food—in a great bonfire. None was more impoverished, Thoreau writes, than the "seemingly wealthy, . . . who have accumulated dross, but know not how to use it, or get rid of it, and have thus forged their own golden or silver fetters."[27]

Thoreau spurned not only the upward spiral of the individualist way of life ("shall we always study to obtain more of these things, and not sometimes to be content with less?") but also those institutions that would set his needs for him. After graduating from Harvard College, Thoreau had tried his hand at teaching but soon gave it up because, as he explains in *Walden*, "my expenses were in proportion, or rather out of proportion, to my income, for I was obliged to dress and train, not to say think and believe, accordingly."[28] Not wishing to have his needs imposed by others, Thoreau cast around for alternative ways of life that would allow him to control both his needs and resources.

The isolated cabin by the shore of Walden Pond was such a place where Thoreau could lead "a life of simplicity," without the temptations and distractions of society. Decreasing his worldly needs enabled Thoreau to devote less time to "making ends meet" and more time to fulfilling spiritual needs. Laboring in the workaday world, Thoreau feared, too often distracted men from cultivating their inner world. He tells, for instance, of having been "terrified" to find that three pieces of limestone adorning his desk required dusting daily while "the furniture of my mind was all undusted still." Having made this terrible discovery, Thoreau promptly hurled the offending pieces of stone "out the window in disgust."[29]

To live simply, for Thoreau, was a prerequisite for a life of true wisdom. It was "life near the bone" that was "sweetest." "None can be an impartial or wise observer of human life," he thought, "but from the vantage ground of what we should call voluntary poverty."[30] For models of exemplary behavior he looked to the ascetics of the East: "The ancient philosophers of Chinese, Hindoo, Persian, and Greek, were a class than which none has been poorer in outward riches, none so rich in inward." Not only was money "not required to buy one necessary of the soul," but the possession of "superfluous wealth" invariably led to a feeding of the baser passions at the expense of starving the human spirit.[31]

"Trade curses every thing it handles"

Thoreau rejects in no uncertain terms the individualist ideal of bidding and bargaining. "Trade," he laments, "curses every thing it handles; and though you trade in messages from heaven, the whole curse of trade attaches to the business." In its place he puts forward the autonomous ideal of self-sufficiency. "Instead of studying how to make it worth men's while to buy my baskets," he explained, "I studied rather how to avoid the necessity of selling them."[32]

At Walden, Thoreau withdrew as much as possible from the coercive networks of "restless city merchants" and "adventurous country traders," trying where possible to "avoid all trade and barter" and to minimize his reliance on "distant and fluctuating markets." For the individualist, bidding and bargaining provides opportunities to increase the size of one's network, to make oneself a "Big Man" even if at the expense of relegating those less fortunate to the status of "Rubbish Men." Thoreau, hermit that he is, refuses to engage in this game of social manipulation and coercion. Rather than "live in this restless, nervous, bustling, trivial Nineteenth Century," he prefers to "stand or sit thoughtfully while it goes by."[33]

Thoreau has only barely disguised contempt for those who do naught but "buying and selling, and spending their lives like serfs."

> I respect not his labors, his farm where every thing has its price; who would carry the landscape, who would carry his God, to market, if he could get any thing for him; ... on whose farm nothing grows free, whose fields bear no crops, whose meadows no flowers, whose trees no fruits, but dollars; who loves not the beauty of his fruits, whose fruits are not ripe for him till they are turned to dollars.

The marketplace, he insists, can never realize the true value of objects. Lakes like Walden, he tells us, "are too pure to have a market value; they contain no muck." Nor can huckleberries yield "their true flavor" to those who would buy or sell them.[34]

Thoreau believed that the marketplace—far from promoting self-reliance, as the individualist would have it—impels men to demean and humiliate themselves. Sellers must lie, flatter, and deceive in order to obtain buyers. "What mean and sneaking lives many of you live," writes Thoreau contemptuously, "contracting yourselves into a nutshell of civility or dilating into an atmosphere of thin and vaporous generosity, that you may persuade your neighbor to let you make his shoes, or his hat, or his coat, or import his groceries for him." How much more healthy for each man to provide for himself, to be dependent on no one but himself for his food, clothing, and shelter.[35] The curse of capitalist exchange relations, as Thoreau sees it, is not that it isolates people (the communitarian complaint), but rather that it dangerously increases people's dependence upon one another.

"Such sweet and beneficent society in Nature"

The distinctiveness of Thoreau's preferred way of life is reflected, too, in his conception of the natural world. For the individualist entrepreneur, nature must be subdued so that her rich resources may be harnessed to serve the needs of men. Nature must be tamed in order to be exploited. Man's relation to nature in this worldview is instrumental in character.

Thoreau totally repudiates this instrumental conception of nature. Rather than using nature, Thoreau attempts to make himself "a part of ... Nature." Nature reciprocates by befriending him. "In the midst of a gentle rain," Thoreau recounts becoming "suddenly sensible of such sweet and beneficent society in Nature, in the very pattering of the drops, and in every sound and sight around my house, an infinite and unaccountable friendliness. . . . Every little pine needle expanded and swelled with sympathy and befriended me." He counted Walden Pond as a "neighbor," enjoyed "the friendship of the seasons," became a "neighbor to the birds," a "friend . . . of the pine tree," and even felt "intimate" with his beans.[36]

Thoreau exhibits little desire to tame or bend nature to his will. Instead he wishes to have his existence swallowed up in nature's "indescribable . . . beneficence." Nature is to be enjoyed, not used; admired, not subdued. "I love the wild," he tells us, "not less than the good." He marvels at the savagery he witnesses when he stumbles upon a ferocious war between colonies of red and black ants. But nature's wildness is more often peaceful and innocent. It is "the peep of the young" or the "turtle-doves fluttering "from bough to bough of the soft white-pines."[37]

His disdain for the entrepreneurial conception of nature as a resource to be exploited is evident from the story he relates of a group of some hundred men who came one winter to cut the ice from Walden Pond. To Thoreau, their enterprising scheme seems patently absurd and not a little unkind to "Squaw Walden." Was it not cruel to take "off the only coat, ay, the skin itself of Walden Pond in the midst of hard winter"? Was it not ridiculous to treat a lake as if it were a "model farm" fit for "ploughing, harrowing, rolling, furrowing"?[38] For Thoreau, Walden Pond is not something to be mastered but an occasion for meditation and reflection.

"Time is but the stream I go a-fishing in"

The hermit possesses a distinctive conception of time. He lives in the present, wishing to be bound neither by obligations from the past nor worries about the future. With the arrival of each day the slate is wiped clean; the sins and deeds of yesterday are erased. Fears for what tomorrow may bring are not allowed to intrude themselves lest they generate the anxious enterprise that would disturb the hermit's deliberate peace.[39]

In this respect, as in so many others, Thoreau shows himself a representative hermit. In Thoreau's view, "all times and places and occasions are now and here. God himself culminates in the present moment, and will never be more divine in the lapse of all the ages." "We should be blessed," Thoreau was persuaded, "if we lived in the present always . . . and did not spend our time in atoning for the neglect of past opportunities, which we call doing our duty." Each day, he believed, should bring with it a fresh beginning, washing clean the sins of the past. His guide is the injunction of Confucius to "renew thyself completely each day; do it again, and again, and forever again."[40]

The past, for Thoreau, is dominated by voices of ignorance and error. Little is worth preserving or remembering. "The old," Thoreau says, "have no very important advice to give the young" for "their lives have been such miserable failures." In "thirty years on this planet," Thoreau adds, "I have yet to hear the first syllable of valuable or even earnest advice from my seniors." "You may say the wisest thing you can old man," he taunts at another point, but "I hear an irresistible voice which invites me away from all that." To build monuments to past foolishness only compounds the folly. "To what end," Thoreau asks, "is so much stone hammered?" It is only the "insane ambition" of nations "to perpetuate the memory of themselves by the amount of hammered stone they leave."[41]

Just as people should not dwell in the past, so should they not worry about the future. Thoreau saw no sense in "spending the best part of one's life earning money in order to enjoy a questionable liberty during the least valuable part of it." What foolishness it is to devote oneself to "laying up treasures which moth and rust will corrupt." Fretting about the future produces only an empty life of "desperate haste." Instead he offers us the opportunity to "spend one day as deliberately as Nature," looking neither backward nor forward in time.[42]

For those with the inner fortitude to shut out both past and future, time becomes timeless. As a model, Thoreau offers us the artist of Kouroo. The artist determines to make a staff "perfect in all respects," laying aside all concern for the time it would take him to do so.

> He proceeded instantly to the forest for wood, being resolved that it should not be made of unsuitable material; and as he searched for and rejected stick after stick, his friends gradually deserted him, for they were old in their works and died, but he grew not older by a moment. His singleness of purpose and resolution, and his elevated piety, endowed him, without his knowledge, with perennial youth. As he made no compromise with Time, Time kept out of his way, and only sighed at a distance because he could not overcome him.[43]

Thoreau believed that the "undisturbed solitude and stillness" of Walden Pond enabled him to realize something of the timeless state achieved by the artist of Kouroo. Sitting "in my sunny doorway from sunrise till noon, rapt in revery, amidst the pines and hickories and sumachs," Thoreau tells us that "for the most part, I minded not how the hours went." "My days were not days of the week, bearing the stamp of any heathen deity, nor were they minced into hours and fretted by the ticking of a clock; for I lived like the Puri Indians, of whom it is said that 'for yesterday, to-day, and to-morrow they have only one word.'"[44]

"Read not the Times, Read the Eternities"

Nowhere is the hermit in Thoreau more evident than in his claim to a transcendent knowledge. The world of appearances is one of mere "shams and delusions." "Petty fears and petty pleasures are but the shadow of the reality." The nation's "so-called internal improvements" are "all external and superficial." Men are "deceived by shows"; they "think that *is* which *appears* to be"; they consider "not what is truly respectable, but what is respected"; they are led more by "a regard for the opinions of men" than by "a true utility"; their "inventions are wont to be pretty toys, which distract our attention from serious things."[45]

It is the hermit's insufferable arrogance that he claims to see what others cannot. Thoreau claims a privileged vantage point from which he can see the world as it "really is before a true gaze." He lays claims to a crystal clear vision capable of penetrating beneath "the mud and slush of opinion, and prejudice, and tradition, and delusion, and appearance" to grasp the "hard bottom and rocks in place, which we can call reality."[46] The hermit does not know a truth but *the* truth; he seeks not a version of reality but *the* reality.

For Thoreau, moreover, truth is not something to be sought in cooperation with one's fellow man (as the egalitarian might have it) or gained through experiential trial and error (as the individualist might have it) or through mastering a received body of technical knowledge (as the hierarchist might have it). Rather it is to be achieved through solitary contemplation. True knowledge comes not to those who formulate and test hypotheses but to those who look within themselves. The highest wisdom, Thoreau explains, "does not inspect, but behold."[47] He advised his readers to be "the . . . Lewis and Clark . . . of your own streams and oceans; explore your own latitudes. . . . Nay, be a Columbus to whole new continents and worlds within you."[48]

Little wonder then that the most remarked upon feature of Thoreau's character was his arrogant self-conceit. Even Thoreau's mentor Ralph Waldo Emerson, in an otherwise laudatory eulogy, could not deny that his friend's virtues "sometimes ran into extremes." In dealing with admirers, Emerson acknowledged, Thoreau "was never affectionate, but superior, didactic, scorning their petty ways." And Thoreau's habit of controverting others, he admitted, was "a little chilling to the social affections." His "dangerous frankness" and "inexorable demand . . . for exact truth," Emerson concluded, perhaps "made this willing hermit more solitary even than he wished."[49]

Thoreau's haughty aloofness, though perhaps amplified by the peculiarities of his personality, is a natural accompaniment of the autonomous way of life. The hermit's transcendence of this world invariably produces a disdain for the multitude, whose "vision does not penetrate the surface of things."[50] Judged against the hermit's truth, society seems hollow, insincere, hypocritical, deceitful. Before "such terrible eyes,"[51] the web of half-truths, ambiguities, and compromises that hold society together are revealed as sordid corruptions of individual principle.

Thoreau felt it his duty to seek the truth and damn the consequences. "Rather than love, than money, than fame," exclaimed Thoreau, "give me truth." This is a noble ideal for the scholar but a questionable guide for citizens, politicians, neighbors, and friends. A nation of incurable truth-tellers would have no means of compromising differences. Ambiguity, evasion, and even deception are the necessary accompaniments of social peace. Secure in his place on the sidelines of social life, however, the hermit can afford to dismiss our "mean and sneaking lives" with a derisive laugh or knowing smile. But for we who must live in this compromised world, the contemptuous voice of the hermit ("Not one of my readers has yet lived a whole human life"; "We are sound asleep nearly half our time"; "It is a fool's life, as they will find when they get to the end of it") may seem an irresponsible if not irrelevant annoyance.[52]

The Hermit's Relationship to the Outside World

The hermit's relationship to the wider world is characterized by a certain ambivalence. From his privileged position, the hermit is often tempted to prescribe how others should live. Yet his withdrawal from the world also inclines him to leave the world and those in it alone. The urge to prescribe behavior is further dimmed by the hermit's belief that no individual should interfere with the proper course of another's life.

Throughout *Walden*, Thoreau wrestles with this dilemma. He wishes to offer advice but without prescribing; he desires to alert his neighbors to the errors of their ways yet disavows any wish to reform their lives or the world. The preaching Thoreau is the Thoreau who brags "as lustily as chanticleer in the morning, standing on his roost, if only to wake my neighbors up." This is the Thoreau who records and publishes his living expenses in luxuriant detail so as to provide a practical model for others to follow. Yet the withdrawn Thoreau is equally in evidence. This is the Thoreau who scathingly attacks reformers and do-gooders: "If I knew for a certainty a man was coming to my house with the conscious design of doing me good, I should run for my life." And it is the Thoreau who cautions his readers, in the best hermit tradition, that each man should "be very careful to find out and pursue his own way."[53] This is also the Thoreau who could write (in a letter to the abolitionist Parker Pillsbury on the eve of the firing on Fort Sumter): "I do not so much regret the present condition of things in this country (provided I regret it at all) as I do that I ever heard of it."[54] Only by renouncing all interest in making the world a better place can the hermit sustain his separation from the world.

The attention paid to the hermit sometimes seems grossly out of proportion to his direct influence on society. By his refusal to organize, the hermit ensures that his commitments will be ineffectual. But his removal from society also carries with it a form of indirect power, for the hermit's uncompromising voice can seem a beacon of sanity and truth. Those disillusioned with the dominant social relationships may feel that the hermit has stripped away the hypocrisy masking societal stupidity and coercion. The tendency to look to the hermit for an indictment of existing social relations explains much of Thoreau's subsequent fame. For egalitarians, Thoreau provides a powerful indictment of the competitive individualist's multiplication of wants and exploitation of nature. For those who have lost out (or refuse ever to join) in the competitive struggle, Thoreau's withdrawal provides an appealing justification for dropping out.

If those in society are sometimes attracted by the hermit's charms, it is also true that the hermit is not immune to the enticements of society. The isolated, spartan life of the hermit is difficult to sustain. Thoreau, after all, lived at Walden Pond only slightly over two years. And even while he lived at the pond he was far from severing all his connections to society: he visited friends and family in Concord on a number of occasions, sold his beans to give variety to his diet, and supplemented his income by hiring himself out as a day laborer.[55]

Yet for all of these limitations on his autonomy, Thoreau did articulate a social vision that offered a stark alternative to the competitive individualist way of life. To say, as so many do, that Thoreau's individualism and independence are "very much in the American grain"[56] only shows the lamentable imprecision of these categories. For those who divide the modern world into two camps—socialistic collectivism and bourgeois individualism—it is a short, unavoidable step from finding that Thoreau is no socialist to the conclusion that he is an enterprising individualist. Alternatively, Thoreau may be dismissed as a nature lover or idiosyncratic crank who lacked a coherent social vision. None of these views does justice to Thoreau's vision. If instead we view Thoreau through the lens of Mary Douglas's cultural categories, his distinctive belief system comes into sharper focus. By increasing the variety of social types, we can avoid having either to treat Thoreau as a singular eccentric or to choose between Thoreau the socialist and Thoreau the capitalist. Viewing Thoreau as a hermit makes sense of his attitudes toward time, nature, knowledge, work, possessions, exchange, competition, cooperation, reform, even friendship. No other analytic scheme can claim as much.

9

Culture, Context, and Consensus

Political conflict in the United States has been and continues to be animated by fundamentally different visions of the good life. Beneath the debate over this or that policy lies a conflict between rival political cultures. At stake in most policy disputes is not merely the best means to an agreed upon end but competing conceptions of the ends worth pursuing. These cultural differences may be obscured by appeals to shared words, or by attempts to sidestep disagreement on ends by seeking agreement on means. That all sides appeal to terms such as equality or democracy or liberty should not conceal from us the fundamentally different meanings these terms have in different political cultures.[1] That politicians sublimate or paper over value disagreements should not hide that it is precisely such disagreements over ultimate ends that make the politician's skills necessary.[2]

These disagreements over forms of life, moreover, are patterned. It is not a case of every individual having his or her own vision of the good life. Visions must be shareable with others else they are merely hallucinations or daydreams. Visions must be livable else they are merely utopias—nice to think about but impossible to live. My claim, following Mary Douglas, is that the value systems that are both shareable and livable can be usefully grouped into five types: individualism, egalitarianism, hierarchy, fatalism, and hermitude.

My first thesis—that political conflict in the United States is rooted in rival political cultures—contradicts the consensus theory, which holds that the United States has been peculiarly innocent of the struggles over rival visions of the good life that have beset other nations. My second thesis—that political cultures can be grouped into five categories—runs directly counter to historians' feeling for the complexity and diversity of human experience. In today's academic climate, people will be only too ready to embrace the first thesis—consensus theory is said to homogenize history, to ignore race, class, and gender, and to exaggerate America's exceptionalism—but the second thesis can expect a far more skeptical hearing. Why five political cultures? Why not ten or twenty or fifty?

My aim in this concluding chapter is not to answer all possible objections to the categorization presented in these pages.[3] Such a task is well beyond my capabilities and would, in any event, be rather tedious. My more modest objective is to compare the theory applied in this book with a range of other synthetic theories of American political culture that have been offered by political scientists. My hope is that doing so will encourage the reader to measure the utility of the Douglas typology against the benchmark of alternative theories, demonstrate the continuities between the grid–group categories and the categories used by other theorists of American political culture,[4] and indicate the ways in which the grid–group framework adds to our understanding of the American past and present.

Before proceeding to analyze these rival theories, all but one of which take off from and amend Louis Hartz in some fashion, I wish to address two objections that might lead some (particularly historians) to ask, Why bother? The first objection is that Hartz and the idea of liberal consensus are irrelevant in the wake of the voluminous literature documenting the prevalence of classical republican ideas in the early republic. The second objection is that grand synthetic theorizing of the sort that Hartz practiced is misguided because it ignores historical context.

Classical Republicanism and Modern Liberalism

Taking issue with Hartz will seem passé to those accustomed to thinking that notions of a Lockean liberal consensus have long since been discredited by "the republican synthesis."[5] But the republican hypothesis, at least as formulated by the likes of Bernard Bailyn, Gordon Wood, and Drew McCoy, argues not that the liberal, individualistic frame of reference was unimportant in the American past but that it established a position of dominance only after the demise of classical republicanism around the close of the eighteenth century. Where Hartz, following Tocqueville, maintains America was *born* liberal, republican revisionism counters that America *became* liberal. The republican hypothesis, in short, disputes less the existence of a liberal consensus than the timing of the emergence of that consensus.

Beneath the often fierce debate over whether particular individuals or groups were more classical republican or liberal individualistic is general agreement about the direction of change in American political development. "The disagreement," as John Murrin comments, "is more about dates than substance. No one denies that America became a liberal society. The question of when remains open." For Gordon Wood, it is the creation of the American Constitution that signals the "end of classical politics" premised on virtue and the birth of a modern politics based on interest; according to Michael Lienesch, "America became a modern nation" sometime between the Peace of Paris in 1783 and the election of Thomas Jefferson in 1800; Stephen Watts identifies the decades from 1790 to 1820 as the critical years in which took place "a massive, multifaceted transformation away from republican traditions and toward modern liberal capitalism in America"; and Daniel Joseph Singal sees "the market capitalist ideology . . . gradually gain[ing] ascendancy over republicanism in the North beginning in the 1820s and 1830s."[6] Even critics of republican revisionism like Joyce Appleby and Isaac Kramnick do not deny the existence of classical republicanism in the American past but instead maintain that by the time of the American Revolution "the world view of liberal individualism was fast pushing aside older paradigms" of thought.[7] What seems agreed upon by all sides is that at some point between the American Revolution and the Age of Jackson the classical republican conception of politics was replaced by a modern individualistic one.[8]

This dichotomy between tradition and modernity is familiar enough to social scientists. Indeed it underlies much of the sociological and political theory of the nineteenth and twentieth centuries, from Herbert Spencer and Max Weber to Talcott Parsons and Gabriel Almond.[9] Modern social science has been understandably attracted by the power and simplicity of the notion that world history can be understood in terms of a unidirectional transition from one form of social and political relations to another. Although a powerful theoretical tool for certain purposes, the dichotomy misleads by neglecting the occurrence of hierarchy in modern social systems and individualism in traditional systems, as well as by leaving out altogether those systems of social relations that are neither hierarchical nor individualistic (i.e., egalitarianism, fatalism, and autonomy).

The consequences of this impoverished notion of cultural variety and cultural change are evident in the debate between Appleby and Lance Banning over how best to describe the ideological character of the Jeffersonians.[10] Appleby describes the transition from classical republicanism to liberal capitalism in late-eighteenth-century America strictly in terms of the change from traditional hierarchy to modern individualism. For Appleby, classical republicanism is defined in terms of its preference for hierarchy, deference, and organicism. Thus Jeffersonians, who wanted to liberate men from formal, institutional restraints, clearly fell on the modern or individualistic side of the classical–modern divide.

If classical political culture is equated with hierarchy, then no doubt Appleby is correct. For there can be little question that Jefferson and the great preponderance of his followers rejected hierarchy.[11] But are these our only choices: organic hierarchy or competitive individualism? By distinguishing between Douglas's "grid" and "group" dimensions (dimensions that are conflated in the tradition–modernity dichotomy), one can see that Banning's republican hypothesis is not that Jeffersonians endorsed a collective social order that established hierachical gradations between groups (high-grid, high-group) but rather that they believed in a community that would encourage widespread participation by all citizens (low-grid, high-group). The argument between Banning and Appleby is thus an argument over whether the Jeffersonians were competitive individualists or egalitarian collectivists. Banning recognizes this as the nub of the issue: "The irreducible difference between a strictly liberal interpretation of Jeffersonian ideology and a republican hypothesis may lie in our understanding of the way in which the Jeffersonians related the public and private spheres of life." For the liberal individualist, the sum of private transactions produced a close approximation of the public good; for classical republicans a commitment to the public weal was necessary to achieve the public good. Classical republicanism taught that the citizenry's commitment to the community would suffer if individuals devoted themselves to private gain; the doctrine of liberalism denied any conflict between the pursuit of private enterprise and the health of society.[12]

My point is not that Banning is correct and Appleby is wrong. Jeffersonian political culture combined elements of both individualism and egalitarianism.

Agreed on the ills of hierarchy, Jeffersonians were divided and ambivalent about the proper relation of the individual to the collectivity. We do not yet know nearly enough to explain the beliefs or circumstances that permitted Jefferson and his followers to reconcile the egalitarian belief in civic participation and "ward" democracy with the individualistic ideal that good government should leave men "free to regulate their own interests."[13] Nor have we a good grasp on the proportion of the party inclined toward egalitarian participation and the proportion favoring individualistic self-regulation. These are among the types of questions that future research needs to address if we are to get beyond the current impasse.[14]

If Jeffersonian republicanism contained two competing cultural traditions so, too, classical republicanism contained at least two rival cultural strains, the first hierarchical, the other egalitarian. Daniel Walker Howe points to this division within classical republicanism when he corrects J. G. A. Pocock's dichotomy between court and country by adding that "there were really *two* country parties: one radical Whig and the other reactionary Tory." Howe warns us against conflating radical commonwealthmen like Price with reactionary Tories like Bolingbroke.[15] It was the egalitarian strain of the commonwealthmen, not the hierarchical doctrines of Bolingbroke, that shaped Jeffersonian political culture.

One can push this line of thought still further. For classical republicanism arguably also contained within it an *individualistic* cultural strain that overlapped at many points with Lockean liberalism, most obviously in its distrust of central power and in its praise of the independent, property-holding citizen.[16] Moreover, core words of the classical republican language became redefined over time in ways that became consistent with liberal individualism. "Virtue," for instance, became defined in terms of economic industriousness rather than selfless participation in civic affairs.[17] Thus the discovery of classical republican ideas or language in the American past does not ipso facto invalidate the Hartzian thesis of a liberal individualistic consensus. If the republican synthesis has driven Hartz's Lockean liberalism from the field, the rout has been achieved in large part by stretching the concept of republicanism to include notions of citizen independence and vigilance against executive encroachment that merge imperceptibly with individualistic liberalism. If we adhere to a narrower definition of republicanism as a privileging of the collectivity over the individual, the public over the private, and pay more attention to the changing meaning of such terms as virtue and corruption, Hartz's liberal consensus begins to look more defensible even if still incomplete. I accept the revisionist position that "the Lockean monolith" posited by Hartz cannot be sustained, but it is even more difficult to make sense of American political culture and political development by shunting liberal individualism off to the sidelines.[18] A satisfactory theory of American political cultures must acknowledge both the pervasive individualism of American life and the centrality of the communitarian or egalitarian challenge.

Cultures in Context

Historians' skepticism of Hartz's consensus thesis has perhaps less to do with the relative importance of classical republicanism and Lockean liberalism and more to do with the sweeping generality of Hartz's thesis. Such a thesis not only seems to wash out the rich variety of life but to commit the error of reading history backward by imposing contemporary categories and definitions upon the past. Each time period, historians tell us again and again, must be understood in its particular context. We must strive "to comprehend segments of the past within their own sockets of time and place." The scholar must "divest himself as far as possible of those preconceptions which have been established only in later times [and understand] the past according to its own intentions."[19] Suspending disbelief, the historian's task is "to climb inside [past] minds . . . and look out at the world through their eyes."[20]

Among the most influential proponents of a contextualist or historicist understanding of political ideas are Quentin Skinner and J. G. A. Pocock. Their contribution to the study of political thought is to insist that ideas be understood in their historical context.[21] To make sense of a text one must know about the people who formulated the ideas: what their motives and purposes were, what kind of societies they lived in, and the constraints of the language they thought in. This historicist understanding of ideas challenges the traditional emphasis in political theory on the timeless truths of classic texts. The history of political thought, in the view of Skinner and Pocock, is to be understood "not as a series of attempts to answer a canonical set of questions, but as a sequence of episodes in which the questions as well as the answers have frequently changed."[22] The only general truth that emerges from the history of ideas is that "there are in fact no such timeless concepts, but only the various different concepts which have gone with various different societies."[23]

Much of the contextualist approach is methodologically consistent with the theory of cultural biases presented in this book. Skinner's contention that "each society places unrecognized constraints upon our imaginations" is perfectly compatible with the Durkheimian program of identifying the social constraints on cognition that underlies Douglas's grid–group typology.[24] The same is true of Pocock's claim that "each [mode of discourse] will present information selectively as relevant to the conduct and character of politics, and it will encourage the definition of political problems and values in certain ways and not in others."[25] In shifting attention to the "variety of viable moral assumptions and political commitments" manifest in classical texts of ethical and political thought,[26] Pocock and Skinner help to bring a much-needed anthropological sensibility to the field of political theory.

Laudable, too, is their admonition to focus on historical context. They are undoubtedly correct that failure to do so can result in a distorted understanding of past actions. This has often been the case, for instance, with the doctrine of laissez-faire in America. When contemporary conservatives approvingly cite Thomas Paine's warnings against government intervention to bolster their individualist cultural bias, they distort the past by taking ideas out of context. For Paine's opposition to governmental intervention in the economy was grounded in his commitment to egalitarianism. Unlike contemporary egalitarians, however, who see the central government as a potential source of redressing inequalities, early American egalitarians like Paine believed that an active government inevitably created greater inequalities.

But if a proper understanding of the context in which historical actors act helps prevent the indiscriminate ransacking of the past, contextualism can be and often is carried too far.[27] The theory of cultural biases presented in this book maintains that, contrary to Pocock and Skinner, it is possible to identify certain modes of thought that recur throughout history. There are a variety of modes of thinking and of organizing, as Skinner and Pocock suggest, but there is not an infinite variety of ways. All people who choose to live together, no matter their level of technology, face certain choices about how to organize their social and political lives. Douglas's theory focuses on two such choices: how much autonomy the individual will have vis-à-vis the community, and how prescribed the individual's life will be. Different people in different places and different times will resolve these questions differently, but none can avoid them altogether. Grid–group theory is thus an attempt to go beyond the contextualist position that society constrains cognition by systematizing the *types* of social constraints that are possible.

The historicist claim that the task of the student of culture is to discover "the languages in which the inhabitants of an era did in fact present their society and cosmos to themselves and to each other" has been further criticized by Isaac Kramnick for its "assumption that there is but one language—one exclusive or even hegemonic paradigm—that characterizes the political discourse of a particular place or moment in time."[28] Appleby, too, criticizes the contextualist position that there is only "a single, shared world view operating within a given society" and that individuals are imprisoned within this "presiding paradigm." Both Appleby and Kramnick point to competing languages or paradigms that allow for "reality testing" and enable individuals to avoid being confined to a single worldview.[29]

The limitations of the contextualist or historicist approach to political ideas parallel in some respects the inadequacies of early anthropological studies of culture. Pioneering works on "national character," such as those offered by anthropologists Ruth Benedict and Margaret Mead, focused upon the unique configuration of values, beliefs, and practices that constituted a tribe's or a nation's culture. The Zuni differed from the Kwakiutl differed from the Dobuans.[30] The Russians differed from the Japanese differed from the Americans.[31] Comparison between nations or tribes was rarely attempted, and variation within was largely ignored.

Cultural analysis needs to break free from both the assumption of consensus and the assumption of noncomparability. The first requires paying more attention to cultural differences within such units as nation-states, ethnic groups, political parties, and historical eras. The second requires creating analytic constructs that allow generalizing about seemingly diverse phenomena. It is much easier to persuade historians of the existence of subcultures than it is to persuade them of the need for overarching general theories or typologies. Indeed it is historians who have been primarily responsible for uncovering the existence of a rich array of subcultures within the United States, thereby discrediting all notions of cultural consensus or a unified national character. But recognizing America's cultural diversity without also endorsing systematic typologies threatens an unrestrained pluralism in which every distinct time and place and race is endowed with a unique culture. Culture is piled upon culture, creating "a kind of old-fashioned ethnographic museum, in which text after text piles up, parts catalogued without a sense of the whole."[32]

Once one accepts that typologizing is necessary to an understanding of political cultures, the question becomes which typology best orders and explains the historical data. Political scientists, who generally need little persuading of the need for some theoretical anchor, have advanced a number of promising candidates. In the next sections, I discuss several of these alternative theories and compare them with the typology of political cultures employed in this book. The first theories I consider are those that attempt to account for cultural conflict from within the confines of the Hartzian consensus thesis.

Conflict Within Consensus: The Theories of Samuel P. Huntington, Seymour Martin Lipset, and J. David Greenstone

The Politics of Creedal Passion[33]

Among the more creative attempts to reconcile conflict with consensus is Samuel P. Huntington's *American Politics: The Promise of Disharmony.*[34] Huntington agrees with Hartz that "in contrast to most European societies, a broad consensus exists and has existed in the United States on basic political values and beliefs,"[35] but he sees this liberal-democratic consensus as the source of America's "disharmonic polity." Because political practice invariably must fail to measure up to the values espoused in the American creed, the consensus stands as a permanent indictment of existing institutions. Periods of bitter disagreement—"creedal passion periods"— stem from periodic attempts to put into practice the nation's values. Everyone believes in the creed, Huntington suggests; some just believe more passionately in its values than do others. What separates Americans is not competing political values but rather the differing intensity with which common values are held.

What differentiated Lincoln from the abolitionist William Lloyd Garrison, Huntington would argue, is not *what* they believed in but *how* they believed it. Garrison simply believed more intensely in the American creed than did Lincoln. An alternative hypothesis is that Garrison and Lincoln disagreed because they believed in different values, and that they meant different things even when they used the same words. Huntington is correct that both Lincoln and Garrison appealed to a common document—the Declaration of Independence—to condemn slavery but each attached a very different meaning to the phrase "all men are created free and equal." For Garrison, like many other abolitionists, slavery was wrong because no man had the right to tell another man what to do. Lincoln, like most Republicans, objected to slavery primarily because it denied individuals the opportunity to compete and improve their condition.[36]

Americans can agree about the desirability of equality only because they mean different things by it. Similarity in language often disguises deep differences in worldviews. Both Ronald Reagan and Ralph Nader, for instance, profess a deep commitment to equality without sharing a common political culture. Equality for Reagan means an equal opportunity to be different; for Nader, equality means equalizing results. There is a world of difference between those who wish to reduce authority so as to create opportunities and promote individual differences and those who reject authority so as to diminish differences among people.

Similarly, although both Reagan and Nader profess a belief in democracy their conceptions of democracy are startlingly different. The two men have radically different conceptions of the desirable relationship between private and public spheres. Reagan wishes to minimize the public sphere so as to maximize individual autonomy in the private sphere. Nader, in contrast, desires "people to spend more time as public citizens," "to go to the village green and improve their polity, their community." Such "hypercitizenship" is what Nader calls "real democracy."[37] What differentiates Reagan from Nader is not the intensity of their belief in the American creed but fundamentally different conceptions of democracy.

The four periods that Huntington identifies as creedal passion periods—the Revolutionary, Jacksonian, and Progressive eras, and the 1960s—were all associated with a rise in egalitarian movements. Huntington himself recognizes that reducing differences among people was a central thrust of each of these periods. "Each was in some measure a period in which distinctions—whether based on status, occupation, knowledge, or position—were denigrated, in which there was a stress on homogenization, on 'the great principle' (in the words of Jacksonians) 'of amalgamating all orders of society.'"[38] But is American society, either now or in the past, characterized by agreement on the proposition that distinctions among people—between men and women, rich and poor, old and young, parents and children—should be reduced? The answer is no. The principle of reducing differences is antithetical to core values of the liberal individualist ethos. It is the principle of differences, not similarities, that animates "Lockean liberalism." That vision is one of a social order based on exchange, implying differences and differentiation rather than similarity and likeness.[39] Lockean political values (i.e., individualism) include equal treatment before the law and expansion of individual opportunities but do not include a collectivist equality dedicated to diminishing differences between people.

Having assumed consensus on radical egalitarianism, Huntington concludes with a startling paradox: "the legitimacy of American government varies inversely with belief in American political ideals."[40] Abandoning the assumption of consensus leaves a more plausible (if less striking) proposition: the legitimacy of American institutions varies inversely with the strength of egalitarianism. The denigration of authority so characteristic of creedal passion periods stems not from the individualistic consensus "coming alive"[41] but from an upsurge of egalitarianism.

Achievement Versus Equality

The tension between individualist and egalitarian values is the focus of Seymour Martin Lipset's *The First New Nation*.[42] Like Huntington, Lipset accepts Hartz's characterization of American society as essentially consensual. All Americans, Lipset insists, "believe strongly" in the "two values that are at the core of the American creed—individualism and egalitarianism." Unlike Huntington, however, Lipset explores what he sees as the "deep contradiction" between the core values of the American creed.[43] From Huntington's hierarchical vantage point, the differences between egalitarianism and individualism are of secondary importance; what is important to him is that both are anti-authority and antigovernment. For Lipset, however, the differences between these two anti-authority values assume paramount importance.

The consensual adherence to a "value system" encompassing both individualism and egalitarianism creates conflict because the two values lead in contradictory directions. Encouraging individual achievement, for instance, tends to undermine economic equality.[44] Political debate in the United States, Lipset explains, "often takes the form of one consensual value opposing the other."

> Liberals and conservatives typically do not take alternative positions on issues of equality and freedom. Instead, each side appeals to one or the other core values, as liberals stress egalitarianism's primacy and the social injustice that flows from unfettered individualism, while conservatives enshrine individual freedom and the social need for mobility and achievement as values "endangered" by the collectivism inherent in liberal nostrums. Both sides treat as their natural constituency the entire American public. In this sense, liberals and conservatives are less opponents than they are competitors, like two department stores on the same block trying to draw the same customers by offering different versions of what everyone wants.[45]

Lipset deftly weaves together themes of consensus and conflict. His reliance on the distinction between liberal and conservative, left and right, is straight out of the Progressive tradition of Charles Beard and Vernon Parrington. But in contrast to Progressive historiography, Lipset stresses that what makes the United States exceptional is that the left does not reject individual freedom and the right does not reject equality. Ideological conflict in America, Lipset suggests, occurs not between clearly differentiated social groupings or classes as in Europe but within the psyche of every American.

Lipset stresses not only value consensus at a given point in time but, like Hartz, continuity over time in "the American value system." The egalitarian theme is traced to the Revolutionary struggle against the hierarchical traditions of the Old World, and the emphasis on achievement is attributed to Puritanism's emphasis on success and hard work.[46] This "basic value system," Lipset argues, was "solidified in the early days of the new nation" and has been passed on essentially unchanged down to the present day. To support this thesis of an "unchanging American character," Lipset cites a study showing that in four separate eras of United States history visitors remarked upon Americans' "belief in equality of all as a fact and as a right."[47] But as the historicist critic would be quick to point out, this begs the question of what Americans of these diverse historical periods meant by "equality." Indeed it does not even tell us whether Americans of a single historical period meant the same thing when they spoke approvingly of equality. If Americans have attached different meanings to the term "equality," as I argue in chapter 3, the conflict begins to look less psychological and more cultural, less within Americans than between Americans. Lipset's abstract value consensus begins to dissolve as one specifies whether equality means equality before the law, equality of opportunity, or equality of results.

Reliance on Parsonian assumptions of a "common value system" and anthropological notions of "national character" leaves Lipset ill-prepared to appreciate fully the cultural divide separating individualists from egalitarians in the United States. Lipset's consensual formulation underestimates the extent to which egalitarians in America, from Samuel Adams to Wendell Phillips to Henry George to Tom Hayden, repudiated individualist values of competition and capitalist accumulation. By the same token, it understates individualists' hostility toward equalizing conditions.

By equating individualist values with conservatism and egalitarian values with liberalism, Lipset's analytic scheme also runs into some of the same difficulties that beset the left–right distinction.[48] Most important among these problems is that the designation "conservative" or "right" fails to discriminate between competitive individualism and hierarchical collectivism. Who is more conservative or further to the right: George Will or Milton Friedman, Fisher Ames or William Graham Sumner? This question cannot be answered within the confines of the liberal–conservative continuum because Will and Ames adhere to a hierarchical variant of conservatism while Friedman and Sumner adhere to an individualistic variant of conservatism. Equating conservatism with individualism leaves no room for the hierarchical culture of Federalists, Whigs, and southern slaveowners. Hierarchy may be weaker in America than in most other nations but, as chapter 6 shows, it has not been completely absent.

In sum, Lipset's bipolar distinction between individualism and egalitarianism captures a central conflict that runs throughout United States history, but it also neglects other enduring cultural strands in the American past. In addition to overlooking the role of hierarchy, it also leaves out those fatalists who see little chance of being able to succeed through their own individual effort and show scant support for the collective action necessary to achieve more equal outcomes. Missing, too, is the voice of the autonomous hermit who rejects both the competitive rat race of individualism and the collectivist thrust of egalitarianism in favor of a simple life of self-sufficiency.

The Permanent Liberal Bipolarity

Another innovative effort to account for conflict within the confines of the consensus paradigm is presented by J. David Greenstone. Greenstone agrees with Hartz that all Americans share a "liberal commitment to freedom, individualism, and pluralism" but suggests that "American liberalism is divided between two liberal outlooks, one of which is Lockean as Hartz maintains, while the other stems from a Protestant mode of thought and social action that [Hartz] largely ignores."[49] These two traditions constitute what Greenstone calls a "permanent liberal bipolarity."

One side of this liberal bipolarity emphasizes a utilitarian politics that seeks to maximize individual rights, interests, and preferences. The other side of the liberal bipolarity involves a politics of humanitarian reform that emphasizes moral self-development. The first is privatistic and materialistic; the second is reformist and benevolent. The former has a procedural conception of democracy; the latter a substantive conception of democracy. Lockean or empiricist liberalism, Greenstone explains, "focuses on formal or procedural criteria for equitable treatment and considers the question of substantive ends as a matter of private judgment." Public ends are the sum of private interests. In contrast to the procedural tradition that treats every individual preference as equally valid, Protestant or substantive liberalism sets up a public interest that is separate and apart from the sum of private interests. This version of liberalism establishes community standards by which the preferences of different individuals can be compared, weighed, and judged.

A shared commitment to liberty, Greenstone argues, cloaks the different meanings that these rival traditions attach to that word. Lockean liberals define liberty as an absence of external restraint, what Isaiah Berlin has termed "negative liberty." Substantive liberals, in contrast, favor a conception of "positive liberty" that frees individuals "to cultivate their own distinctively human faculties—and help others do so as well." Freedom, in this understanding of the term, means "the exercise and development of these [human] faculties rather than absence of restraint in the pursuit of privately determined ends."[50]

Greenstone cites Stephen Douglas as an exemplar of negative, Lockean liberty, and Daniel Webster as a proponent of positive liberty. For Douglas, a Jacksonian Democrat, liberty meant freedom from "the political community's moralistic meddling in private affairs." Webster, a conservative Whig, taught that to do exclusively as one pleased endangered the unity and harmony of the collectivity and was not true liberty. Freedom, for Webster, meant not the liberty to act on one's every passing impulse but rather the liberty to cultivate one's higher self and thus avoid becoming a slave to one's baser passions.[51]

Greenstone's most important contribution is to uncouple liberalism from Lockeanism. In Hartz's account the two are conflated so that every non-Lockean notion is by definition nonliberal. Greenstone's analysis enables us to make sense of an American political culture that is clearly non-Lockean in its suspicion of the unregulated marketplace, as well as in its privileging of the public interest over private interests and community morality over individual desires. In contrast to Huntington, moreover, Greenstone recognizes that attacks on capitalism in America stem not from an overly literal application of individualistic values but rather from a rival set of values.[52]

My main reservation with Greenstone's formulation is that substantive or Protestant liberalism encompasses too much. It includes, for instance, both radical abolitionists and conservative Whigs. Both William Lloyd Garrison and Daniel Webster rejected untrammeled, Lockean individualism, but they differed radically over the proper role of authority. Both wished to privilege the public over the private, the collectivity over the individual, but Webster wanted that collectivity to be hierarchical while Garrison desired collective life to be egalitarian. Webster preferred a social order in which the many would defer to the few and the few would display a "paternal sympathy"[53] for the many; Garrison opted for a life in which all would have an obligation to participate and no individual would have the right to exercise authority over another.

Greenstone is too keen a student of American political culture not to notice this profound difference between conservative Whigs and radical abolitionists. He observes, for instance, that "rather than celebrate the Protestant conscience,

[Webster] berated antislavery 'agitators' for their moralism, their 'belief that nothing is good but what is perfect.'"[54] But what Greenstone the historian notices, Greenstone the theorist cannot account for. Missing is a way to distinguish between hierarchical collectivism and egalitarian collectivism. Greenstone's dichotomy between procedural and substantive liberalism successfully distinguishes between those political cultures that elevate the individual over the collectivity and those that subordinate the individual to the collectivity, but it fails to sort out the difference between those collectivities structured around formal authority and those that reject all authority as a form of inequality. In this respect, if in no other, grid–group theory is a more powerful, more discriminating theoretical tool.

Explaining the Growth of Government: James Morone and the Democratic Wish

The essence of American Lockean liberalism, as conceived by Hartz, is the placing of limits upon the state to keep it from interfering with the pursuits of private persons. Locke's own position had been more ambiguous for it had entailed both "a defense of the state that is implicit and a limitation of the state that is explicit." But American Lockeans could be more consistent in opposing a powerful central government than could Locke, for whom the state had been a weapon to destroy the preexisting feudal order. The absence of feudalism in America meant that Americans "were not conscious of having already done anything to fortify the state, but were conscious only that they were about to limit it. One side of Locke became virtually the whole of him."[55]

Hartz's interpretation of American political culture as overwhelmingly antistatist raises an obvious question. If Americans are concerned only with limiting government, how does one account for the tremendous growth of government in America over the past half-century? Hartz recognizes the difficulty and addresses this question in a chapter on the New Deal. In it, he argues that the New Deal succeeded in expanding government's role by submerging classical Lockean liberalism in pragmatic problem-solving. By defining reform in terms of problems to be solved rather than in terms of the clash of rival philosophies, Roosevelt successfully "hid the departure from Locke," a necessary step since even "the normal Liberal Reform departures from Locke cannot be tolerated consciously by the absolute mind of the nation."[56]

Whatever the validity of this interpretation of the New Deal, Hartz's thesis demonstrably fails to account for subsequent expansion of the state, especially since the Great Society. The New Deal may have been legitimated by its claim to be solving problems and saving capitalism from itself, but the Great Society programs were enacted not during a crisis of capitalism but during a period of economic prosperity. It and subsequent expansions of government have been undertaken not in the name of pragmatic problem-solving but rather in the name of social justice, equity, fairness, and democracy.

A more compelling answer to the question of how a nation with a dread of public power keeps getting bigger and bigger government is offered by James

Morone in *The Democratic Wish*. A student of Greenstone's, Morone argues that Americans periodically overcome their "antistatist trepidations by pursuing their democratic wish." The dread of public power as a threat to liberty is overridden by their yearning for "direct, communal democracy." By promising to empower the people, "reformers—oppressed groups, public officials, policy entrepreneurs—have repeatedly overcome the checks and balances of a polity biased against the expansion of government. Calling for community," Morone explains, "is the acceptable form in which to cast collectivist sentiment in a society of state bashers." American political development is thus marked by the striking irony that "the search for more direct democracy builds up the bureaucracy."[57]

More clearly than any other analyst of American political culture discussed in this chapter, Morone grasps the critical distinction between communitarianism and liberalism, between participatory egalitarianism and competitive individualism. According to Morone, the participatory and communitarian vision of the democratic wish presents "a direct counter to every aspect of liberalism." Where American liberalism promotes "individuals, representation (in a limited government full of checks and balances), private interest, and individualism," the democratic wish responds with "the people, participation, common good, and community."[58]

The problem with the democratic wish, Morone argues, is that it increases the size and extends the scope of government without bolstering government's legitimacy, coherence, or autonomy. It gives us bigger but not better government. The democratic wish provides the main spur to public action in the American polity, but it is not any more supportive of central governmental authority than "its liberal antithesis." Both liberal and communitarian traditions, Morone points out, "rest on a suspicion of government." The one rejects centralized political authority in favor of "private choices," the other in favor of participatory community.[59]

Morone's criticism of the democratic wish emerges not from a libertarian perspective but from the perspective of one who believes that in an increasingly complex, interdependent world, "Americans need a more powerful political center," "a competent, coherent, independent state."[60] Morone sees the current United States government as "chaotic," "fragmented," "sprawling," "unwieldy," "poorly coordinated," "narrow," "incoherent," and woefully inadequate to cope with the complex problems of the contemporary world.[61] "Government-bashing individualists pressing their own self-interest with only the loosest coordination from the center," Morone insists, offer "few solutions" to the complex problems facing contemporary America.[62]

From Morone's social democratic perspective, the democratic wish thus poses possibilities as well as perils. Its "call to community" offers Americans "a counter to both their individualism and their institutions," as well as providing "an ideological key for redefining and reconstructing the state." Morone wonders whether "the search for the people can be reconfigured" in a way "that embraces the ministers Americans have always revolted against." He admits it is "an unlikely aspiration" but concludes that "the American future may depend on it."[63]

Whether the American future depends on marrying the democratic wish with an acceptance of the state is questionable, but the viability of social democracy in America certainly does depend on the consummation of this marriage.[64] Morone's book can be read as a perceptive analysis of the unintended effects of egalitarianism in the absence of a strong hierarchical tradition. In the absence of a strong state, Morone suggests, the democratic wish "leaves behind the underlying conditions it found: a political economy of self-seeking interests pushing ahead within a complex welter of political rules that advantage some citizens, disadvantage others, and seem almost invisible to all." The end result of the democratic wish, "when the marches, strikes, and late-night meetings fade away," is the inclusion of more groups within the system, but the system remains fundamentally the same. Embarking with "communal hopes," reformers end up "reinforc[ing] liberal institutions."[65] The embrace of Rousseau ultimately leads America back to Locke.

This thesis is provocative but problematic in view of Morone's insistence that the mismatch between demands for state action and the authority of the state to undertake such actions has created a fundamental regime crisis for America's liberal institutions and beliefs. If the democratic wish generates demands for state action that far outstrip support for state authority, how can the democratic wish be said to reinforce liberal individualism? For individualism is maintained, as Aaron Wildavsky argues, by balancing low support for authority with low demands for authority.[66] If Morone is correct that the democratic wish (i.e., egalitarianism) generates demands for state action, as I believe he is, then he cannot be right that the democratic wish simply functions to shore up the individualistic system.

Calling on government to do more while simultaneously challenging its authority and expertise ("the best way to build government," Ralph Nader tells us, "is to attack government")[67] does hinder the emergence of a social democratic alliance of hierarchy and egalitarianism. This much Morone has exactly right. But it does not follow from this that liberal individualistic institutions are therefore recreated and reinforced by the democratic wish. On the contrary, to the extent that government grows larger and more intrusive as a result of the democratic wish, there is less scope for self-regulation (what Morone calls "private choices"), the defining characteristic of a liberal individualistic regime. Morone's description of an American state unable to meet the demands of its people is a sign not of the strength of liberal individualism but of its weakness. It is egalitarianism, Wildavsky reminds us, that "demands both an increase in bureaucracy and a decrease in authority"[68]— an increase in bureaucracy because only government is deemed able to regulate the ill effects of competitive individualism; a decrease in authority because authority and expertise are violations of equal participation.

Morone argues that only a drastic increase in state capacity (away from individualism and toward hierarchy) can solve America's chronic problems; individualists would counter that America's problems are best solved by reducing demands upon the state (away from egalitarianism and toward individualism). Although these two sides differ over whether to seek the remedy on the demand side or the supply side, both agree that the current imbalance between the demands Americans currently make of government and the support they give that government goes a long way toward explaining government's failures as well as the citizenry's disillusionment and cynicism.

The source of this current imbalance, I would argue, is to be found not in the recurrence of past patterns but in the breakdown of those past patterns. In contrast to contemporary egalitarians, most egalitarians of the eighteenth and nineteenth centuries did not believe that the cause of social justice would be served by increasing the power of the federal government. On the contrary, as I showed in chapter 3, most viewed government intervention as the primary source of economic and social inequality. Egalitarian radicals like Tom Paine harbored a faith in the unregulated marketplace as an engine of relative equality and held a view of commerce as an instrument of noncoercive sociality that is inconceivable to the contemporary egalitarian. Their view of the inverse relationship between government intervention and equality made it possible for nineteenth-century egalitarians to make common ground with individualists in keeping the size of government relatively small.[69]

This alliance with individualism has been foreclosed to most contemporary egalitarians, who believe government intervention is essential to redress inequalities caused by the marketplace. At the same time, an alliance with hierarchy aimed at strengthening the authority of government in the name of equality is difficult in view of hierarchy's traditional weakness in America. Thus Morone's justifiable pessimism. Having broken their alliance with individualism, which structured much of nineteenth-century American political life, contemporary egalitarians find themselves lacking an obvious ally. They look with hope to the fatalists, to the disaffected masses who if they could only be coaxed to the polls would help usher in a new egalitarian majority.[70] But the masses have so far proven distinctly unreceptive to egalitarian overtures. If it is unlikely, as Morone reluctantly concludes, that the American egalitarian will build up "a competent, coherent, independent state,"[71] it seems even less likely that egalitarians can create a governing coalition by mobilizing the powerless.

The Subcultures of Daniel Elazar: Individualism, Traditionalism, and Moralism[72]

The theories proposed by Morone, Greenstone, Lipset, and Huntington modify and improve upon Hartz by taking into account modes of thought and action that do not fit the pattern of aggressive, atomistic capitalism described by Hartz. Each of these authors grasps that there is more to American political culture than competitive individualism. But none breaks radically enough from Hartz. Morone comes the closest to doing so, describing the communitarian, democratic wish and liberal individualism as "almost mirror opposites."[73] Yet Morone also describes a liberal individualistic system that repeatedly absorbs and coopts communitarian opposition. The communitarian counter to Lockean liberalism turns out to be not a sustainable political culture but a periodic outburst that inevitably fails to transform the dominant liberal institutions in a communitarian direction. We are left with a more inclusive but still fundamentally liberal political regime.

A dramatic departure from the idea of a consensual national political culture is proposed by political scientist Daniel Elazar. Elazar identifies three distinct political subcultures within the United States: individualism, traditionalism, and moralism. The individualistic subculture emphasizes a "conception of the democratic order as a marketplace [and] places a premium on limiting community intervention—whether governmental or nongovernmental—into private activities to the minimum necessary to keep the marketplace in proper working order."[74] Elazar's individualistic subculture corresponds closely to Douglas's competitive individualist way of life and Hartz's Lockean liberalism.

The traditionalistic subculture, Elazar tells us, is characterized by "an ambivalent attitude toward the marketplace coupled with a paternalistic and elitist conception of the commonwealth." This political culture "reflects an older, precommercial attitude that accepts a substantially hierarchical society as part of the ordered nature of things, authorizing and expecting those at the top of the social structure to take a special and dominant role in government."[75] The traditionalistic subculture obviously bears a strong resemblance to the political culture I term hierarchical. "Traditionalism," however, is a poor label for this way of life because it suggests that at its core is a commitment to doing things the way they have been done in the past, regardless of the substance of those past behaviors and beliefs. This is unsatisfactory because it is not parallel with "individualism," which suggests a commitment to a particular set of substantive beliefs, and because it does not correspond to the description offered by Elazar, which focuses on the culture's ambivalence toward the self-regulation and competition of markets as well as its distrust of popular participation. The designation "hierarchical" more accurately denotes the way of life (as defined by Elazar) that its adherents believe in.

The moralistic political culture is the most ambiguous of Elazar's three categories. The confusion begins with the label "moralistic," which suggests that moralism is the peculiar province of one political culture. The question is not which culture is moralistic, but what is to count as morality (and moralism) in different cultures. Is the individualist who lectures the poor on how they have no one to blame but themselves for their plight any less moralistic than the egalitarian who lectures the rich that they are responsible for the oppression of the poor? Every culture is held together by moralizing about how not to behave.

The "moralism" label is also misleading because it does not adequately convey the characteristics that Elazar offers as the defining features of this way of life. Elazar argues that the political cultures of moralism and individualism represent "two contrasting conceptions of the American political order." In contrast to the individualistic political culture, where "the political order is conceived as a marketplace in which the primary public relationships are products of bargaining among individuals and groups acting out of self-interest," Elazar tells us, the moralistic culture conceives of the political order as "a commonwealth—a state in which the whole people have an undivided interest—in which the citizens cooperate in an effort to create and maintain the best government in order to implement certain shared moral principles."[76] This description suggests that Elazar conceives of these two cultures as existing at opposite poles of a single continuum based on the degree to which individuals believe political life should be oriented toward a collective public good. This dimension might be seen as roughly coincident with Douglas's group dimension. Rather than use the designation "moralism" (which seems to denote a *style* of expressing one's beliefs in contrast to individualism, which refers to the *content* of those beliefs), it would better capture Elazar's meaning to refer to this culture as "communitarian" or "communalism."[77]

How then does moralism (or communalism) relate to traditionalism? Both, according to Elazar, believe in the commonwealth. The difference is that while the traditionalistic culture harbors a deferential conception of the commonwealth, the communitarian culture adheres to a participatory vision of the collective. Politics from the communitarian perspective, Elazar tells us, is "ideally a matter of concern for every citizen, not just for those who are professionally committed to political careers. Indeed, it is the duty of every citizen to participate in the political affairs of his commonwealth."[78] In short, this is the difference between hierarchical collectivism and egalitarian communalism.

If one adheres to this definition of a moralistic political culture as both participatory and communitarian, it becomes evident that much of what Elazar classifies as moralism is actually better described by the traditionalistic category (deferential and collectivist). Consider the case of Massachusetts, the state Elazar identifies as the seedbed of the moralistic culture. The political culture of eighteenth- and early nineteenth-century Massachusetts was, as Elazar posits, strongly collectivistic. But that collectivity was predominantly deferential, not participatory. In no other state in the Union (except perhaps Delaware) was Federalism, the party of hierarchy, more powerful than in Massachusetts. If distrust of widespread political participation and suspicion of an unregulated free market are the hallmarks of a traditionalistic political culture, then the Federalist party of Massachusetts was the embodiment of the traditionalistic (i.e., hierarchical) culture.[79]

The limitations of Elazar's moralistic category (as with Greenstone's category of substantive liberalism) can be seen by comparing radical abolitionism with conservative Whiggery. Both movements drew their greatest strength from New England, yet culturally they were worlds apart. Like their Federalist predecessors, conservative New England Whigs upheld a politics of deference and noblesse oblige. They defended authority, whether in the person of teacher, husband, parent, or political leader. Many relied upon an analogy between the human body and the social and political system. Just as there was a need in the body for the higher faculties of conscience and reason to restrain the lower animal passions, so in society it was necessary for the more responsible at the upper end of the social stratum to discipline and regulate the profligate at the lower end.[80]

Radical abolitionists, in contrast, were committed to an egalitarian form of social organization. Their commitment to egalitarianism led many abolitionists to call into question, for instance, traditional gender roles. "Our object," William Lloyd Garrison revealed, "is *universal* emancipation—to redeem woman as well as man from a servile to an equal condition." A member of the Garrison-led Boston Clique, Lydia Maria Child, explained that she and her husband "despised the idea of any distinction in the appropriate spheres of human beings." Garrison and his followers were highly critical of familial authority relations, whether between husband and wife or parent and child. One radical Garrisonian went so far as to liken "every family" to "a little embryo plantation." Another explained that she was always "very conscientious not to use the least worldly authority over her child." The abolitionist marriage that most closely approximated the egalitarian ideal was that of Garrisonian followers Abigail Kelley and Stephen Foster. Both agreed that either could withdraw from the marriage whenever they chose, and the farm they bought after their marriage was deeded to Foster and Kelley jointly. While Abigail was out lecturing, moreover, Stephen would often stay home and take care of the farm and child. To one admiring abolitionist, the Foster home was a place in which there could be "seen the beauty and the possibility of a permanent partnership of equals."[81]

The Whig attitude toward the family was radically different. Whigs not only defended the husband's control over the wife and parents' authority over children, but they generalized from these familial relationships to the ideal political relationships. The patriarchal family was identified by a number of conservative Whigs as the ultimate origin of the state.[82] Leaders, in the Whig view, should assume the same attitude toward the citizenry that "a parent holds to a child, or a guardian to a ward." By the same token, the citizenry was obligated to show the same deference toward political leaders that children owed to their parents, and wives to their husbands. The government ought to be "the parent of the people," exerting "a beneficent, paternal, fostering usefulness upon the Industry and Prosperity of the People" as well as protecting and guiding "the weak and disabled and . . . more dependent members of society." The nation, in the words of Daniel Webster, was a "family concern."[83]

Where Whigs saw each member of society, from the mighty to the meek, as an integral if unequal part of the whole, abolitionists tried to separate themselves from the wider society that they believed to be shot through with corruption. In the view of Garrisonian abolitionists, historian Lawrence Friedman explains, black slavery was "only the worst example of American reliance on force—on man oppressing his fellow man rather than partaking in mutual love. Oppression of man by man in all its forms, not simply Southern racial bondage, made up the American slave system." Using this expansive definition of slavery, Garrison could proclaim "that Pennsylvania is as really a slave-holding State as Georgia."[84] Believing that the "slaveholding spirit" suffused government, Garrisonians refused to vote or hold political office. Participation in the system could only result in cooptation and defilement.

The different ways of life adhered to by New England Whigs and abolitionists also led them, despite a common Puritan heritage, to radically different views of religious institutions. While Garrisonian abolitionists employed religious conscience as a standard by which to question authority, Whigs looked to religion to uphold authority. The predominantly Whig Congregational Association of Massachusetts, for instance, defended "deference and subordination [as] essential to the happiness of society, and particularly so in the relation of a people to their pastor."[85] Garrisonians, by contrast, attacked Protestant churches for being "the bulwarks of American slavery" and indicted the clergy for its "truckling subservience to power, . . . clinging with mendicant sycophancy to the skirts of wealth and influence." The churches' corruption stemmed, in their view, from the "desire among clergy to assert their authority." Faced with a conflict between their preferred way of life and church membership, many Garrisonians withdrew from their church, thus escaping what they perceived as their "spiritual bondage."[86]

Whigs and immediatist abolitionists also differed in their views of human nature. Consistent with their hierarchical predisposition, Whigs had a pessimistic view of human nature. "The weakness, the follies, and the vices of human nature" were unquestioned axioms of Whig thought. Because human nature was "so strongly inclined to go astray," Whigs believed it was safer to rely on institutions to "keep it on a path approximate to the parallel of rectitude, than to give it unlimited freedom to go right or wrong."[87] Garrisonian abolitionists believed in the ultimate perfectibility of man. By renouncing membership in the corrupt (i.e., hierarchical) institutions of this world and joining with other like-minded brethren, they believed they could recreate heaven on earth and man's basic goodness would emerge.[88]

Elazar deserves credit for recognizing that abolitionism was not simply a variant of the dominant culture of competitive individualism, merely capitalists with a conscience, but instead represented a distinct cultural type. So, too, he has made a valuable contribution in pointing out that the collectivist orientation of Whig (as well as Federalist) political culture distinguished it from competitive individualism. But as Elazar is right to point out that individualism cannot encompass the variety of the American experience, his moralistic category is inadequate because it tends to run together two distinct ways of life: egalitarianism and hierarchy. It will not do to throw the radical abolitionism of William Lloyd Garrison and the hierarchical Whiggery of Daniel Webster, the patriarchal mormonism of Joseph Smith and the egalitarian communalism of Hopedale, the leveling of the Populists and the paternalism of the Mugwumps, into the same cultural melting pot.

Elazar's neglect of these cultural differences is all the more puzzling in view of his analytic statement distinguishing between participatory and deferential cultures. The explanation, I believe, lies in a failure to distinguish the particular historical manifestation of a type from the type itself. The traditionalistic culture, for instance, is made virtually synonymous with the political culture of the southern states. The result is to overlook the occurrence of hierarchy (i.e., collectivism combined with deference) among northern Yankees. Cultural variation within the South is also underestimated. Alabama and Virginia, for instance, are both identified as traditionalistic cultures by Elazar, yet V. O. Key's seminal study of southern politics suggests that

> the political distance from Virginia to Alabama must be measured in light years. Virginian deference to the upper orders and the Byrd machine's restraint of popular aberrations give Virginia politics a tone and a reality radically different from the tumult of Alabama. There a wholesome contempt for authority and a spirit of rebellion akin to that of the Populist days resist the efforts of the big farmers and "big mules"—the local term for Birmingham industrialists and financiers—to control the state. Alabamians retain a sort of frontier independence, with an inclination to defend liberty and to bait the interests.[89]

The fundamental weakness of Elazar's categorization (as well as Greenstone's and Lipset's) is that the categories are not derived from dimensions.[90] Instead Elazar's types are derived inductively from regional variations in the United States: traditionalism is drawn from the plantation-centered system of the South, and moralism from New England Puritanism. As a result the categories are neither mutually exclusive nor exhaustive. Nor do they travel well across time or space. Whatever their deficiencies, however, Elazar's categories are a welcome relief for those unwilling to accept that every state or region is culturally unique or alike.

The American Jeremiad: From Consensus to Hegemony

Rather than modify Hartz to take ideological conflict into account, as Huntington, Lipset, Greenstone, and Morone have done, or reject Hartz in favor of "the republican synthesis," as Pocock, Bailyn, and Banning have done, literary historian Sacvan Bercovitch reasserts the Hartzian theory with a vengeance. In a sense, as David Harlan suggests, "Bercovitch has rewritten Hartz's *The Liberal Tradition in America*, transferring to American Puritanism all the intellectually stultifying traits Hartz had attributed to American [Lockean] liberalism."[91] Like Hartz, Bercovitch insists on

> a sweeping qualitative distinction between America and all other modern countries. In England (and the Old World generally), capitalism was an economic system that evolved dialectically, through conflict with earlier and persistent ways of life and belief. Basically New England bypassed the conflict. . . . In all fundamental ideological aspects, New England was from the start an outpost of the modern world. It evolved from its own origins, as it were, into a middle-class culture.[92]

For Bercovitch, the Puritans provide the key to the "astonishing cultural hegemony" of competitive individualist values—free enterprise, laissez-faire, self-reliance, social mobility—in America. Not only was there an "elective affinity," as Weber and others have argued, between Puritanism and capitalism, but the Puritans bequeathed a rhetorical form—the jeremiad—that functioned to sustain the cultural hegemony of competitive capitalism. It was the Puritan jeremiad, Bercovitch argues, that "gave contract the sanctity of covenant, free enterprise the halo of grace, [and] progress the assurance of the chiliad."[93]

Bercovitch accents the ways in which the American Puritans transformed the traditional European jeremiad.[94] The European jeremiad operated "within a static hierarchical order: the lessons it taught, about historical recurrence and the vanity of human wishes, amounted to a massive ritual reinforcement of tradition." God's wrath was invoked to instill resignation to one's lot in life and obedience to constituted authority. The American Puritans transformed this message by infusing it with "affirmation and exultation," inverting "the doctrine of vengeance into a promise of ultimate success."[95] The American jeremiad is characterized by "affliction *and* promise," "lament *and* celebration." It "both laments an apostasy and heralds a restoration." "Thundering denunciations of a backsliding people" are joined to "the promise of the millennium." "Cries of declension and doom" are accompanied by promises of "a second paradise, a Canaan abounding in blessings beyond anything they had had or imagined."[96]

Where the European jeremiad "used fear and trembling to teach acceptance of fixed social norms," the American Puritan jeremiad used anxiety to assure a continual striving for improvement and progress. "New England's Jeremiahs," Bercovitch writes, "set out to provide the sense of insecurity that would ensure . . . their promised future," a future defined in terms of the values of free enterprise and self-reliance.[97] The European jeremiad sustained traditional hierarchy; the American jeremiad sustained modern individualism. It is the American jeremiad, by fusing the sacred and the profane and "sanctifying an errand of entrepreneurs,"[98] that accounts for America's almost religious devotion to free enterprise capitalism.

To test Bercovitch's thesis, one might ask, Which groups in contemporary American society speak the language of impending catastrophe and future redemption? Which groups issue warnings of cataclysm while holding out hope of a millennial future? Which groups denounce selfish materialism and worldly ambition and envision a future world of friendship and mutual love? Where, in short, do we hear the echoes of the American jeremiad? My hands-down choice would be radical environmental groups (particularly those like Friends of the Earth and Greenpeace) and antinuclear groups (the Clamshell Alliance or Abalone Alliance, for instance).

Chicken Little–like warnings of impending doom pervade the environmental movement. "The world is dissolving. . . . The world is going to tumble around its ears if the Sierra Club—or someone—doesn't do a job in the next five years." "The ocean can die, these horrors could happen. And there would be no place to hide." "There will soon be no such things as fresh air for us to breathe [if we] continue to pollute the air in our major cities."[99] If we do not act immediately to remedy "the greenhouse effect," the earth will become uninhabitable. Three Mile Island and Chernobyl are signs of the terrible devastation that awaits us if we continue down the nuclear path. For the secular environmental movement, as Douglas and Wildavsky note, nature substitutes for God: "Either God will punish or nature will punish; the jeremiad is the same and the sins are the same: worldly ambition, lust after material things, large organization."[100]

As Stephen Cotgrove points out, "warnings of an impending crisis from exceeding the earth's capacity is only half the message." The other half is a message of salvation. "A return to small-scale decentralized communities" will enable the saved to construct "a web of political life that will have no place for exploitative values, destructive technologies, and dehumanized relationships." Alongside its message of impending doom and destruction, Douglas and Wildavsky observe, groups like Friends of the Earth maintain "an almost utopian vision of future society in which all forms of life will exist harmoniously without political, economic, and technological restraints."[101] Redemption is possible for the elect who change their lifestyles in an egalitarian direction.

To what end does the environmental movement invoke the rhetoric of the jeremiad? What, to put the question somewhat differently, is the function of the jeremiad? Do the laments of doom and promises of salvation serve, as Bercovitch would have it, to bolster free enterprise? Hardly. Far from functioning to sustain competitive individualism, the environmentalist's jeremiad attacks the very foundations of the competitive individualist worldview. Justifying individualism's bold —often reckless—experimentation is a view of nature as resilient; no matter how much you knock nature about, it will bounce back. Environmentalists wish to substitute a radically different view of nature as fragile and terrifyingly unforgiving in order to support a radically different way of life.[102] For if nature is fragile, the polity is justified in reining in and regulating the acquisitive entrepreneur. Salvation, environmentalists insist, will come only to those who jettison the manipulation and waste of competitive individualism in favor of the noncoercive, decentralized, small-scale egalitarian way.

Bercovitch's response to such criticisms has been to counter that dissent in America only seems to challenge "the liberal American Way" but that it in fact "re-present[s] the strategies of a triumphant middle-class hegemony." "The catastrophic alternative" is just part of "a strategy for channeling revolution into the service of society." Radicalism is "socialized into an affirmation of order" and redefined as an "affirmation of cultural values." The symbol of America transforms "what might have been a search for moral or social alternatives into a call for cultural revitalization." Invoking the "true America," no matter that the ideal America is used to score the actually existing America, serves only to reaffirm the hegemony of "the American ideology."[103]

This formulation assumes that "America" is a symbol with a fixed meaning over time and across space. Bercovitch speaks as if the meaning of the "myth of America" or "the symbol of America" was self-evident and meant the same to all—free enterprise, laissez-faire, and upward mobility. But the definition of America is part-and-parcel of the conflict between rival political cultures.[104] To denounce American practice in the name of the American ideal, therefore, is not necessarily to reaffirm competitive capitalism. It depends, obviously, on the meaning that one attaches to "America."

Bercovitch strains to fit even the most bitter critics of America within the confines of a "pervasive middle-class hegemony." Even the total rejection of America becomes, in Bercovitch's scheme, quintessentially American. Endorsing Loren Baritz's view that Melville's work "became so purely American because of the depths of his rejection of America," Bercovitch argues that even the "antijeremiad" is "not so much a rejection of the culture as it is a variation on a central cultural theme." The antijeremiad remains "locked within the same symbolic structure," whereby America is either the "world's fairest hope" or "man's foulest crime."[105]

No matter whether "the symbol of America" is invoked to rail against American society or whether it is rejected altogether; either case is evidence of an "extraordinary cultural hegemony."[106] What, one begins to wonder, would count as evidence against the hegemony thesis? Bercovitch's theory appears to have many of the same qualities he ascribes to "the American ideology." Both seem impervious to criticism, turning every dissent into affirmation. That may be a strength for a culture, but it is a debilitating weakness in a theory.

Although dressed in the radical chic of hegemony and hermeneutics, Bercovitch's theory bears a strong resemblance in certain respects to the now discredited structural functionalism of the 1950s and early 1960s. Any action not resulting in the total transformation of society becomes identified as functional for society. Structural functionalists argued, for instance, that the persistence of deviance meant that deviance must be functional for society. Any behavior, no matter how bizarre or seemingly antisocial (up to and including suicide) could be shown to have its functions. More and more inventive functional explanations were dreamed up for every conceivable behavioral pattern.[107] Eventually, however, struc-

tural functionalism collapsed under the realization that short of total societal collapse, everything could be interpreted as functional for society. One had no way
of determining whether society continued to exist despite rather than because of
these practices.[108]

My objection to Bercovitch's analysis is not that it is functionalism but that it
is crude functionalism.[109] Lacking a discriminating conception of system change
short of total revolution, every action—no matter how subversive on its face—can
be made to seem functional for the hegemonic system.[110] Bercovitch is correct,
I believe, in understanding the jeremiad as a form of social control, but his
impoverished understanding of cultural variation obscures the question of what
behaviors the jeremiad was designed to control and in the name of what set of
social relations. In a country in which all roads lead to competitive capitalism,
social control can raise only one question: How does "the dominant culture" circumscribe, coopt, or preempt potential dissent and resistance? But once one sees
American society as consisting of competing cultures, the question of social control becomes more complex and more interesting.

Bercovitch's premise is that in Puritan New England there was "no competing
order—no alternative set of values except the outmoded Old World order they rapidly discarded."[111] If hierarchy was absent or rapidly vanishing, then competitive
individualism was hegemonic. But Bercovitch slights the possibility of a Puritan
culture in early America that resembled neither medieval hierarchy nor modern
capitalism—a Puritan culture that as I argued in chapter 1 combined strongly
bounded groups with individual voluntarism. The tradition–modernity dichotomy
thus leads Bercovitch to overlook perhaps the most self-evident function of
the Puritan jeremiad: to control individual behavior that might undermine group
cohesion.[112] The Puritan jeremiad, after all, repeatedly singles out untrammeled
individualism, materialism, selfishness, and the pursuit of profit as threats to the
collectivity.

It is strongly bounded groups, not freewheeling unattached entrepreneurs, who
have characteristically joined millenarian hopes with warnings of approaching
calamity. The jeremiad's social source is not free enterprise capitalism but rather
the problem of exit in strongly bounded groups. For the Puritans, the jeremiad's
function was to warn against the seductions of the frontier, land speculation, and
commercial ventures that would pull apart the tightly bound group. Similarly, the
jeremiad functions within the contemporary environmental movement to hold
together the group in the face of the opportunities for gain offered by individualism. The wrath of nature, like the wrath of God, is invoked to resist the lures of
individualism. Far from being an affirmation of competitive individualism, the
American jeremiad makes a claim on the individual in the name of the group.

Bercovitch insists that the American jeremiad "serves to blight, and ultimately
to preclude, the possibility of fundamental social change," but I think David Harlan
is far closer to the truth when he says that it is Bercovitch's "impoverished notion
of hegemony that blights whatever it touches."[113] What else are we to make of a
theory that construes John Cotton's stern warnings against profits to be an affirmation of free enterprise, that interprets Walt Whitman's denunciation of America's
"materialistic advancement" as "canker'd, crude [and] saturated in corruption" to

be just another voice in the capitalist chorus, or that sees the social withdrawal of Thoreau's *Walden* as simply one more embodiment of "the myth of American laissez-faire individualism"?[114] American political culture, I would suggest, is not nearly as impoverished as some of the theories used to explain it.

Culture: A Prism, Yes; A Prison, No

The consensus thesis has sometimes been criticized by those on the left for its celebration of consensus and its conservative implications. This is to seriously misunderstand the thesis, at least as it is formulated by Hartz (and also Berco-vitch).[115] True, Hartz did not hold out much hope for a radical transformation of America's liberal tradition.[116] But Hartz's quiescence is born of dark pessimism, not complacent satisfaction. Far from glorying in American consensus, Hartz anguishes over the debilitating consequences of America's "liberal absolutism."[117] From this "absolute and irrational attachment" to Locke comes the frenzy of McCarthyism, the obsession with all things un-American, and an isolationist dis-trust of the larger world.[118] Americans, in Hartz's model, are imprisoned within "the moral unanimity of a liberal society" that affords few if any opportunities for "that grain of relative insight" that is the path to "an understanding of self and an understanding of others."[119] In the grips of an "irrational Lockianism,"[120] learning from errors and from others becomes virtually impossible. Although Hartz did hold out some hope that the United States might transcend its "national blindness" through contact with other nations in an increasingly interdependent world, he remained less than sanguine about the chances of America ever breaking free of its liberal prison.[121]

America's unchallenged liberal political culture, according to Hartz, is "a submerged faith."[122] It constitutes a set of unquestioned axioms or unarticulated premises that remain safely tucked away beneath the surface of partisan rhetoric.[123] Liberalism in America is thus not so much a preference as a prerational identifi-cation. Because attachment to a political culture rests deep within the subconscious or unconscious mind of America, the attachment cannot be "tested" by reality in any meaningful way. Individuals thus have little possibility of transcending their inherited political culture. Indeed individuals are incapable of recognizing let alone modifying the culture that constrains them.[124]

Hartz's pessimistic view of the relation between the individual and culture follows from his premise that only a single culture exists within the United States. In the absence of conflict between rival cultures no social force exists to raise fundamental beliefs to the surface where they can be examined, probed, and tested. If there is only one tradition available to us—the liberal tradition—and there is no way of thinking outside of some tradition—as there most assuredly is not—then we are indeed, as Hartz grimly suggests, hopelessly trapped inside our Lockean culture.

If, however, one allows for competing cultural biases within American soci-ety, there is no need to acquiesce in Hartz's "metaphysical pathos."[125] Culture no longer seems an inescapable prison. If Americans inhabit a social universe characterized by rival cultural biases, then they have ample opportunity to compare their bias with other biases. If alternative ways of looking at the world are constantly clashing, then there is significant scope for learning about, testing, and even abandoning cultural attachments. Culture is a prism that biases the way one experiences the world, not a prison that shuts one completely off from that world.

Cultural biases may be likened to scientific theories.[126] Both are resistant to change; anomalies are explained away, pigeonholed, ignored, or just not seen. Neither life nor science can stand still while each piece of evidence that might contradict an accepted idea is tested. Were every surprise or disappointment to send us scrambling for an alternative theory, both science and life would lack the necessary stability. Science would lose its cumulative character, and social relations would be characterized by a permanent state of anomie.

But cultures, like theories, cannot exclude reality altogether. Although cultures (like poorly formulated theories) build in lots of self-protection, reality can and does intrude. They may predict consequences that prove false, create blind spots that lead to disaster, or generate expectations that go unfulfilled. As evidence builds up against theories, or cultures fail to pay off for their adherents, doubts build up, followed by defections. A persistent pattern of surprises forces individuals to cast around for alternative cultures (or theories) that can provide a more satisfying fit with the world as it is.

Conceiving culture in this way prevents one from turning the individual into a pawn in the hands of disembodied languages or norms. Cultures are not passively received and internalized but instead are continually negotiated and renegotiated by individuals. Culture, as Michael Thompson and Aaron Wildavsky write, "is transmitted from generation to generation . . . [b]ut it is not transmitted unchanged, nor is it transmitted without question. Cultural transmission is absolutely not a game of pass-the-parcel."[127]

This conception of culture rejects both the Parsonian view that sees individuals passively conforming to consensual norms and the Gramscian view that portrays individuals mindlessly consenting to the hegemonic worldview of the dominant group.[128] Both make cultural reproduction too easy, as if collective values mysteriously and inexorably seep into the consciousness of unsuspecting individuals. As with Hartz, the individual is left little choice but to reproduce the existing social system. Barring some exogenous disturbance to the system, the values and institutional commitments of a culture remain the same. If, however, we allow for rival cultural biases within a society, the active, negotiating individual is restored by giving the individual competing norms and values over which to negotiate. No longer is the individual faced with only a grim choice between conformity and deviance, as Parsons would have it, or between submission and revolution, as Gramsci would have it, or between the "national blindness" of an irrational Lockeanism and impotent marginality, as Hartz would have it. In a world of clashing biases, culture is a prism, not a prison.

Notes

Preface

1. David Riesman, *Constraint and Variety in American Education* (Lincoln: University of Nebraska Press, 1956), 78–79. Lest it be thought that this view of historians is held only by nonhistorians, consider Edmund S. Morgan's admission that "most historians don't much like generalizations. Indeed they make a trade of showing that this or that generalization about the past will not work here or there or then" ("Mothers of Us All," *New York Review of Books*, February 1, 1990, 18).

2. Clifford Geertz, "Politics Past, Politics Present: Some Notes on the Uses of Anthropology in Understanding the New States," *European Journal of Sociology* 8 (1967), 4. Reprinted as chapter 12 in Geertz, *The Interpretation of Cultures* (New York: Basic Books, 1973).

3. Clifford Geertz, *Local Knowledge: Further Essays in Interpretive Anthropology* (New York: Basic Books, 1983), 4. This trend in Geertz's thought, which is already apparent in the introductory essay to *The Interpretation of Cultures*, becomes more fully manifest in *Negara: The Theatre State in Nineteenth-Century Bali* (Princeton: Princeton University Press, 1980) and *Local Knowledge*. Also see the profile of Geertz by John Horgan, "Ethnography as Art," *Scientific American* 261 (July 1989), 28–31. The contrast between the early and later Geertz is brought out well by Jeffrey Alexander in *Twenty Lectures: Sociological Theory Since World War II* (New York: Columbia University Press, 1989), 302–29.

4. Historian Ronald G. Walters agrees that "Geertz's influence has sunk historians deeper in old [antitheoretical] habits of thought. . . . The tendency of thick description . . . is to reinforce the impulse to burrow in and not to try to connect the dots" ("Signs of the Times: Clifford Geertz and Historians," *Social Research* 47 [Autumn 1980], 551, 556). Contrast this with Richard R. Beeman's hope that Geertz's "semiotic approach" to culture would help historians "begin the task of fashioning some coherent interpretive pattern from the scores of local studies on which historians have been so busily engaged" ("The New Social History and the Search for 'Community' in Colonial America," *American Quarterly* 29 [Fall 1977], 443).

5. John Higham, "Introduction," in John Higham and Paul K. Conkin, eds., *New Directions in American Intellectual History* (Baltimore: Johns Hopkins University Press, 1979), xvi. Daniel Joseph Singal comments that Geertz's "writings have been cited so often by

historians that it has become something of a professional embarrassment" ("Beyond Consensus: Richard Hofstadter and American Historiography," *American Historical Review* 89 [October 1984], 998). More recently, Robert F. Berkhofer, Jr., suggested that "following the transition from cultural unity to social division as the clue to interpreting culture has been the passing of the patron sainthood of cultural studies from Clifford Geertz to Raymond Williams" ("A New Context for a New American Studies," *American Quarterly* 41 [December 1989], 607n4). This is not quite right, for Geertz's conception of knowledge as ineluctably local is perfectly consistent with an emphasis on social division, and Williams's call for a "theory of social totality" as well as his use of the concept of hegemony has a great deal to say about cultural unity. Perhaps what Williams provides that Geertz does not is an anticapitalist ideological bias more compatible with the views of many younger historians and students of American studies.

    6. Walters, "Signs of the Times," 551. Also see Thomas Bender's "Wholes and Parts: The Need for Synthesis in American History," *Journal of American History* 73 (June 1986), 120–36. Bender complains that "as ever deeper explorations into the interior meaning shared by groups achieve greater and greater ethnographic integrity, they become more and more self-contained," thus setting back "the cause of synthesis" (129).

    7. The question of synthesis is often confused with the quite different question of accessibility to the general reading public (see, e.g., Bender, "Wholes and Parts"; and Eric H. Monkkonen, "The Dangers of Synthesis," *American Historical Review* 91 [December 1986], 1146–57). A synthesis need not be any more accessible or less technical than the most narrow monograph. Louis Hartz's *The Liberal Tradition in America* (New York: Harcourt, 1955) is a case in point of a synthesis that is quite inaccessible to the general public.

    8. Geertz, *Interpretation of Cultures*, 10, 12.

    9. See Mary Douglas, *Natural Symbols: Explorations in Cosmology* (London: Barrie & Rockliff, 1970); and "Cultural Bias," in Douglas, *In the Active Voice* (London: Routledge & Kegan Paul, 1982). These categories are further elaborated in Michael Thompson, Richard Ellis, and Aaron Wildavsky, *Cultural Theory* (Boulder: Westview Press, 1990).

    10. The phrase "a wilderness of detail" comes from Tocqueville. "The Deity," Tocqueville observed, "has no need of general ideas. . . . With one glance He sees every human being separately and sees in each the resemblances that make him like his fellows and the differences which isolate him from them. . . . It is not like that with man. If a human intelligence tried to examine and judge all the particular cases that came his way individually he would soon be lost in a wilderness of detail and not be able to see anything at all" (Alexis de Tocqueville, *Democracy in America* [Garden City, N.Y.: Anchor, 1969], 437).

    11. Carl Degler, "Remaking American History," *Journal of American History* 67 (June 1980), 17. Degler is approvingly paraphrasing Frances FitzGerald. Similarly, Jack P. Greene and J. R. Pole lament that "as scholars have concentrated more and more upon smaller and smaller units in their laudable efforts to recover the context and texture of colonial life in as much detail as the sources and scholarly ingenuity will permit . . . one paradoxical result [has] been a signal loss of overall coherence until we are now less clear than ever before about precisely what the central themes and the larger questions are in the field as a whole" ("Reconstructing British-American Colonial History: An Introduction," in Greene and Pole, eds., *Colonial British America: Essays in the New History of the Early Modern Era* [Baltimore: Johns Hopkins University Press, 1984], 7).

    12. Kristen Luker uses the phrase "the tip of the iceberg" in the context of analyzing the contemporary battle over abortion. "Each side of the abortion debate," she writes, "has an internally coherent and mutually shared view of the world that is tacit, never fully articulated, and, most importantly, completely at odds with the world view held by their oppo-

nents" (*Abortion and the Politics of Motherhood* [Berkeley: University of California Press, 1984], 159).

13. T. H. Marshall, *Sociology at the Crossroads and Other Essays* (London: Heinemann, 1963), 36–37.

Chapter 1

1. See, e.g., Max Weber, *The Protestant Ethic and the Spirit of Capitalism* (New York: Scribner's, 1958), 222n22; Arthur O. Lovejoy, *The Great Chain of Being* (Cambridge: Harvard University Press, 1957), 5–6; Koenraad W. Swart, "'Individualism' in the Mid-Nineteenth Century," *Journal of the History of Ideas* 23 (January–March 1962), 77; Steven Lukes, *Individualism* (New York: Harper & Row, 1973), ix; Robert N. Bellah, Richard Madsen, William M. Sullivan, Ann Swidler, and Steven M. Tipton, *Habits of the Heart: Individualism and Commitment in American Life* (New York: Harper & Row, 1985), 142.

2. Lukes, *Individualism*. Lukes, "Types of Individualism," in Philip P. Wiener, ed., *Dictionary of the History of Ideas: Studies of Selected Pivotal Ideas*, 4 vols. (New York: Scribner's, 1973), 2: 594–604.

3. In addition to the sources cited in note 8 of the preface, see Mary Douglas, "Introduction to Grid/Group Analysis," in Douglas, ed., *Essays in the Sociology of Perception* (London: Routledge & Kegan Paul, 1982); and Mary Douglas, "Converging on Autonomy: Anthropology and Institutional Economics," in Oliver E. Williamson, ed., *Organization Theory: From Chester Barnard to the Present and Beyond* (New York: Oxford University Press, 1990), 98–115.

4. Emile Durkheim, *Suicide: A Study in Sociology* (Glencoe: Free Press, 1951), esp. chap. 5.

5. Douglas, "Introduction to Grid/Group Analysis," 3. Mary Douglas, "Cultural Bias," in Douglas, *In the Active Voice* (London: Routledge & Kegan Paul, 1982), 205. Casey Hayden's description of her relationship to the civil rights movement provides a good illustration of what Douglas has in mind by "strong group." "The movement," Hayden writes, "today is commonly known as the civil rights movement, but it was considerably more than that. To me, it was everything: home and family, food and work, love and reason to live. When I was no longer there, and then when it was no longer there at all, it was hard to go on" (Casey Hayden, preface to Mary King, *Freedom Song: A Personal Song of the 1960s Civil Rights Movement* [New York: Morrow, 1987], 7).

6. Douglas, "Introduction to Grid/Group Analysis," 3.

7. "How can there be only four ways of life?" is a question often raised about this approach. This is the wrong question, as a hypothetical example will help to show. If we selected two variables, say height and weight, no one would find it surprising or improbable that we could place the entire population in the fourfold typology that these two variables would generate. Because all human beings have a height and a weight, all human beings could be placed at some point (and only one point) in the typology. The real question is, What is gained by analyzing the world in terms of height and weight, lumping together all those relatively short and heavy people, and contrasting them with tall and heavy people or with tall and skinny people? For those interested in political phenomena the answer would be very little (the answer might differ if one was interested in heart attack rates). The proper question, then, for the grid–group typology, is, What is gained by analyzing the world in terms of grid and group, lumping together all people who prefer (or live in) a hierarchical collectivity, and contrasting these people with, say, those who prefer

(or live in) an egalitarian community? As Robert Brown puts the point, "*Any* criterion will organize data—will order items in clases—but only some classifications will be scientifically useful" ("Structural-Functional Analysis: Some Problems and Misunderstandings," *American Sociological Review* 21 [April 1956], 129). My aim in this book is to demonstrate that the classifications derived from the grid–group typology are "scientifically useful," by which I mean nothing more than that they help us to understand or explain important aspects of social life.

8. Bernard Bailyn, *Education in the Forming of American Society* (Chapel Hill: University of North Carolina Press, 1960), 15–16, 49. Also see Richard Brown, *Modernization: The Transformation of American Life, 1600–1865* (New York: Hill & Wang, 1976); James A. Henretta, *The Evolution of American Society, 1700–1815: An Interdisciplinary Analysis* (Lexington: Heath, 1973), esp. 206–14; James A. Henretta, "Reply" to James T. Lemon, *William and Mary Quarterly* 37 (October 1980), 698–99; James A. Henretta, *The Origins of American Capitalism: Collected Essays* (Boston: Northeastern University Press, 1991). The influence of the tradition–modernity dichotomy on historians is documented in Thomas Bender's excellent *Community and Social Change in America* (New Brunswick: Rutgers University Press, 1978), esp. chap. 3. Also see Joyce Appleby, "Value and Society," in Jack P. Greene and J. R. Pole, eds., *Colonial British America: Essays in the New History of the Early Modern Era* (Baltimore: Johns Hopkins University Press, 1984), 290–316.

9. Alexis de Tocqueville, *Democracy in America*, trans. George Lawrence (Garden City, N.Y.: Anchor, 1969), 509. Also see Roger Boesche, "The Prison: Tocqueville's Model for Despotism," *Western Political Quarterly* 33 (December 1980), 550–63.

10. A survey a class of mine recently carried out of nine groups—abortion rights activists, prolife activists, Catholic priests, professors, students, Democratic precinct workers, car salesmen, homeless people, and prisoners—found little evidence of fatalistic attitudes. Using a six-item scale—There is no use in doing things for people, you only get it in the neck in the long run; The future is too uncertain for a person to make serious plans; No matter how hard people try, fate largely determines the outcome; A person is better off if he or she does not trust anyone; Failure is mostly a matter of bad luck; People like me don't have any say about what government does—less than 4 percent of the nonprison population answered more than two of the fatalist items in the affirmative. Among Democratic party activists, Catholic priests, abortion rights activists, and prolife activists, no respondent answered more than one of the fatalist items in the affirmative, and over 9 in 10 of these respondents rejected all six of the statements. Among prisoners (all of whom were repeat offenders), by contrast, roughly three quarters answered three or more of the fatalist items in the affirmative, and over half assented to four or more of the statements.

11. I take up the question of hierarchy's relative strength or weakness in the United States in chapter 6.

12. Louis Hartz, *The Liberal Tradition in America* (New York: Harcourt, 1955). Daniel J. Boorstin, *The Genius of American Politics* (Chicago: University of Chicago Press, 1953).

13. Martin Taylor, "In Search of 'Omens, Myths, Heroes' for 1990s," *Willamette Collegian*, October 19, 1990, 2. Linda MacRae-Campbell, quoted in Mark Satin in *New Options*, May 29, 1989, and reprinted in the *Utne Reader* (September/October 1990), 81–82. Parker J. Palmer, quoted in Don Wycliff, "Critic of Academia Wins Applause on Campus," *New York Times*, September 12, 1990, B9. John Taylor, "Are You Politically Correct?" *New York,* January 21, 1991, 35. Robert Booth Fowler agrees that "rampant individualism is the favorite theme and the constant complaint among many American intellectuals. . . . One cannot read American political intellectuals today without recognizing that capitalism receives scant sympathy" (*The Dance with Community: The Contemporary Debate in American Political Thought* [Lawrence: University Press of Kansas, 1991], 11, 16).

14. Swart, "'Individualism' in the Mid-Nineteenth Century," 78. Christopher J. Berry, *The Idea of a Democratic Community* (New York: St. Martin's Press, 1989), esp. 2. There are, of course, also conservative or hierarchical critics of individualism. One thinks here of such political thinkers as Robert Nisbet, Russell Kirk, Alan Bloom, and Alasdair MacIntyre (see Fowler, *Dance with Community*, esp. chap. 6). Generally, though, the communitarian critique of individualism within the academy has come from the egalitarian left. Robert Fowler, although attentive to the great diversity of approaches among community-oriented intellectuals, is "struck by the shared assumptions behind much communitarian thinking today." Chief among these is "the attraction of many communitarians to substantive equality in numerous, even all, possible areas of life. Community is often taken to imply equality without question" (149). For a recent attempt to redirect the communitarian critique of liberalism in a more hierarchical direction, see Clarke E. Cochran, "The Thin Theory of Community: The Communitarians and Their Critics," *Political Studies* 37 (September 1989), 422–35. The difference between Cochran's communitarian vision and the communitarianism put forward by an egalitarian like Benjamin Barber becomes evident when Cochran specifies what he has in mind by a "thicker" conception of community. Cochran writes, "Authority is one of the constituent features of community, so community cannot be understood without unpacking the meaning of authority. . . . Certain concepts travel with authority, concepts such as loyalty, commitment, obedience, law and coercion, all ideas that seem to demand sacrifice of individuality to diminish agency. There are other concepts as well that make community 'thick': ritual, tradition, common good and common action. Communities cannot exist without these. . . . Acceptance of authority, loyalty to ideals and commitment to an historical community, though they do require sacrifice and closure of options, are the very stuff of character-building. Accepting sacrifice for a community forms identity" (434).

15. Bellah et al., *Habits of the Heart*, 277. Also see Robert N. Bellah, Richard Madsen, William M. Sullivan, Ann Swidler, and Steven M. Tipton, *The Good Society* (New York: Knopf, 1991), and the review of both books by Andrew Greeley, "Habits of the Head," *Society* 29 (May/June 1992), 74–81. A preference for communitarianism without the hierarchical restrictions on individual freedom characteristic of past ages is evident, too, in Ralph Ketcham's question: "How can the intricate interplay of market forces, however preferable to guild monopoly . . . , be itself a ground of genuine community? . . . How does equality before the law, however corrosive of formal hierarchy and privilege, get beyond the adversarial style of the courtroom to social harmony and the equivalent of what Quakers call 'the sense of the meeting'?" (Ralph Ketcham, *Individualism and Public Life* [New York: Basil Blackwell, 1987], 30–31).

16. Benjamin R. Barber, *Strong Democracy: Participatory Politics for a New Age* (Berkeley: University of California Press, 1984), 71–72, 231. The quotation from Robert Nisbet is from *The Twilight of Authority* (New York: Oxford University Press, 1975), 217.

17. Christopher Lasch, *The True and Only Heaven: Progress and Its Critics* (New York: Norton, 1991), 172; also 15. Also see Don Herzog, "Some Questions for Republicans," *Political Theory* 14 (August 1986), 474, 476; Thomas L. Pangle, *The Spirit of Modern Republicanism: The Moral Vision of the American Founders and the Philosophy of Locke* (Chicago: University of Chicago Press, 1988), 28; and Fowler, *Dance with Community*, 27, 63–79.

18. William M. Sullivan, *Reconstructing Public Philosophy* (Berkeley: University of California Press, 1982), 55. Also see Fowler, *Dance with Community*, esp. 63–64, 69–70.

19. See Aaron Wildavsky with Brendon Swedlow, "Is Egalitarianism Really on the Rise?" in Aaron Wildavsky, *The Rise of Radical Egalitarianism* (Washington, D.C.: American University Press, 1991), 86–94; and Sidney Verba and Gary Orren, *Equality in America: The View from the Top* (Cambridge: Harvard University Press, 1985).

20. Jean Bethke Elshtain, *Public Man, Private Woman: Women in Social and Political Thought* (Princeton: Princeton University Press, 1981), 246. Ann Ferguson and Nancy Folbre, quoted in Hester Eisenstein, *Contemporary Feminist Thought* (Boston: G. K. Hall, 1983), 144.

21. Elinor Lenz and Barbara Myerhoff, *The Feminization of America* (Los Angeles: Jeremy P. Tarcher, 1985), 37, 38, 230, 10.

22. Kathy E. Ferguson, *The Feminist Case Against Bureaucracy* (Philadelphia: Temple University Press, 1984), 198.

23. Ann Ferguson, *Blood at the Root: Motherhood, Sexuality and Male Dominance* (London: Pandora, 1989), 155, 230, 228. The consciousness-raising group favored by so many feminists typifies the egalitarian combination of high group and low grid. Such consciousness-raising groups are supposed to be characterized by "an ethic of openness, honesty, and self-awareness." The aim is to discuss intimate subjects in a confidential, non-hierarchical atmosphere, and to confront "sources of inequality on the basis of which members feel subordinated or excluded" (Catherine MacKinnon, quoted in Stephen Macedo, "Review Essay: Justice, Sex, & Doing the Dishes," *Polity* 24 [Spring 1992], 519). "Setting up a structure that is too confining may destroy the atmosphere of free association and, as a result, someone's story may be lost" (Patricia Cain, quoted in Michael Weiss, "Feminist Pedagogy in the Law Schools," *Academic Questions* 5 [Summer 1992], 79; also see Christina Hoff Sommers, "Sister Soldiers," *New Republic*, October 5, 1992, 29–33.) Men are excluded from these groups on the grounds that only then can women be open and share in a truly egalitarian fashion (Macedo, "Justice, Sex, & Doing the Dishes," 521; also see Weiss, "Feminist Pedagogy," 83).

24. Barbara Katz Rothman, *Recreating Motherhood: Ideology and Technology in a Patriarchal Society* (New York: Norton, 1989), 253, 59, 221, 251.

25. Martha A. Ackelsberg and Kathryn Pyne Addelson, "Anarchist Alternatives to Competition," in Valerie Miner and Helen E. Longino, eds., *Competition: A Feminist Taboo?* (New York: Feminist Press, 1987), 230–31, 223.

26. Wildavsky with Swedlow, "Is Egalitarianism Really on the Rise?"

27. Tocqueville, *Democracy in America*, 35–36, 279, 39.

28. David Hackett Fischer, *Albion's Seed: Four British Folkways in America* (New York: Oxford University Press, 1989), 25–31, 174–80, esp. 27–28, 177–80. Also see Stephen Foster, *Their Solitary Way: The Puritan Social Ethic in the First Century of Settlement in New England* (New Haven: Yale University Press, 1971), 32–33; and Kenneth A. Lockridge, *Settlement and Unsettlement: The Crisis of Political Legitimacy Before the Revolution* (Cambridge: Cambridge University Press, 1981), 20.

29. Fischer, *Albion's Seed*, 118, 101, 180. Also see Foster, *Solitary Way*, 28.

30. John Winthrop, "A Model of Christian Charity," in Perry Miller and Thomas Johnson, eds., *The Puritans* (New York: Harper & Row, 1963), 195; spelling and punctuation have been modernized throughout. Norman Fiering, *Jonathan Edwards' Moral Thought and Its British Context* (Chapel Hill: University of North Carolina Press, 1981), 131. Also see William Perkins, "A Treatise on the Vocations or Callings of Men," in Edmund S. Morgan, ed., *Puritan Political Ideas, 1558–1794* (Indianapolis: Bobbs-Merrill, 1965), 39. For further evidence of hierarchical rhetoric and practices within Puritanism, see Dean Hammer, "Puritanism in the Making of a Nation," typescript, 1992, esp. chap. 1; Melvin Yazawa, *From Colonies to Commonwealth: Familial Ideology and the Beginnings of the American Republic* (Baltimore: Johns Hopkins University Press, 1985), esp. 9–27; and Philip Greven, *The Protestant Temperament: Patterns of Child-Rearing, Religious Experience, and the Self in Early America* (Chicago: University of Chicago Press, 1977), esp. 194–98. Also see Perry Miller, *Nature's Nation* (Cambridge: Harvard University Press, 1967), 43.

31. Michael Walzer, *The Revolution of the Saints* (Cambridge: Harvard University Press, 1965), 169.

32. Perry Miller, *Errand into the Wilderness* (New York: Harper & Row, 1956), 147. Also see Harry S. Stout, *The New England Soul: Preaching and Religious Culture in Colonial New England* (New York: Oxford University Press, 1986), 18–19.

33. Joshua Miller, "Direct Democracy and the Puritan Theory of Membership," *Journal of Politics* 53 (February 1991), 65. Peter N. Carroll, *Puritanism and the Wilderness: The Intellectual Significance of the New England Frontier, 1629–1700* (New York: Columbia University Press, 1969), 132. Also see Miller, *Errand into the Wilderness*, 148.

34. Fischer, *Albion's Seed*, 78. Also see Edmund S. Morgan, *The Puritan Family: Religion and Domestic Relations in Seventeenth-Century New England* (New York: Harper & Row, 1966), 83–86.

35. Miller, *Errand into the Wilderness*, 148.

36. Walzer, *Revolution of the Saints*, chap. 5, esp. 151–53.

37. Michael Zuckerman, "The Fabrication of Identity in Early America," *William and Mary Quarterly* 34 (April 1977), 202.

38. John Cotton, "Swine and Goats," in Miller and Johnson, *The Puritans*, 314. Also see David D. Hall, *Worlds of Wonder, Days of Judgment: Popular Religious Belief in Early New England* (New York: Knopf, 1989), 117.

39. In the area of food, too, the Puritans were among the earliest Americans to "associate plain cooking with piety, and vegetables with virtue" (*Fisher, Albion's Seed*, 135).

40. Fischer, *Albion's Seed*, 141–43.

41. Fischer, *Albion's Seed*, 172, 166–68, 174.

42. Hartz, *Liberal Tradition*, 62.

43. Winthrop, "A Model of Christian Charity," in Miller and Johnson, *The Puritans*, 198.

44. Fischer, *Albion's Seed*, 24.

45. "Bradford's History of Plymouth Plantation, 1606–1646," quoted in Alan Simpson, *Puritanism in Old and New England* (Chicago: University of Chicago Press, 1955), 33.

46. Miller, *Errand in the Wilderness*, 143.

47. Richard Baxter, quoted in Walzer, *Revolution of the Saints*, 170.

48. Fischer, *Albion's Seed*, 73, 181. Also see Carroll, *Puritanism and the Wilderness*, 141–47.

49. Carroll, *Puritanism and the Wilderness*, 137, 79, 150, 127. Also see Dennis J. Coyle, "Breaking Through the Hedge: Puritan Attitudes Toward Expansion into the Wilderness," typescript, University of California, Berkeley, 1983.

50. Fischer, *Albion's Seed*, 72, 186.

51. Fischer, *Albion's Seed*, 156–58.

52. Ralph Barton Perry, *Puritanism and Democracy* (New York: Vanguard Press, 1944), 297. Also see Sacvan Bercovitch, *The American Jeremiad* (Madison: University of Wisconsin Press, 1978) and, of course, Max Weber, *The Protestant Ethic and the Spirit of Capitalism*. A critique of Weber along these lines is elaborated in Michael Thompson, Richard Ellis, and Aaron Wildavsky, *Cultural Theory* (Boulder: Westview Press, 1990), chap. 9. A seminal article on this point is Michael Walzer, "Puritanism as a Revolutionary Ideology," *History and Theory* 3 (1963), 59–90.

53. Miller and Johnson, "Introduction," in Miller and Johnson, *The Puritans*, 5. Fischer, *Albion's Seed*, 153. Increase Mather, "An Earnest Exhortation to the Inhabitants of New England," in Richard Slotkin and James F. Folsom, *So Dreadful a Judgment; Puritan Responses to King Phillip's War, 1676–1677* (Middletown: Wesleyan University Press, 1978), 179–80.

54. Perry Miller, *The New England Mind: The Seventeenth Century* (Cambridge: Harvard University Press, 1954), 462. Walzer, *Revolution of the Saints*, 170. Rupert Wilkinson, *The Pursuit of National Character* (New York: Harper & Row, 1988), 56. Zuckerman, "The Fabrication of Identity," 205. Also see Charles H. George and Katherine George, *The Protestant Mind of the English Reformation, 1570–1640* (Princeton: Princeton University Press, 1961), 101–4; Charles E. Hambrick-Stowe, *The Practice of Piety: Puritan Devotional Disciplines in Seventeenth-Century New England* (Chapel Hill: University of North Carolina Press, 1982), esp. 39–50; and Lasch, *The True and Only Heaven*, 551.

55. So, for instance, Sacvan Bercovitch writes that the Puritan thought of John Cotton and John Winthrop reflects "an earlier ideal of class deference while foreshadowing later developments toward free enterprise" (*American Jeremiad*, 22), and Perry Miller describes the Puritans' "crablike progress" from an "aristocratic" order to "a middle-class empirical enterprising society" (quoted in *American Jeremiad*, 27).

56. Puritanism, of course, was not all of a piece. As David D. Hall recently pointed out, Puritanism was "a movement that wore several different faces at any point in time" ("On Common Ground: The Coherence of American Puritan Studies," *William and Mary Quarterly* 44 [April 1987], 195). Kenneth Lockridge, for instance, contrasts an "intensely localistic version of Puritanism" that resisted hierarchical authority with an "establishment" version of Puritanism that believed that "delegation of authority to aristocratic leaders was . . . a precondition of a valid state" (*Settlement and Unsettlement*, 37, 48). The more radical or egalitarian brand of Puritanism is well analyzed in Philip F. Gura, *A Glimpse of Sion's Glory: Puritan Radicalism in New England, 1620–1660* (Middletown: Wesleyan University Press, 1984).

57. It must be conceded that the Puritans' support of deference and authority in certain areas of life fits ambiguously at best into the tradition of egalitarian community identified in this chapter. Puritanism, Robert Fowler points out, has generally not proven "to be an attractive area of study for intellectuals in search of community in our history, in part because . . . Puritans do not fit neatly with the secular, egalitarian, and often participatory democratic model of community favored by other communitarians at present" (Fowler, *Dance of Community*, 31). A recent exception that does look at the Puritans for a model of egalitarian community is Miller, "Direct Democracy and the Puritan Theory of Membership." More complex, but in a similar vein to Miller's analysis, is John H. Schaar, "Liberty/Authority/Community in the Political Thought of John Winthrop," *Political Theory* 19 (November 1991), 493–518. My own view is that Puritanism is Janus-faced: one side anticipates the hierarchy of New England Federalism and Whiggery, the other looks toward the egalitarianism of radical abolitionism.

58. Mercy Warren, "History of the Rise, Progress and Termination of the American Revolution" (1805) in Herbert Storing, ed., *The Complete Anti-Federalist*, 7 vols. (Chicago: University of Chicago Press, 1981), 6:216–17, 230, 206.

59. Gordon S. Wood, *The Creation of the American Republic, 1776–1787* (Chapel Hill: University of North Carolina Press, 1969), 68.

60. Isaac Kramnick, "The 'Great National Discussion': The Discourse of Politics in 1787," *William and Mary Quarterly* 45 (January 1988), 4–5.

61. J. G. A. Pocock, "Virtue and Commerce in the Eighteenth Century," *Journal of Interdisciplinary History* 3 (Summer 1972), 121.

62. Forrest McDonald, *Novus Ordo Seclorum: The Intellectual Origins of the Constitution* (Lawrence: University Press of Kansas, 1985), 70–71. Much the same tension is captured in J. G. A. Pocock's observation that classical republicanism contained "an image of the human personality, at once intensely autonomous and intensely participatory" ("Virtue and Commerce," 134).

63. J. G. A. Pocock, *The Machiavellian Moment: Florentine Political Thought and the Atlantic Republican Tradition* (Princeton: Princeton University Press, 1975), 546.

64. Gordon Wood comments that "no other leader took classical republican values quite as seriously as Adams did" (Gordon S. Wood, "Interests and Disinterestedness in the Making of the Constitution," in Richard Beeman, Stephen Botein, and Edward C. Carter II, *Beyond Confederation: Origins of the Constitution and American National Identity* [Chapel Hill: University of North Carolina Press, 1987], 82). Also see Pauline Maier, *The Old Revolutionaries: Political Lives in the Age of Samuel Adams* (New York: Knopf, 1980), 32.

65. Kramnick, "The 'Great National Discussion,'" 15. Also quoted in Wood, *Creation of the American Republic*, 61, 421.

66. Vernon L. Parrington, *Main Currents in American Thought: The Colonial Mind, 1620–1800* (New York: Harcourt, 1927), 251.

67. Wood, *Creation of the American Republic*, 114–18. Drew R. McCoy, *The Elusive Republic: Political Economy in Jeffersonian America* (Chapel Hill: University of North Carolina Press, 1980), 23, 70–75. McDonald, *Novus Ordo Seclorum*, 87–89.

68. Baron de Montesquieu, *The Spirit of the Laws*, trans. Thomas Nugent (New York: Hafner, 1949), bk. 5, chaps. 2 and 3; bk. 7, chap. 2, pp. 40–41, 96.

69. Wood, *Creation of the American Republic*, 65. Also see McCoy, *The Elusive Republic*, 72–73.

70. Wood, *Creation of the American Republic*, 65.

71. Montesquieu, *Spirit of the Laws*, bk. 5, chap. 5, p. 43.

72. Wood, *Creation of the American Republic*, 63–64, 419. Pauline Maier, *From Resistance to Revolution: Colonial Radicals and the Development of Opposition to Britain, 1765–1776* (New York: Random House, 1972), 137–38. McDonald, *Novus Ordo Seclorum*, 88–90.

73. McCoy, *Elusive Republic*, 100–101. Also see Wood, *Creation of the American Republic*, 418; and Herbert Storing, *What the Anti-Federalists Were For* (Chicago: University of Chicago Press, 1981), 21.

74. Wood, *Creation of the American Republic*, 230–31, 227–28.

75. Eric Foner, *Tom Paine and Revolutionary America* (New York: Oxford University Press, 1976), 152, 169–70, 194, and chap. 5 passim. Also see McDonald, *Novus Ordo Seclorum*, 89–90.

76. Letters of Cato III, in *Complete Anti-Federalist*, 2:110; the quotation from Montesquieu is from *Spirit of the Laws*, bk. 8, chap. 16.

77. Kramnick, "The 'Great National Discussion,'" 12. Storing, *What the Anti-Federalists Were For*, 20.

78. "Essays of Brutus," October 18, 1787, in *Complete Anti-Federalist*, 2:369.

79. "The Impartial Examiner," February 27, 1788, in *Complete Anti-Federalist*, 5:180.

80. Letters of Cato III, in *Complete Anti-Federalist*, 2:110. *The Federalist*, ed. Jacob E. Cooke (Middletown: Wesleyan University Press, 1961), No. 51, p. 349. Also see Wood, *Creation of the American Republic*, 500.

81. "A Federal Republican," in *Complete Anti-Federalist*, 3:76. Also see Kramnick, "The 'Great National Discussion,'" 11.

82. Letters of Cato V, in *Complete Anti-Federalist*, 2:117.

83. Storing, *What the Anti-Federalists Were For*, 30–31. Patrick Henry, Speech in the Virginia Ratifying Convention, June 5, 1788, in *Complete Anti-Federalist*, 5:227.

84. Charles Turner, Speech in the Massachusetts Ratifying Convention, January 17, 1788, in *Complete Anti-Federalist*, 4:219.

85. "The Impartial Examiner," March 5, 1788, in *Complete Anti-Federalist*, 5:188.

86. Wood, *Creation of the American Republic*, 499. Wood, "Interests and Disinterestedness," 101.

87. William G. McLoughlin, *Revivals, Awakenings, and Reform* (Chicago: University of Chicago Press, 1978), 128. William G. McLoughlin, "Charles Grandison Finney," in David Brion Davis, ed., *Ante-Bellum Reform* (New York: Harper & Row, 1967), 105. The quotation from Christ is at Matthew 5:48.

88. John L. Thomas, "Romantic Reform in America, 1815–1866," *American Quarterly* 17 (Winter 1965), 656–81; the quotation from Emerson is at 656. The impact of perfectionism on antebellum reform is also documented in Gilbert Barnes, *The Anti-Slavery Impulse, 1830–1844* (New York: Harcourt, 1964), and Whitney R. Cross, *The Burned-Over District* (Ithaca: Cornell University Press, 1950).

89. McLoughlin, *Revivals, Awakenings, and Reform*, 129.

90. John L. Thomas, "Antislavery and Utopia," in Martin Duberman, ed., *The Antislavery Vanguard: New Essays on the Abolitionists* (Princeton: Princeton University Press, 1965), 247–48. Lewis Perry, *Radical Abolitionism: Anarchy and the Government of God in Antislavery Thought* (Ithaca: Cornell University Press, 1973), 63.

91. See Thomas, "Antislavery and Utopia," 240–45. The quote from Finney is in McLoughlin, *Revivals, Awakenings, and Reform*, 129.

92. This theme is developed throughout the corpus of Friedrich A. Hayek. See, for instance, *The Constitution of Liberty* (Chicago: University of Chicago Press, 1960) and *Law, Legislation and Liberty* (Chicago: University of Chicago Press, 1973). Also see Thomas Sowell, *A Conflict of Visions* (New York: William Morrow, 1987), 26.

93. Adam Smith, *The Wealth of Nations*, ed. Edwin Cannon, (New York: Modern Library, 1937), 14.

94. McLoughlin, *Revivals, Awakenings, and Reform*, 114. The connection between views of human nature and ways of life is explored at greater length in Thompson, Ellis, and Wildavsky, *Cultural Theory*, chap. 1.

95. Adin Ballou quoted in Thomas, "Antislavery and Utopia," 252.

96. Jonathon A. Glickstein, "'Poverty Is Not Slavery': American Abolitionists and the Competitive Labor Market," in Lewis Perry and Michael Fellman, eds., *Antislavery Reconsidered: New Perspectives on the Abolitionists* (Baton Rouge: Louisiana State University Press, 1979), 198n6.

97. This argument is elaborated in Richard Ellis and Aaron Wildavsky, "A Cultural Analysis of the Role of Abolitionists in the Coming of the Civil War," *Comparative Studies in Society and History* 31 (January 1990), 89–116. Also see Daniel J. McInerney, "'A State of Commerce': Market Power and Slave Power in Abolitionist Political Economy," *Civil War History* 37 (June 1991), 101–19. "In the eyes of [abolitionist] reformers," McInerney shows, "the slavepower built a world on the model of the market: it merchandised mankind, encoded accumulation, and systematized selfishness" (107). "The slavepower," in the abolitionists' indictment, "dominated economic relations, internalized the mercenary standards of the marketplace, [made] the market the measure of all things, [and] hoped to subordinate the polity and society completely to the principles of economic transaction" (102–3).

98. Lawrence J. Friedman, *Gregarious Saints: Self and Community in American Abolitionism, 1830–1870* (Cambridge: Cambridge University Press, 1982), 97.

99. Friedman, *Gregarious Saints*, 4. Lawrence J. Friedman, "'Pious Fellowship' and Modernity: A Psychosocial Modernity," in Alan M. Kraut, ed., *Crusaders and Compromisers: Essays on the Relationship of the Antislavery Struggle to the Antebellum Party System* (Westport, Conn.: Greenwood Press, 1983), 241. A decade after the Civil War,

Wendell Phillips recalled fondly that "large and loving group [the Boston Clique] that lived and worked together [and had been] all the world to each other" (Friedman, *Gregarious Saints*, 66).

100. Friedman, *Gregarious Saints*, 68.

101. Thomas, "Antislavery and Utopia," 247.

102. Charles Follen, quoted in Friedman, *Gregarious Saints*, 44.

103. Ronald G. Walters, *The Antislavery Appeal: American Abolitionism After 1830* (Baltimore: Johns Hopkins University Press, 1976), 111, 119, 113. After the Civil War, Phillips would go much further and call for "the overthrow of the whole profit-making system" (quoted in Richard Hofstadter, *The American Political Tradition* [New York: Vintage, 1973], 202).

104. Walters, *Antislavery Appeal*, 112, 148.

105. Quoted in Friedman, "'Pious Fellowship' and Modernity," 242.

106. Thomas, "Antislavery and Utopia," 263.

107. Quoted in Ronald G. Walters, *American Reformers, 1815–1860* (New York: Hill & Wang, 1978), 37.

108. Walters, *Antislavery Appeal*, 119.

109. Walters, *Antislavery Appeal*, 127.

110. Thomas, "Antislavery and Utopia," 249–52. Walters, *American Reformers*, 49–50.

111. John L. Thomas, *The Liberator: William Lloyd Garrison* (Boston: Little, Brown, 1963), 312–13. Thomas, "Antislavery and Utopia," 254–57.

112. Thomas, "Antislavery and Utopia," 260–63; quotation at 263.

113. Thomas, "Romantic Reform," 678.

114. Friedman, *Gregarious Saints*, 67.

115. Thomas, *The Liberator*, 315. Thomas, "Antislavery and Utopia," 251.

116. An anonymous contributor to the *Liberator*, quoted in Arthur Bestor, *Backwoods Utopias*, 2nd ed. (Philadelphia: University of Pennsylvania Press, 1970), 19.

117. Quoted in Walters, *American Reformers*, 77. Also see Aileen S. Kraditor, *Means and Ends in American Abolitionism: Garrison and His Critics on Strategy and Tactics, 1834–1850* (New York: Pantheon, 1969), 252.

118. Hartz, *Liberal Tradition*, 229–30, 239.

119. See especially Daniel T. Rodgers, "In Search of Progressivism," *Reviews in American History* 10 (December 1982), 113–31. Also see Peter G. Filene, "An Obituary for 'The Progressive Movement,'" *American Quarterly* 22 (Spring 1970), 20–34; John D. Buenker, "Progressive Era: A Search for a Synthesis," *Mid-America* 51 (October 1969), 175–93; and David M. Kennedy, "Progressivism: An Overview," *The Historian* 37 (May 1975), 453–68.

120. David B. Danbom, *"The World of Hope": Progressives and the Struggle for an Ethical Public Life* (Philadelphia: Temple University Press, 1987), 38–40. Henry F. May, *Protestant Churches and Industrial America* (New York: Harper, 1949), 209–10.

121. David P. Thelen, *The New Citizenship: Origins of Progressivism in Wisconsin, 1885–1900* (Columbia: University of Missouri Press, 1972), 82–83.

122. Danbom, *"World of Hope,"* 94.

123. Danbom, *"World of Hope,"* 74–75. Also see May, *Protestant Churches and Industrial America*, 226.

124. Danbom, *"World of Hope,"* 46.

125. Thelen, *New Citizenship*, 109. Also see David P. Thelen, *Robert M. La Follette and the Insurgent Spirit* (Boston: Little, Brown, 1976), 25.

126. Danbom, *"World of Hope,"* 95.

127. Thelen, *The New Citizenship*, 109.

128. David E. Price, "Community and Control: Critical Democratic Theory in the Progressive Period," *American Political Science Review* 68 (December 1974), 1667. Jean B. Quandt, *From the Small Town to the Great Community: The Social Thought of Progressive Intellectuals* (New Brunswick: Rutgers University Press, 1970), 26, 28, 46–47.

129. Quandt, *Small Town to Great Community*, 144, 45; the words are Quandt's, not Follett's.

130. Quandt, *Small Town to Great Community*, 39, 42, 29, 141; also see 40.

131. Robert M. Crumden, *Ministers of Reform: The Progressives' Achievement in American Civilization, 1889–1920* (New York: Basic Books, 1982), 68. Also see Allen F. Davis, *Spearheads for Reform: The Social Settlements and the Progressive Movement, 1890–1914* (New York: Oxford University Press, 1967), esp. 27, 29; and May, *Protestant Churches and Industrial America*, 226–27.

132. Quoted in Daniel Levine, *Varieties of Reform Thought* (Madison: State Historical Society of Wisconsin, 1964), 20.

133. J. David Greenstone, "Dorothea Dix and Jane Addams: From Transcendentalism to Pragmatism in American Social Reform," *Social Service Review* 53 (December 1979), 535. Quandt, *Small Town to Great Community*, 46, 48.

134. Levine, *Reform Thought*, 20–21.

135. John Dewey, *The School and Society* (1899), quoted in Louis S. Feuer, "John Dewey and the Back to the People Movement in American Thought," *Journal of the History of Ideas* 20 (October–December 1959), 564.

136. Quoted in Quandt, *Small Town to Great Community*, 48, 98.

137. Katherine Camp Mayhew and Anna Camp Edwards, *The Dewey School: The Laboratory School of the University of Chicago, 1896–1903* (1936), quoted in Feuer, "John Dewey," 559.

138. Herbert Croly, *The Promise of American Life* (Indianapolis: Bobbs-Merrill, 1965; originally published 1909), 106.

139. Danbom, *"World of Hope,"* 85.

140. Crumden, *Ministers of Reform*, 49.

141. Price, "Community and Control," 1671.

142. Danbom, *"World of Hope,"* 106. Quandt, *Small Town to Great Community*, 24. Similar language was employed by an Appleton editor who wrote, "We live in common wealth or common suffering" (Thelen, *New Citizenship*, 82).

143. Danbom, *"World of Hope,"* ix.

144. Robert H. Wiebe, *Businessmen and Reform: A Study of the Progressive Movement* (Cambridge: Harvard University Press, 1962), 9. The phrase "fraternal equality" is used approvingly by William Allen White (Quandt, *Small Town to Great Community*, 152).

145. Danbom, *"World of Hope,"* 102. Quandt, *Small Town to Great Community*, 139.

146. Richard Hofstadter, *The American Political Tradition* (New York: Vintage, 1973), xxxv.

147. Hofstadter, *American Political Tradition*, xxxv.

148. Quandt, *Small Town to Great Community*, 151, and passim.

148. Quandt, *Small Town to Great Community*, 151–52; also see 46.

150. Mary Parker Follett, quoted in Quandt, *Small Town to Great Community*, 39. Also see Danbom, *"World of Hope,"* passim.

151. Danbom, *"World of Hope,"* 109.

152. Danbom, *"World of Hope,"* 101. Also see Roy Lobove, "The Twentieth Century City: The Progressive as Municipal Reformer," in Blaine A. Brownell and Warren E. Stickle, eds., *Bosses and Reformers: Urban Politics in America, 1880–1920* (Boston: Houghton

Mifflin, 1973), 82–96. Further evidence of the communitarian content of the progressive vision may be found in Casey Nelson Blake, *Beloved Community: The Cultural Criticism of Randolph Bourne, Van Wyck Brooks, Waldo Frank, and Lewis Mumford* (Chapel Hill: University of North Carolina Press, 1990); and Richard H. Pells, *Radical Visions and American Dreams: Culture and Social Thought in the Depression Years* (Middletown: Wesleyan University Press, 1973).

153. The centrality of this tension in the formation of the American identity is suggested in Bellah et al., *Habits of the Heart*, 256; Wilkinson, *Pursuit of American Character*, chap. 4; Rowland Berthoff, "Peasants and Artisans, Puritans and Republicans: Personal Liberty and Communal Equality in American History," *Journal of American History* 69 (December 1982), 579–92; Kammen, *People of Paradox*, 112, 115–16, 173, 179; and especially Zuckerman, "The Fabrication of Identity."

154. Richard Hofstadter, *The Age of Reform: From Bryan to F.D.R* (New York: Vintage, 1955), 15–16. Seymour Martin Lipset, "Why No Socialism in the United States?" in Seweryn Bialer and Sophia Sluzur, eds., *Sources of Contemporary Radicalism* (Boulder: Westview Press, 1977), 125. Grant McConnell, *Private Power and American Democracy* (New York: Knopf, 1966), 22. Samuel P. Huntington, *American Politics: The Promise of Disharmony* (Cambridge: Harvard University Press, 1981), 91–93.

155. Walter Lippmann, quoted in Huntington, *American Politics*, 143.

156. Sidney E. Mead, "The 'Nation with the Soul of a Church,'" *Church History* 36 (September 1967), 262–83.

157. Michael Kammen, *People of Paradox: An Inquiry Concerning the Origins of American Civilization* (New York: Oxford University Press, 1972), 116, 165, 175, 178. Hartz, who shares Kammen's fondness for paradox, also draws attention to Americans' contradictory commitments to "pragmatism and absolutism . . . materialism and idealism" (*Liberal Tradition*, 63).

158. *People of Paradox*, 98, 109–10.

159. *People of Paradox*, 99; emphasis added.

160. Hartz, *Liberal Tradition*, 40, 38. Boorstin, *Genius of American Politics*, 94.

161. Gordon S. Wood, "Rhetoric and Reality in the American Revolution," *William and Mary Quarterly* 23 (January 1966), 25, 26. Also see Bernard Bailyn, *The Ideological Origins of the American Revolution* (Cambridge: Harvard University Press, 1967).

162. Huntington, *American Politics*, 154, 160, 91. Daniel J. Elazar also sees Puritanism as the source of what he labels the "moralistic" political subculture in the United States (*American Federalism: A View from the States,* 2nd ed. [New York: Crowell, 1972], 108). I discuss Elazar's categories at greater length in chapter 9.

163. Hartz, *Liberal Tradition*, 250. Huntington, *American Politics*, 89, 154. Hartz concedes that "there are aspects of our original life in the Puritan colonies" that do not fit into the category of liberal "in the classic Lockian sense" (*Liberal Tradition*, 4) but seems to believe that by the eighteenth century these aspects of Puritanism had ceased to have a significant impact on American political culture. Boorstin takes a similar tack, conceding that Puritan "dogma" was out of step with "the pragmatic spirit" of America while insisting that Puritanism "was gradually eroded by the American climate" (*Genius of American Politics*, 37, 53).

164. See Douglas, "Cultural Bias," esp. 183–90.

165. This distinction is briefly touched on in Merle Curti's preface to *The Course of American Thought* (New Brunswick: Transaction Publishers, 1982).

166. Mary Douglas, *Natural Symbols: Explorations in Cosmology* (London: Barrie & Rockliff, 1970), 119, 125. Also see Mary Douglas, *Purity and Danger: An Analysis of the Concepts of Pollution and Taboo* (London: Routledge & Kegan Paul, 1966); and Mary

Douglas and Aaron Wildavsky, *Risk and Culture: An Essay on the Selection of Technical and Environmental Dangers* (Berkeley: University of California Press, 1982).

167. Douglas, *Natural Symbols*, 25. Mary Douglas, *How Institutions Think* (Syracuse: Syracuse University Press, 1986), 38–41.

168. This is not to suggest that these movements occupy *identical* positions along the grid and group dimensions. Group boundaries, for instance, are significantly more pronounced for the Puritans than they are for the other movements considered in this chapter. The Puritans are also significantly higher on the grid dimension than are the others.

169. Karl Marx, quoted in Jon Elster, *Making Sense of Marx* (Cambridge: Cambridge University Press, 1985), 18. Also see Douglas, *How Institutions Think*, 42; and Douglas, "Introduction to Grid/Group Analysis," 7.

Chapter 2

1. See, for instance, Donald J. Devine, *The Political Culture of the United States* (Boston: Little, Brown, 1972), esp. 52–53; and William C. Mitchell, *The American Polity: A Social and Cultural Interpretation* (New York: Free Press, 1962), esp. 107–8.

2. This literature is ably reviewed in Seymour Martin Lipset, "Why No Socialism in the United States?" in Seweryn Bialer and Sophia Sluzur, eds., *Sources of Contemporary Radicalism* (Boulder: Westview Press, 1977), 31–149.

3. See Leon Fink, "The New Labor History and the Powers of Historical Pessimism: Consensus, Hegemony, and the Case of the Knights of Labor," *Journal of American History* 75 (June 1988), 115–35, esp. 123–24.

4. Larry Siedentop, "Two Liberal Traditions," in Alan Ryan, ed., *The Idea of Freedom* (New York: Oxford University Press, 1979), 153.

5. Sheldon S. Wolin points to a similar distinction when he criticizes those who "have lumped together two distinct traditions of political thought: democratic radicalism and liberalism" (*Politics and Vision: Continuity and Innovation in Western Political Thought* [Boston: Little, Brown, 1960], 293). The confusion that the term "liberalism" invites is suggested by the fact that Roberto M. Unger, a radical egalitarian, can label his vision of politics a kind of "superliberalism" (*False Necessity: Anti-Necessitarian Social Theory in the Service of Radical Democracy* [Cambridge: Cambridge University Press, 1987], 588; see the superb review essay by Bernard Yack, "Toward a Free Marketplace of Social Institutions: Roberto Unger's 'Super-Liberal' Theory of Emancipation," *Harvard Law Review* 101 [June 1988], 1961–77), as can exemplary competitive individualists like Friedrich Hayek and Ludwig von Mises.

6. Richard Hofstadter, *The American Political Tradition* (New York: Vintage, 1973), xxxvii.

7. See the suggestive remarks in Keith Joseph and Jonathan Sumption, *Equality* (London: John Murray, 1979), 8–11.

8. Judith N. Shklar, *Men and Citizens: A Study of Rousseau's Social Theory* (Cambridge: Cambridge University Press, 1969), 49.

9. Compare, e.g., Neal Wood, *John Locke and Agrarian Capitalism* (Berkeley: University of California Press, 1984) or C. B. Macpherson, *The Political Theory of Possessive Individualism: Hobbes to Locke* (London: Oxford University Press, 1962) with James Tully, *A Discourse on Property: John Locke and His Adversaries* (Cambridge: Cambridge University Press, 1980) or Richard Ashcraft, *Revolutionary Politics and Locke's Two Treatises of Government* (Princeton: Princeton University Press, 1986) and idem, *Locke's Two Treatises of Government* (London: Allen & Unwin, 1987). The quotation is from Ashcraft,

Revolutionary Politics, 272. See the thoughtful review of Ashcraft's work by Jeffrey Friedman in "Locke as Politician," *Critical Review* 1 (Summer 1988), 64–101. Also see John Dunn, *The Political Thought of John Locke: An Historical Account of the "Two Treatises of Government"* (Cambridge: Cambridge University Press, 1969); Nathan Tarcov, *Locke's Education for Liberty* (Chicago: University of Chicago Press, 1984); and Edward J. Harpham, ed., *John Locke's Two Treatises of Government: New Interpretations* (Lawrence: University Press of Kansas, 1992).

10. Louis Hartz, *The Liberal Tradition in America* (New York: Harcourt, 1955), 6.

11. Hartz, *Liberal Tradition in America*, 8. I am, of course, mindful of the extensive historical scholarship arguing for a civic republican alternative to Locke (in fact I explicitly take up this debate in chapter 9). My aim here, however, is to argue that even if one concedes to Hartz that Lockean precepts have been almost universally accepted in the United States, his conclusion of a competitive individualist value consensus would still not be justified.

12. This ambiguity in Locke's thought is pointed out in, among other places, Eric Foner, *Tom Paine and Revolutionary America* (New York: Oxford University Press, 1976), 40; Rush Welter, *The Mind of America, 1820–1860* (New York: Columbia University Press, 1975), 408; Richard Schlatter, *Private Property: The History of an Idea* (London: George Allen & Unwin, 1951), 157; and Alan P. Grimes, *American Political Thought* (New York: Henry Holt, 1955), 333. Also see Daniel T. Rodgers, *Contested Truths: Keywords in American Politics Since Independence* (New York: Basic Books, 1987), 129.

13. John Locke, *Two Treatises of Government*, ed. Peter Laslett (Cambridge: Cambridge University Press, 1960), 332 (2d tr. [treatise], chap. 5, par. 31).

14. Locke, *Two Treatises*, 376, 188 (2d tr., chap. 11, par. 135; 1st tr., chap. 4, par. 42). Also see Forrest McDonald, *Novus Ordo Seclorum: The Intellectual Origins of the Constitution* (Lawrence: University Press of Kansas, 1985), 65–66; and Peter Laslett's introduction to Locke, *Two Treatises,* 105.

15. Locke, *Two Treatises*, 306 (2d tr., chap. 5, par. 27).

16. Schlatter, *Private Property*, 169.

17. David Freeman Hawke, *Paine* (New York: Harper & Row, 1974), 50.

18. Hartz, *Liberal Tradition in America*, 73.

19. Foner, *Paine and Revolutionary America*, 40.

20. Foner, *Paine and Revolutionary America*, 39, 102.

21. Thomas Paine, *Agrarian Justice*, in *The Thomas Paine Reader*, ed. Michael Foot and Isaac Kramnick (Harmondsworth: Penguin, 1987), 476.

22. Paine, *Agrarian Justice*, 477–78.

23. Locke, *Two Treatises*, 305–6 (2d tr., chap. 5, par. 27).

24. Paine, *Agrarian Justice*, 485; emphasis added.

25. Paine, *Agrarian Justice*, 485.

26. Throughout *Agrarian Justice*, Paine insists that "it is justice, and not charity," that requires restitution for the dispossessed (483; also 477, 482). For Locke, in contrast, it is "Charity [that] gives every Man a Title to so much out of another's Plenty, as will keep him from extream want" (188; 1st tr., chap. 4, par. 42).

27. Paine, *Agrarian Justice*, 478, 482, 484.

28. Schlatter, *Private Property*, 175n3. Foner, *Paine and Revolutionary America*, 251.

29. Isaac Kramnick, "Editor's Introduction," in Thomas Paine, *Common Sense* (Harmondsworth: Penguin, 1986), 49.

30. Paine, *Agrarian Justice*, 482.

31. Paine, *Agrarian Justice*, 485.

32. Paine's egalitarianism is accented in Gregory Claeys, *Thomas Paine: Social and*

Political Thought (Boston: Unwin Hyman, 1989); Thomas A. Horne, *Property Rights and Poverty: Political Argument in Britain, 1605–1834* (Chapel Hill: University of North Carolina Press, 1990), 203–9; and John Seaman, "Thomas Paine: Ransom, Civil Peace, and the Natural Right to Welfare," *Political Theory* 16 (February 1988), 120–42.

33. Locke, *Two Treatises*, 188 (1st tr., chap. 4, par. 42).

34. Welter, *Mind of America*, 132.

35. Charles J. Ingersoll, a Pennsylvania Democrat, quoted in John Ashworth, *"Agrarians" and "Aristocrats": Party Political Ideology in the United States, 1837–1846* (Cambridge: Cambridge University Press, 1987), 89.

36. Ashworth, *"Agrarians" and "Aristocrats,"* 97–98.

37. Horne, *Property Rights and Poverty*, 224. Rodgers, *Contested Truths*, 74.

38. Horne, *Property Rights and Poverty*, 225. Although Skidmore accepted the possessive individualist postulate that "the personal exertions of each individual of the human race are exclusively and unalienably his own" (Rodgers, *Contested Truths*, 75), he clearly departed from Locke in rejecting the notion that the mixing of one's labor with raw materials gives one a just title to property. When a savage makes a bow, Skidmore argued, "the material, of which the bow is made, is the property of mankind. It is property, too, which, previous to the existence of government, has never been alienated to anyone. If it has not been alienated, it cannot belong to another . . . what right then had that other, to bestow his labor upon it. . . . Instead of acquiring a right, thereby, to the bow, he has rather committed a trespass upon the great community of which he is a member" (Horne, *Property Rights and Poverty*, 225).

39. Sean Wilentz, *Chants Democratic: New York City and the Rise of the American Working Class* (New York: Oxford University Press, 1984), 183. Hartz dismisses Skidmore (whom he mistakenly identifies as Stephen Skidmore [on the previous page he mistakenly identifies Thomas Sedgwick as Ellery Sedgwick) as little more than an advocate of "small propertied individualism" who "wanted to guarantee inalienable the right of acquiring wealth" (*Liberal Tradition*, 123–24).

40. Thomas Skidmore, "A Plan for Equalizing Property," in Joseph L. Blau, *ed., Social Theories of Jacksonian Democracy* (Indianapolis: Bobbs-Merrill, 1954), 356.

41. William Leggett, "Objects of the Evening Post," in Blau, *Social Theories of Jacksonian Democracy*, 72.

42. William Leggett, "True Functions of Government," in Blau, *Social Theories of Jacksonian Democracy*, 77.

43. William Leggett, "Associated Effort," in Blau, *Social Theories of Jacksonian Democracy*, 82.

44. William Leggett, "Rich and Poor," in Blau, *Social Theories of Jacksonian Democracy*, 70–71.

45. *Democratic Review*, January 1845, quoted in Ashworth, *"Agrarians" and "Aristocrats,"* 23. Also see Edward Pessen, *Most Uncommon Jacksonians: The Radical Leaders of the Early Labor Movement* (Albany: State University of New York Press, 1967), 174–75.

46. George Henry Evans, quoted in Pessen, *Uncommon Jacksonians*, 175. Also see Jonathan Glickstein, *Concepts of Free Labor in Antebellum America* (New Haven: Yale University Press, 1991), esp. 7, 260–61, 319–20; and Lawrence Frederick Kohl, *The Politics of Individualism: Parties and the American Character in the Jacksonian Era* (New York: Oxford University Press, 1989).

47. George Bancroft, quoted in Hartz, *Liberal Tradition*, 127.

48. Ashworth, *"Agrarians" and "Aristocrats,"* 23. Also see Pessen, *Uncommon Jacksonians*, 105.

49. William G. Shade, *Banks or No Banks: The Money Issue in Western Politics, 1832–1864* (Detroit: Wayne State University Press, 1972), 124. Arthur M. Schlesinger, Jr., *The Age of Jackson* (Boston: Little, Brown, 1945), 153.

50. Vernon L. Parrington, *Main Currents in American Thought*, 3 vols. (New York: Harcourt, 1927), 2: 247.

51. Stephen Simpson, a Philadelphia labor leader, quoted in Pessen, *Uncommon Jacksonians*, 175.

52. Fred Somkin, *Unquiet Eagle: Memory and Desire in the Idea of American Freedom, 1815–1860* (Ithaca: Cornell University Press, 1967), 196, 202.

53. George Bancroft, *The Necessity, the Reality, and the Promise of the Progress of the Human Race* (1854), quoted in Arthur M. Schlesinger, Jr., "Affirmative Government and the American Economy," in Schlesinger, *The Cycles of American History* (Boston: Houghton Mifflin, 1986), 230.

54. Hartz, *Liberal Tradition*, chap. 5.

55. James Whitcomb, governor of Indiana, quoted in Shade, *Banks or No Banks*, 117.

56. See Arthur M. Schlesinger, Jr., "The Ages of Jackson," *New York Review of Books* 35 (December 7, 1989), 49.

57. Hartz, *Liberal Tradition*, 156. See also Samuel P. Huntington, *American Politics: The Promise of Disharmony* (Cambridge: Harvard University Press, 1981), 111.

58. Eric Foner, "Thaddeus Stevens, Confiscation, and Reconstruction," in Stanley Elkins and Eric McKitrick, eds., *The Hofstadter Aegis: A Memorial* (New York: Knopf, 1974), 164. Also see James Brewer Stewart, *Wendell Phillips: Liberty's Hero* (Baton Rouge: Louisiana State University Press, 1986), 247.

59. James M. McPherson, *The Struggle for Equality: Abolitionists and the Negro in the Civil War and Reconstruction* (Princeton: Princeton University Press, 1964), 252.

60. Patrick W. Riddleberger, "George W. Julian: Abolitionist Land Reformer," *Agricultural History* 29 (July 1955), 109.

61. McPherson, *Struggle for Equality*, 249.

62. Foner, "Thaddeus Stevens," 164.

63. Riddleberger, "George Julian," 108–9.

64. McPherson, *Struggle for Equality*, 250.

65. Riddleberger, "George Julian," 109.

66. Staughton Lynd, *Intellectual Origins of American Radicalism* (New York: Pantheon, 1968), 89.

67. LaWanda Cox, "The Promise of Land for the Freedmen," *Mississippi Valley Historical Review* 45 (December 1958), 421.

68. Riddleberger, "George Julian," 109.

69. Thomas F. Woodley, *Great Leveler: The Life of Thaddeus Stevens* (New York: Stackpole, 1937), 375; emphasis added.

70. Foner, "Thaddeus Stevens," 176.

71. Quoted in Richard Hofstadter, William Miller, and Daniel Aaron, *The United States: The History of a Republic* (Englewood Cliffs: Prentice-Hall, 1957), 395.

72. McPherson, *Struggle for Equality*, 407.

73. McPherson, *Struggle for Equality*, 411.

74. McPherson, *Struggle for Equality*, 252.

75. Cox, "Promise of Land," 431.

76. Foner, "Thaddeus Stevens," 162.

77. Foner, "Thaddeus Stevens," 161.

78. McPherson, *Struggle for Equality*, 411.

79. McPherson, *Struggle for Equality*, 411.

80. McPherson, *Struggle for Equality*, 187, 251, 411–12.

81. Norman Pollack, *The Just Polity: Populism, Law, and Human Welfare* (Urbana: University of Illinois Press, 1987), 8, 21, 44.

82. W. Scott Morgan, *History of the Wheel and Alliance* (1889); and The Omaha Platform of 1892, in Norman Pollack, ed., *The Populist Mind* (Indianapolis: Bobbs-Merrill, 1967), 249, 63.

83. *Farmer's Alliance*, September 26, 1890; and James B. Weaver, *A Call to Action* (1892), in Pollack, *Populist Mind*, 17, 110.

84. S. M. Smith, secretary of the Illinois State Farmers' Association, in Ari Hoogenboom and Olive Hoogenboom, eds., *The Gilded Age* (Englewood Cliffs: Prentice-Hall, 1967), 61. In a similar vein, the Omaha Platform of 1892 stated that "the fruits of the toil of millions are boldly stolen to build up colossal fortunes for a few" (Pollack, *Populist Mind*, 60).

85. Morgan, *History of the Wheel*; and *Farmer's Alliance*, October 22, 1891, in Pollack, *Populist Mind*, 249, 20.

86. Morgan, *History of the Wheel*, in Pollack, *Populist Mind*, 246–47.

87. William A. Peffer, *The Farmer's Side* (1891), in Pollack, *Populist Mind*, 82; emphasis added.

88. Inaugural address, January 9, 1893, in Pollack, *Populist Mind*, 52.

89. Quoted in Gary Lee Malecha, "Understanding Agrarian Fundamentalism: A Cultural Interpretation of American Populism," paper prepared for delivery at the 1988 Annual Meeting of the American Political Science Association, Washington, D.C., September 1–4, 1988, pp. 34–35. Also see James H. Davis, *A Political Revelation* (1894), in Pollack, *Populist Mind*, 217.

90. Thomas L. Nugent, in Pollack, *Populist Mind*, 305.

91. Omaha Platform, in Pollack, *Populist Mind*, 64.

92. Pollack, *Populist Mind*, 12–13.

93. *Farmer's Alliance*, October 22, 1891, in Pollack, *Populist Mind*, 21.

94. Weaver, *Call to Action*, in Pollack, *Populist Mind*, 111.

95. Davis, *Political Revelation,* in Pollack, *Populist Mind*, 217–18.

96. Leon Fink, *Workingmen's Democracy: The Knights of Labor and American Politics* (Urbana: University of Illinois Press, 1983).

97. Davis, *Political Revelation*, in Pollack, *Populist Mind*, 218; emphasis added.

98. Ignatious Donnelly, *Caesar's Column* (1891), in George Brown Tindall, ed., *A Populist Reader: Selections from the Works of American Populist Leaders* (New York: Harper & Row, 1966), 107.

99. Henry George, *Progress and Poverty* (New York: Robert Schalkenbach Foundation, 1990; originally published 1879), 445–46.

100. John L. Thomas, *Alternative America: Henry George, Edward Bellamy, Henry Demarest Lloyd and the Adversary Tradition* (Cambridge: Harvard University Press, 1983), 50.

101. George, *Progress and Poverty*, 11, 436, 334.

102. George, *Progress and Poverty*, 336, 337, 545, 358, 336, 335, 370.

103. George, *Progress and Poverty*, 464, 552, 353, 545.

104. See, e.g., Francis Fukuyama, "The End of History?" *National Interest* 16 (Summer 1989), 3–18.

105. Fukuyama's glowing optimism is the mirror image of Hartz's brooding pessimism. Where Fukuyama revels in the final triumph of liberalism, Hartz is anguished over the debilitating consequences of America's "liberal absolutism" (Hartz, *Liberal Tradition*, 302). For Fukuyama, the disappearance of socialism from the world scene signals an end

to history; for Hartz, the absence of socialism (and feudalism) in the United States has left Americans without "that grain of relative insight" that would enable them to understand their own history as well as the history of others (287, 308). Liberal consensus, for Fukuyama, represents a national triumph; for Hartz, liberal consensus means "national blindness" (287). From their radically different vantage points, Fukuyama and Hartz converge on a common point: a modern world without socialism is a world lacking in fundamental conflict of visions over the good life.

106. George, *Progress and Poverty*, 465. A good contemporary example is the Marxist left's reading of John Rawls's *A Theory of Justice* as at best "a philosophical apologia for an egalitarian brand of welfare-state capitalism" (Robert Paul Wolff, *Understanding Rawls: A Reconstruction and Critique of "A Theory of Justice"* [Princeton: Princeton University Press, 1977], 195; also see C. B. Macpherson, *Democratic Theory: Essays in Retrieval* [Oxford: Oxford University Press, 1973], 87–94). The radical egalitarianism of Rawls's vision, in spite of his acceptance of a "property-owning democracy" and market pricing mechanisms, is persuasively argued by Arthur DiQuattro, "Rawls and Left Criticism," *Political Theory* 11 (February 1983), 53–78; and Richard Krouse and Michael McPherson, "Capitalism, 'Property-Owning Democracy,' and the Welfare State," in Amy Gutmann, ed., *Democracy and the Welfare State* (Princeton: Princeton University Press, 1988).

107. Herbert McClosky and John Zaller, *The American Ethos: Public Attitudes Toward Capitalism and Democracy* (Cambridge: Harvard University Press, 1984), 140–41, 231. Also see S. Robert Lichter and Stanley Rothman, "Media and Business Elites," *Public Opinion* 4 (October/November 1981), 44; and Linda L. Bennett and Stephen Earl Bennett, *Living with Leviathan: Americans Coming to Terms with Big Government* (Lawrence: University Press of Kansas, 1990), 106–7.

108. A 1958 survey found 89 percent of the mass public and 96 percent of political influentials disagreeing with the statement "Some form of socialism would certainly be better than the system we have now" (McClosky and Zaller, *The American Ethos*, 135). A somewhat different picture is presented in a recent poll of Christian religious elites, which found that 31 percent of Catholic religious leaders and 25 percent of mainline Protestant leaders believe that the "U.S. would be better off if it moved toward socialism" (Robert Lerner, Stanley Rothman, and S. Robert Lichter, "Christian Religious Elites," *Public Opinion* 11 [March/April 1989], 57).

109. Nor does a preference for private property translate into a belief that individuals should be able to dispose of their property in any way they see fit. One survey conducted in the mid-1970s found that although only 14 percent of "opinion leaders" agreed that "private ownership of property has often done mankind more harm than good," only 36 percent of that same sample believed that "the way property is used should mainly be decided by the individuals who owned it." And although only 17 percent of opinion leaders thought that public ownership of large industry would be a good idea, 71 percent of those same people believed that "a lumber company that spends millions for a piece of forest land should, nevertheless, be limited by law in the number of trees it can cut" (McClosky and Zaller, *The American Ethos*, 140, 143; also see Dennis Coyle, "Environmental Cultures, Regulation and the Constitution," paper prepared for delivery at the 1991 Annual Meeting of the Western Political Science Association, Seattle, Washington, March 21–23, 1991).

110. Sidney Verba and Gary R. Orren, *Equality in America: The View from the Top* (Cambridge: Harvard University Press, 1985), 78–79. Lerner, Rothman, and Lichter, "Christian Religious Elites," 57. Lichter and Rothman, "Media and Business Elites," 44.

111. Verba and Orren, *Equality in America*, 74.

112. George, *Progress and Poverty*, 453. Jennifer L. Hochschild, *What's Fair? American Beliefs About Distributive Justice* (Cambridge: Harvard University Press, 1981), 211, 123. Also see Christopher Lasch, *The True and Only Heaven: Progress and Its Critics* (New York: Norton, 1991), 486. Contrast this with the individualistic respondent, who expresses unbounded admiration for the wealthy "for their courage, foresight, and accomplishing certain endeavors of such sizable proportions to propel them into the elevations that they stand in" (Hochschild, *What's Fair?*, 205).

113. Ashworth, *"Agrarians" and "Aristocrats,"* 24. Wilentz, *Chants Democratic*, 187. George, *Progress and Poverty*, 452–53.

114. Verba and Orren, *Equality in America*, 72, 158, 155, 79.

115. Locke, *Two Treatises*, 188 (1st tr., chap. 4, par. 42).

Chapter 3

1. George Wilson Pierson, *Tocqueville in America* (Garden City, N.Y.: Anchor, 1959), 449. Unfortunately, in writing about equality and democracy, Tocqueville often failed to heed his own advice. Pierson estimates that Tocqueville used the term *democratie* in no less than "seven or eight different senses" (459).

2 See, e.g., John C. Schaar, "Equality of Opportunity and Beyond," in Roland Pennock and John Chapman, eds., *Equality*, (New York: Atherton, 1967), 228; and George C. Catlin, "Equality and What We Mean by It," ibid., 99.

3. James Fitzjames Stephen, *Liberty, Equality, Fraternity* (1873), quoted in Hugo Adam Bedau, "Egalitarianism and the Idea of Equality," in Pennock and Chapman, *Equality*, 4.

4. The best analysis is Douglas Rae, *Equalities* (Cambridge: Harvard University Press, 1981).

5. I take this distinction between equality as result and equality as process from Thomas Sowell, *A Conflict of Visions: Ideological Origins of Political Struggles* (New York: Quill, 1987), chap. 6.

6. Sowell, *Conflict of Visions*, 122.

7. This is the assumption made by Sidney Verba and Gary R. Orren, for instance, who write that "the parties' differing views on equality are bounded by the American ideology of equal opportunity. . . . Americans have never striven to ensure that all people live alike. Rather, they have *always* followed the ideal of an equal start in the race so that those with greater ability and drive are allowed, and encouraged, to come out ahead" (*Equality in America: The View from the Top* [Cambridge: Harvard University Press, 1985], 124, 71; emphasis added). But this formulation neglects the very different meanings given to the term "equal opportunity" in the equal-results and equal-process vision. In the equal-process vision, the emphasis is on the opportunity to compete; in the equal-results vision, the emphasis is on equalizing the resources with which to compete. Interpreted literally, equality of opportunity becomes not an invitation to excel in the race of life, but rather an insistence that at every step of the way advantages should be redistributed to ensure that everyone has an equal chance. Interpreted in this way, equal opportunity becomes indistinguishable from equal results.

I conducted a survey of bankers and university teachers and administrators that replicated Verba and Orren's finding that overwhelming majorities opted for equality of opportunity ("giving each person an equal chance for a good education and to develop his or her ability") over equality of results ("giving each person a relatively equal income regardless of his or her education and ability"). But a follow-up question asking respondents whether they thought of equality of opportunity more in terms of "the opportunity to compete" or "equalizing the resources with which to compete" found that 62 percent of the university teachers and administrators opted for the "equalizing resources" definition, while only 25 percent of the bankers opted for that definition of equal opportunity. The samples in the survey were small and unrepresentative, so what is important is not the precise percentages but the existence of different conceptions of what equal opportunity entails.

8. Frederick Jackson Turner, *The Frontier in American History* (New York: Holt, 1962). David M. Potter, *People of Plenty: Economic Abundance and the American Character* (Chicago: University of Chicago Press, 1954). Louis Hartz, *The Liberal Tradition in America* (New York: Harcourt, 1955). Amy Bridges, *A City in the Republic: Antebellum New York and the Origins of Machine Politics* (Ithaca: Cornell University Press, 1987). For recent work exploring the old question of American exceptionalism, see Byron Shafer, ed., *Is America Different? A New Look at American Exceptionalism* (New York: Oxford University Press, 1991); Dorothy Ross, *The Origins of American Social Science* (Cambridge: Cambridge University Press, 1991); Richard Rose, "How Exceptional Is the American Political Economy?," *Political Science Quarterly* 104 (Spring 1989), 91–115. Also see Ian Tyrrell, "American Exceptionalism in an Age of International History," *American Historical Review* 96 (October 1991), 1031–55, and the convincing rebuttal by Michael McGerr, "The Price of the 'New Transnational History,'" ibid., 1056–67.

9. This thesis is argued in Aaron Wildavsky, "Resolved, that Individualism and Egalitarianism Be Made Compatible in America: Political-Cultural Roots of Exceptionalism," in Shafer, *Is America Different?*, 116–37. The essay also appears in Aaron Wildavsky, *The Rise of Radical Egalitarianism* (Washington, D.C.: American University Press, 1991), 29–48. Also see Richard Ellis and Aaron Wildavsky, *Dilemmas of Presidential Leadership: From Washington Through Lincoln* (New Brunswick: Transaction Publishers, 1989).

10. Jonathan Glickstein, *Concepts of Free Labor in Antebellum America* (New Haven: Yale University Press, 1991), 402n8.

11. Skidmore believed that "the Creator of the Universe" had "ordered [the earth] to be distributed equally" (Thomas A. Horne, *Property Rights and Poverty: Political Argument in Britain, 1605–1834* [Chapel Hill: University of North Carolina Press, 1990], 225) and insisted that the rights of labor and the poor demanded that "we rip all up, and make a full General Division" of property (Sean Wilentz, *Chants Democratic: New York City and the Rise of the American Working Class* [New York: Oxford University Press, 1984], 183), yet he also held it to be "a truth not to be controverted, that each individual knows better how to apply his own industry, his own faculties, advantages, opportunities, property, etc. than government can possibly do" (Horne, *Property Rights and Poverty*, 225). Abolishing inheritance (and redistributing it equitably to persons who had reached their majority), Skidmore believed, would usher in a world not only without lenders and borrowers, rich and poor, landlords and tenants, but also with the least possible government (Wilentz, *Chants Democratic*, 187).

12. David Ramsay, quoted in Gordon S. Wood, *The Creation of the American Republic, 1776–1787* (Chapel Hill: University of North Carolina Press, 1969), 70; also see 425.

13. Wood, *Creation of the American Republic*, 70.

14. Wood, *Creation of the American Republic*, 478.

15. Thomas M. Doerflinger, *A Vigorous Spirit of Enterprise: Merchants and Economic Development in Revolutionary Philadelphia* (Chapel Hill: University of North Carolina Press, 1986).

16. Eric Foner, *Tom Paine and Revolutionary America* (New York: Oxford University Press, 1976), 194, 129, 133.

17. Wood, *Creation of the American Republic*, 86, 441.

18. Doerflinger, *Vigorous Spirit of Enterprise*, 16, 42–43.

19. Foner, *Paine and Revolutionary America*, 196, 125. Wood, *Creation of the American Republic*, 402.

20. Foner, *Paine and Revolutionary America*, chap. 5; quotation at 170.

21. John Ashworth, *"Agrarians" and "Aristocrats": Party Political Ideology in the United States, 1837–1846* (Cambridge: Cambridge University Press, 1987), 11.

22. Eric Foner, *Free Soil, Free Labor, Free Men: The Ideology of the Republican Party Before the Civil War* (New York: Oxford University Press, 1970), 38.

23. Foner, *Free Soil*, 38, 16.

24. Foner, *Free Soil*, 48, 16, 25; emphasis added.

25. On the Whigs, see Lawrence Frederick Kohl, *The Politics of Individualism: Parties and the American Character in the Jacksonian Era* (New York: Oxford University Press, 1989), 66, 188–89.

26. Ashworth, *"Agrarians" and "Aristocrats,"* 94, 46, 41. Also see Kohl, *Politics of Individualism*, 206; and Harry L. Watson, *Jacksonian Politics and Community Conflict: The Emergence of the Second American Party System in Cumberland County, North Carolina* (Baton Rouge: Louisiana State University Press, 1981), 172.

27. Ashworth, *"Agrarians" and "Aristocrats,"* 27, 46, 26.

28. Ashworth, *"Agrarians" and "Aristocrats,"* 100. Also see Arthur M. Schlesinger, Jr., *The Age of Jackson* (Boston: Little, Brown, 1945), 313.

29. Edward Pessen, *Most Uncommon Jacksonians: The Radical Leaders of the Early Labor Movement* (Albany: State University of New York Press, 1967), 104.

30. Ashworth, *"Agrarians" and "Aristocrats,"* 24.

31. John A. Garraty, *The New Commonwealth, 1877–1890* (New York: Harper & Row, 1968), 325. Also see Samuel P. Huntington, *American Politics: The Promise of Disharmony* (Cambridge: Harvard University Press, 1981), esp. 64.

32. The phraseology is Douglas Rae's in *Equalities*, 2.

33. William Graham Sumner, *What Social Classes Owe to Each Other* (Caldwell, Iowa: Caxton, 1966; originally published 1883), 14, 36; emphasis in original.

34. Quoted in Gene Clanton, "Populism, Progressivism, and Equality: The Kansas Paradigm," *Agricultural History* 51 (July 1977), 579.

35. Robert Green McCloskey, *American Conservatism in the Age of Enterprise, 1865–1910* (Cambridge: Harvard University Press, 1951), 136.

36. Yehoshua Arieli, *Individualism and Nationalism in American Ideology* (Cambridge: Harvard University Press, 1964), 335.

37. Mary Lease, quoted in Gary Lee Malecha, "Understanding Agrarian Fundamentalism: A Cultural Interpretation of American Populism," paper prepared for delivery at the 1988 Annual Meeting of the American Political Science Association, Washington, D.C., September 1–4, 1988, 33.

38. John L. Thomas, *Alternative America: Henry George, Edward Bellamy, Henry Demarest Lloyd and the Adversary Tradition* (Cambridge: Harvard University Press, 1983), 43.

39. Frank Doster, quoted in Norman Pollack, *The Just Polity: Populism, Law, and Human Welfare* (Urbana: University of Illinois Press, 1987), 137.

40. Malecha, "Understanding Agrarian Fundamentalism," 35, 41, 34.

41. The most famous expression of this view was by French novelist and social critic Anatole France: "The law, in its majestic equality, forbids the rich as well as the poor to sleep under bridges, to beg in the streets, and to steal bread."

42. Quoted in Lawrence Goodwyn, *The Populist Moment: A Short History of the Agrarian Revolt in America* (New York: Oxford University Press, 1978), 53.

43. W. Scott Morgan, *History of the Wheel and Alliance* (1889), in Norman Pollack, ed., *The Populist Mind* (Indianapolis: Bobbs-Merrill, 1967), 270.

44. Pollack, *Populist Mind*, 53.

45. Pollack, *Populist Mind*, 21.

46. Wood, *Creation of the American Republic*, 72. In a similar vein, Joyce Appleby has underscored the importance of the Jeffersonian belief that "economic growth in a democracy would level the rich and raise up the poor, causing both 'to approach that middle point, at which the love of order, of industry, of justice and reason naturally establish themselves'" ("The Radical *Double-Entendre* in the Right to Self-Government," in Margaret Jacob and James Jacob, eds., *The Origins of Anglo-American Radicalism* [London: George Allen & Unwin, 1984], 282). Rowland Berthoff and John M. Murrin add that for early nineteenth-century Americans, "the new, liberal principle of laissez-faire was supposed to be egalitarian as well.... Americans could not really face the possibility that their liberty—their freedom to compete—was undermining their equality" ("Feudalism, Communalism, and the Yeoman Freeholder: The American Revolution Considered as a Social Accident," in Stephen G. Kurtz and James H. Hutson, eds., *Essays on the American Revolution* [Chapel Hill: University of North Carolina Press, 1973], 283–84).

47. Thomas Paine, "The Case of the Officers of Excise" (1772), in *The Thomas Paine Reader*, ed., Michael Foote and Isaac Kramnick (Harmondsworth: Penguin, 1987), 45, 41.

48. Foner, *Paine and Revolutionary America*, 95.

49. Foner, *Paine and Revolutionary America*, 181.

50. Foner, *Paine and Revolutionary America*, esp. 93–95.

51. *The Rights of Man*, quoted in "Editor's Introduction: The Life, Ideology and Legacy of Thomas Paine," *Thomas Paine Reader*, 24, 26.

52. "Dissertation on First Principles of Government" (1795), *Thomas Paine Reader*, 462.

53. The prevalence of this view within the Jeffersonian party is documented in Drew R. McCoy, *The Elusive Republic: Political Economy in Jeffersonian America* (Chapel Hill: University of North Carolina Press, 1980).

54. Eric Foner points out that in part 2 of *The Rights of Man*, "Paine asserted for the first time that to do away with poverty in Europe, more was required than a simple transition to republican government. Paine outlined an economic program as close to a welfare state as could be imagined in the eighteenth century. The basis of taxation would be changed from poor rates and regressive levies on articles of consumption to direct, progressive taxes on property, especially land. From the proceeds, every poor family would receive a direct allocation of money to allow it to educate its children; a system of social security would enable all workers to retire at age sixty; [and] public jobs and unemployment relief would be awarded to 'the casual poor'" (*Paine and Revolutionary America*, 218; also see Gary Kates, "From Liberalism to Radicalism: Tom Paine's *Rights of Man*," *Journal of the History of Ideas* 50 [October–December 1989], 569–87).

55. Ellis and Wildavsky, *Dilemmas of Presidential Leadership*, chap. 6.

56. Robert V. Remini, *Andrew Jackson and the Course of American Freedom, 1822–1832* (New York: Harper & Row, 1981), 298.

57. Richard Hofstadter, *The American Political Tradition* (New York: Knopf, 1948). Bray Hammond, *Banks and Politics in America from the Revolution to the Civil War* (Princeton: Princeton University Press, 1957). Also see John Patrick Diggins, *The Lost Soul of American Politics: Virtue, Self-Interest, and the Foundations of Liberalism* (Chicago: University of Chicago Press, 1988), 105–18.

58. Schlesinger, *Age of Jackson*.

59. Ashworth, *"Agrarians" and "Aristocrats,"* 11, 41.

60. *Globe*, September 1840, quoted in Ashworth, *"Agrarians" and "Aristocrats,"* 25; also see 26.

61. Lawrence Frederick Kohl, *The Politics of Individualism: Parties and the American Character in the Jacksonian Era* (New York: Oxford University Press, 1989), 206.

62. Ashworth, *"Agrarians" and "Aristocrats,"* 29, 40.

63. Kohl, *Politics of Individualism*, 208; emphasis added.

64. Ashworth, *"Agrarians" and "Aristocrats,"* 40.

65. Kohl, *Politics of Individualism*, 203.

66. "An Introductory Statement of the Democratic Principle," in Joseph L. Blau, ed., *Social Theories of Jacksonian Democracy* (Indianapolis: Bobbs-Merrill, 1954), 27.

67. Kohl, *Politics of Individualism*, 203.

68. Kohl, *Politics of Individualism*, 206, 208.

69. *Farmer's Alliance*, February 28, 1891, in Pollack, *Populist Mind*, 18–19. A few years later, the newspaper reiterated that "a reigning plutocracy with the masses enslaved, is the natural development and end of individualism" (Norman Pollack, *The Populist Response to Industrial America* [Cambridge: Harvard University Press, 1962], 19).

70. Morgan, *History of the Wheel*, in Pollack, *Populist Mind*, 247–48. Also see Pollack, *The Just Polity*, 89–91.

71. Pollack, *Populist Mind*, 499–500.

72. Lester Ward, "Plutocracy and Paternalism," in Ari Hoogenboom and Olive Hoogenboom, eds., *The Gilded Age* (Englewood Cliffs: Prentice-Hall, 1967), 40.

73. Pollack, *The Just Polity*, 136–37. Kansas governor Lorenzo Lewelling, a close political associate of Doster, voiced the same view in his 1893 inaugural address, which began: "The survival of the fittest is the government of brutes and reptiles, and such philosophy must give place to a government which recognizes human brotherhood. It is the province of government to protect the weak" (Pollack, *Populist Mind*, 51). And, too, see the analysis of Ignatius Donnelly's writings in Pollack, *The Just Polity*, esp. 248–49.

74. Doster, quoted in Pollack, *Populist Mind*, 12. Malecha, "Understanding Agrarian Fundamentalism," 37.

75. Thomas Paine, *Common Sense*, ed. Isaac Kramnick (Harmondsworth: Penguin, 1986), 65.

76. *Democratic Review* (October 1837), quoted in Blau, *Social Theories of Jacksonian Democracy*, 27.

77. Pollack, *Populist Mind*, 62.

78. Pollack, *Populist Mind*, 59.

79. Topeka *Advocate*, quoted in Malecha, "Understanding Agrarian Fundamentalism," 43.

80. Malecha, "Understanding Agrarian Fundamentalism," 42.

81. Ignatius Donnelly, *Caesar's Column* (1890), in Pollack, *Populist Mind*, 481.

82. Walter K. Nugent, *The Tolerant Populists: Kansas Populism and Nativism* (Chicago: University of Chicago Press, 1963), 98.

83. Pollack, *Populist Mind*, 70.

84. James H. Davis, *A Political Revelation* (1894), quoted in Pollack, *Populist Mind*, 28–29.

85. Pollack, *Just Polity*, 181. Also see Richard Hofstadter, *The Age of Reform: From Bryan to F.D.R.* (New York: Vintage, 1955), 63.

86. Henry George, *Progress and Poverty* (New York: Robert Schalkenbach Foundation, 1990; originally published in 1879), xxx, 525, 320, 319, 441, 442, 433, 454, 441.

87. George, *Progress and Poverty*, 440, 300, 303. George's Jacksonian beliefs are evident in his comment that "it needs no argument to show to what abuses and demoralization grants of public money or credit would lead" (322), as well as in his conviction

"that anything that tends to make government simple and inexpensive tends to put it under control of the people and to bring questions of real importance to the front" (303) and that "the differences of natural power are no greater than the differences of stature or of physical strength" (469).

88. Whether the Single Tax could in fact have preserved this alliance of limited government and equal results is doubtful. George's scheme planned to tax the value of land but not the value of improvements, and although George denied this was difficult to do in practice (*Progress and Poverty*, 425), it seems likely that an extensive governmental apparatus would be necessary to determine what constituted the value of the land and what constituted the value of improvements. George also believed that the Single Tax would produce a withering away of government as a "directing and repressive power" (456), because eliminating "from society the thieves, swindlers, and other classes of criminals who spring from the unequal distribution of wealth" would greatly reduce the numbers of policemen, judges, bailiffs, clerks, prison keepers, and prisons (455). Government would no longer have to adjudicate disputes over land and titles; standing armies, public debts, tariffs, and subsidies would also disappear. From this wildly utopian fantasy, George concluded that "all this simplification and abrogation of the present functions of government would make possible the assumption of certain other functions which are now pressing for recognition. Government could take upon itself the transmission of messages by telegraph, as well as by mail; of building and operating railroads, as well as of opening and maintaining common roads." With the revenues gained from the tax on land values, "we could establish public baths, museums, libraries, gardens, lecture rooms, music and dancing halls, theaters, universities, technical schools, shooting galleries, play grounds, gymnasiums, etc. Heat, light, and motive power, as well as water, might be conducted through our streets at public expense; our roads be lined with fruit trees; discoverers and inventors rewarded, scientific investigations supported; and in a thousand ways the public revenues made to foster efforts for the public benefit" (456).

This passage reveals just how far even George, born and raised a Jacksonian, had in fact strayed from Jacksonian orthodoxy. Beneath George's deceptive rhetoric about the Single Tax ushering in "the ideal of Jeffersonian democracy, the promised land of Herbert Spencer, the abolition of government" (455–56) lurked the vision and means to greatly expand government's role in society. Government under the Single Tax, George (sounding eerily like Marx) reassured his readers, "would change its character, and would become the administration of a great co-operative society. It would become merely the agency by which the common property was administered for the common benefit" (457). The promise was the withering away of the state; the reality seemed to be a marked expansion of governmental responsibilities.

89. Typical were William A. Peffer's complaint that "money controls our legislation, it colors our judicial decisions, it manipulates parties, it controls policies" (*The Farmer's Side*, in Pollack, *Populist Mind*, 86), and James B. Weaver's lament that "a bold and aggressive plutocracy has usurped the Government and is using it as a policeman to enforce its insolent degrees" (*A Call to Action*, in Pollack, *Populist Mind*, 110).

90. William Peffer, quoted in Nugent, *Tolerant Populists*, 98; and Pollack, *Populist Mind*, 106.

91. Thomas, *Alternative America*, 144; also see 80–81.

92. Pollack, *The Just Polity*, 130.

93. J. David Greenstone, *Labor in American Politics* (New York: Knopf, 1969). Michael Harrington, *Socialism* (New York: Saturday Review Press, 1972). Seymour Martin Lipset, "Why No Socialism in the United States?" in Seweryn Bialer and Sophia Sluzar, eds.,

Radicalism in the Contemporary Age (Boulder: Westview Press, 1977), 140–45. Also see George E. Mowry, "Social Democracy, 1900–1918," in C. Vann Woodward, ed., *The Comparative Approach to American History* (New York: Basic Books, 1968).

94. Carl Oglesby, "Trapped in a System," speech delivered at antiwar march in Washington, D.C., October 27, 1965, in Massimo Teodori, ed. *The New Left: A Documentary History* (Indianapolis: Bobbs-Merrill, 1969), 186.

95. Quoted in James Miller, *"Democracy is in the Streets": From Port Huron to the Siege of Chicago* (New York: Simon & Schuster, 1987), 265; emphasis added.

96. Kirkpatrick Sale, *Human Scale* (New York: Coward, 1980), 66–68.

97. Todd Gitlin, "Power and the Myth of Progress," in Teodori, *New Left*, 188–91. Sober second thoughts can be found in Todd Gitlin, *The Sixties: Years of Hope, Days of Rage* (New York: Bantam, 1987).

98. Tom Hayden, "The Politics of the Movement," in Teodori, *New Left*, 208. Also see Staughton Lynd, "Coalition Politics or Nonviolent Revolution?" in Teodori, *New Left*, 197–202.

99. Swarthmore College Chapter of SDS, "SDS 1964 Convention Draft Statement," quoted in Miller, *"Democracy is in the Streets,"* 196.

100. Swarthmore College Chapter of SDS, quoted in Miller, *"Democracy is in the Streets,"* 196.

101. Miller, *"Democracy is in the Streets,"* 207.

102. Staughton Lynd, "The New Radicals and 'Participatory Democracy,'" in Teodori, *New Left*, 233.

103. Lynd, "The New Radicals and 'Participatory Democracy,'" 233.

104. Miller, *"Democracy is in the Streets,"* 239. Also see Richard Flacks, "Some Problems, Issues, Proposals," in Paul Jacobs and Saul Landau, eds., *The New Radicals* (New York: Vintage, 1966), 162–65.

105. Report by the Newark Economic Research and Action Project (1964), quoted in Miller, *"Democracy is in the Streets,"* 263.

106. Miller, *"Democracy is in the Streets,"* 263. Todd Gitlin, "The Radical Potential of the Poor," in Teodori, *New Left*, 137. Also see Teodori, *New Left*, 209–17.

107. Gitlin, "The Radical Potential of the Poor," 143.

108. Miller, *"Democracy is in the Streets,"* 263.

109. Gitlin, "The Radical Potential of the Poor," 137.

110. Gitlin, "The Radical Potential of the Poor."

111. Jane Stembridge, quoted in Gitlin, *The Sixties*, 164–65.

112. Richard Flacks, quoted in Miller, *"Democracy is in the Streets,"* 172.

113. Gitlin, *The Sixties*, 185.

114. Carl Wittman and Thomas Hayden, "An Interracial Movement of the Poor?" in Mitchell Cohen and Dennis Hale, eds., *The New Student Left: An Anthology* (Boston: Beacon Press, 1966), 185; Carl Wittman, "Students and Economic Action," in Teodori, *New Left*, 128.

115. Michael Levin, "Comparable Worth: The Feminist Road to Socialism," *Commentary* 78 (September 1984), 13–14.

116. Levin, "Comparable Worth," 18. The subsequent example in the text is drawn from Levin, "Comparable Worth," 17.

117. Abigail M. Thernstrom, *Whose Votes Count? Affirmative Action and Minority Voting Rights* (Cambridge: Harvard University Press, 1987), 13.

118. Thernstrom, *Whose Votes Count?*, 5, 27, 25.

119. See Herman Belz, *Equality Transformed: A Quarter-Century of Affirmative Action* (New Brunswick: Transaction Publishers, 1991).

120. Charles Frankel, "Equality of Opportunity," *Ethics* 81 (April 1971), 191–211, quotation on 192. Similarly, Peter Westen writes that "the language of equal opportunity misleads the people who use it by creating the impression of consensus which does not, in fact, exist. . . . 'Equal opportunity' signifies different things to different people, enveloping all its uses with favorable connotations" (*Speaking of Equality: An Analysis of the Rhetorical Force of 'Equality' in Moral and Legal Discourse* [Princeton: Princeton University Press, 1990], 164).

121. Some egalitarians reject equality of opportunity and the "race" metaphor as inherently antiegalitarian and antidemocratic. See, e.g., Schaar, "Equality of Opportunity and Beyond."

122. Christopher Jencks, "What Must Be Equal for Opportunity to Be Equal?" in Norman E. Bowie, ed., *Equal Opportunity* (Boulder: Westview Press, 1988), 48.

123. Frankel, "Equality of Opportunity," 202.

124. James S. Coleman, "Inequality, Sociology, and Moral Philosophy," *American Journal of Sociology* 80 (November 1974), 751.

Chapter 4

1. Gouverneur Morris, quoted in Alfred Young, "Conservatives, the Constitution, and the 'Spirit of Accommodation,'" in Robert A. Goldwin and William A. Schambra, eds., *How Democratic Is the Constitution?* (Washington, D.C.: American Enterprise Institute, 1980), 119.

2. Jennifer Nedelsky, "Confining Democratic Politics: Anti-Federalists, Federalists, and the Constitution," *Harvard Law Review* 96 (November 1982), 340. The egalitarian, participatory, and/or communitarian character of the Anti-Federalists is also highlighted in Wilson Carey McWilliams, "Democracy and the Citizen: Community, Dignity, and the Crisis of Contemporary Politics in America," in Goldwin and Schambra, *How Democratic Is the Constitution?*, 79–101; Gordon S. Wood, *The Creation of the American Republic, 1776–1787* (Chapel Hill: University of North Carolina Press, 1969), esp. 516; and Joshua Miller, *The Rise and Fall of Democracy in Early America, 1630–1789: The Legacy for Contemporary Politics* (University Park: Pennsylvania State University Press, 1991).

3. Joseph A. Schumpeter, *Capitalism, Socialism and Democracy* (New York: Harper & Row, 1942), chap. 22.

4. "X.Y.Z.," quoted in Richard E. Ellis, *The Jeffersonian Crisis: Courts and Politics in the Young Republic* (New York: Oxford University Press, 1971), 125.

5. *The Federalist*, ed. Jacob E. Cooke (Middletown: Wesleyan University Press, 1961), 62.

6. Herbert J. Storing, *What the Anti-Federalists Were For* (Chicago: University of Chicago Press, 1981), 43.

7. The Address and Reasons of Dissent of the Minority of the Convention of Pennsylvania to Their Constituents, December 18, 1787, in Herbert J. Storing, ed., *The Complete Anti-Federalist* 7 vols. (Chicago: University of Chicago Press, 1981), 3:158; emphasis added. This formulation follows almost exactly the Federal Farmer's statement that "a full and equal representation, is that which possesses the same interests, feelings, opinions, and views the people themselves would were they all assembled" (Letters from the Federal Farmer, October 9, 1787, in ibid., 2:230).

8. Isaac Kramnick, "The 'Great National Discussion': The Discourse of Politics in 1787," *William and Mary Quarterly* 45 (January 1988), 14.

9. Speeches by Melancton Smith, June 21, 1788, in *Complete Anti-Federalist*, 6:157.

10. Essays of Brutus, November 15, 1787, in *Complete Anti-Federalist*, 2:379.

11. The Federal Farmer, January 12, 1788, in *Complete Anti-Federalist*, 2:298.

12. The Federal Farmer, January 12, 1788, in *Complete Anti-Federalist*, 2:300.

13. Melancton Smith, June 21, 1788, in *Complete Anti-Federalist*, 6:160.

14. Essay by Cornelius, December 18, 1787, in *Complete Anti-Federalist*, 4:141.

15. The Fallacies of the Freeman Detected by a [Pennsylvania] Farmer, in *Complete Anti-Federalist*, 3:184.

16. Brutus, October 18, 1787, in *Complete Anti-Federalist*, 2:369.

17. Speech of George Mason in the Virginia State Ratifying Convention, June 4, 1788, in *Complete Anti-Federalist*, 5:257.

18. Address by John Francis Mercer to the Members of the Conventions of New York and Virginia, in *Complete Anti-Federalist*, 5:105. Also see Rosemarie Zagarri, *The Politics of Size: Representation in the United States, 1776–1850* (Ithaca: Cornell University Press, 1987), 91, 102.

19. Essays of An Old Whig I, in *Complete Anti-Federalist*, 3:20–21.

20. Wood, *Creation of the American Republic*, 508. Also see *Federalist* No. 35.

21. A Friend of the Republic, in *Complete Anti-Federalist*, 4:244. Also see the Federal Farmer, January 18, 1788, in *Complete Anti-Federalist*, 2:320.

22. Wood, *Creation of the American Republic*, 508.

23. Letters of Cato III, in *Complete Anti-Federalist*, 2:111. Also see Patrick Henry, quoted in Storing, *What the Anti-Federalists Were For*, 54.

24. A [Pennsylvania] Farmer, in *Complete Anti-Federalist*, 3:184.

25. Storing, *What the Anti-Federalists Were For*, 17. Ellis, *Jeffersonian Crisis*, 126.

26. A Friend of the Republic, in *Complete Anti-Federalist*, 4:244.

27. Letters of Centinel, October 5, 1787, in *Complete Anti-Federalist*, 2:137.

28. The Federal Farmer, January 18, 1788, in *Complete Anti-Federalist*, 2:320.

29. Centinel, October 5, 1787, in *Complete Anti-Federalist*, 2:136–37. Also see Saul Cornell, "Aristocracy Assailed: The Ideology of Backcountry Anti-Federalism," *Journal of American History* 76 (March 1990), 1148–72, esp. 1154.

30. William Findley, quoted in Merrill Jensen, ed., *The Documentary History of the Ratification of the Constitution*, 2 vols. (Madison: State Historical Society of Wisconsin, 1976), 2: 668.

31. Melancton Smith, June 21, 1788, in *Complete Anti-Federalist*, 6:157–58.

32. Melancton Smith, June 21, 1788, in Jonathan Elliot, ed., *The Debates in the Several State Conventions on the Adoption of the Federal Constitution* 5 vols. (Salem, N.H.: Ayer, 1987), 2: 260. Also see Gordon S. Wood, "Interests and Disinterestedness in the Making of the Constitution," in Richard Beeman, Stephen Botein, and Edward C. Carter II, eds., *Beyond Confederation: Origins of the Constitution and American National Identity* (Chapel Hill: University of North Carolina Press, 1988), esp. 100–101.

33. Brutus, October 18, 1787, in *Complete Anti-Federalist*, 2:369.

34. Centinel, October 5, 1787, in *Complete Anti-Federalist*, 2:139. Also see the Federal Farmer, October 13, 1787, in ibid., 2:298.

35. A [Pennsylvania] Farmer, in *Complete Anti-Federalist*, 3:185.

36. A [Pennsylvania] Farmer, in *Complete Anti-Federalist*, 3:184. Brutus, October 18, 1787, in ibid., 2:369.

37. Cato III, in *Complete Anti-Federalist*, 2:112.

38. Melancton Smith, June 21, 1788, in *Complete Anti-Federalist*, 6:160. Also see An Old Whig VIII, in ibid., 3:50.

39. Melancton Smith, June 21, 1788, in *Complete Anti-Federalist*, 6:161.

40. Typical is Joshua Miller's charge that "the distance between the citizenry and power

has produced a pervasive feeling of helplessness to change the course of public life" ("Democracy and the Politics of Experience in Antifederalist Thought," paper delivered at 1984 Annual Meeting of the Western Political Science Association, Sacramento, California, April 12–14, 1984, 3). Also see McWilliams, "Democracy and the Citizen," esp. 96–101; and Carole Pateman, *Participation and Democratic Theory* (Cambridge: Cambridge University Press, 1970).

41. Miller, "Democracy and the Politics of Experience in Antifederalist Thought," 20. Also see Miller, *Rise and Fall of Democracy*, 129, 148.

42. Quoted in John Ashworth, *"Agrarians" and "Aristocrats": Party Political Ideology in the United States, 1837–1846* (Cambridge: Cambridge University Press, 1987), 57–58.

43. Major L. Wilson, "Of Time and Union: Webster and His Critics in the Crisis of 1850," *Civil War History* 14 (December 1968), 299.

44. Quoted in Ashworth, *"Agrarians" and "Aristocrats,"* 57.

45. Ashworth, *"Agrarians" and "Aristocrats,"* 57. In a similar vein, Zachary Taylor wrote to his son that, if chosen to occupy the White House, "I must be untrammeled and unpledged, so as to be president of the nation and not of the party" (K. Jack Bauer, *Zachary Taylor: Soldier, Planter, Statesman of the Old Southwest* [Baton Rouge: Louisiana State University Press, 1985], 232).

46. Ashworth, *"Agrarians" and "Aristocrats,"* 19.

47. Ashworth, *"Agrarians" and "Aristocrats,"* 14.

48. John William Ward, *Andrew Jackson: Symbol for an Age* (New York: Oxford University Press, 1953), 55–56, quotation on 55.

49. M. J. Heale, *The Presidential Quest: Candidates and Images in American Political Culture, 1787–1852* (New York: Longman, 1982), 185.

50. Ashworth, *"Agrarians" and "Aristocrats,"* 13.

51. Ashworth, *"Agrarians" and "Aristocrats,"* 14.

52. Unidentified Democrat, quoted in Ashworth, *"Agrarians" and "Aristocrats,"* 14.

53. Ashworth, *"Agrarians" and "Aristocrats,"* 57.

54. Ashworth, *"Agrarians" and "Aristocrats,"* 14

55. On "delegates" and "trustees" as role orientations in the legislative process, see Heinz Eulau, John C. Wahlke, William Buchanan, and Leroy C. Ferguson, "The Role of the Representative: Some Empirical Observations on the Theory of Edmund Burke," *American Political Science Review* 53 (September 1959), 742–56. Also see Hanna Fenichel Pitkin, *The Concept of Representation* (Berkeley: University of California Press, 1967).

56. James D. Richardson, ed., *A Compilation of the Messages and Papers of the Presidents, 1789–1902* 9 vols. (New York: Bureau of National Literature, 1904), 2: 449.

57. Ashworth, *"Agrarians" and "Aristocrats,"* 13.

58. Walt Whitman, "The True Office of Government," in Joseph L. Blau, ed., *Social Theories of Jacksonian Democracy* (Indianapolis: Bobbs-Merrill, 1954), 134.

59. Ashworth, *"Agrarians" and "Aristocrats,"* 53–54.

60. Amos Kendall, quoted in Ashworth, *"Agrarians" and "Aristocrats,"* 15.

61. Schumpeter, *Capitalism, Socialism and Democracy*, 269.

62. Lawrence Goodwyn, *The Populist Moment: A Short History of the Agrarian Revolt in America* (New York: Oxford University Press, 1978), xx.

63. Benjamin R. Barber, "The Undemocratic Party System: Citizenship in an Elite/ Mass Society," in Robert A. Goldwin, ed., *Political Parties in the Eighties* (Washington, D.C.: American Enterprise Institute, 1980), 34; emphasis added.

64. Massimo Teodori, ed., *The New Left: A Documentary History* (Indianapolis: Bobbs-Merrill, 1969), 165; emphasis added.

65. James Miller, *"Democracy is in the Streets": From Port Huron to the Siege of Chicago* (New York: Touchstone, 1987), 298.

66. Kirkpatrick Sale, *Human Scale* (New York: Coward, 1980), 492–93.

67. Benjamin R. Barber, *Strong Democracy: Participatory Politics for a New Age* (Berkeley: University of California Press, 1984), 151.

68. Quoted in Miller, *"Democracy is in the Streets,"* 192.

69. Miller, *"Democracy is in the Streets,"* 144. Also see 253.

70. Barber, "Undemocratic Party System," 35–36.

71. This is Barber's translation of Rousseau's comment in *The Social Contract* ("Undemocratic Party System," 37). The same passage from Rousseau is also quoted approvingly in Sale, *Human Scale*, 493.

72. Barber, "Undemocratic Party System," 35.

73. Robert J. Pranger, *The Eclipse of Citizenship* (New York: Holt, Rinehart and Winston, 1968), 102.

74. Barber, "Undemocratic Party System," 45. The most influential presentation of this view is Carole Pateman, "Political Culture, Political Structure and Political Change," *British Journal of Political Science* 1 (July 1971), 291–305.

75. Sharon Jeffrey, quoted in Miller, *"Democracy is in the Streets,"* 144. As this quotation illustrates, Benjamin Barber's claim that "the language of consent is *me* language: 'I agree' or 'I disagree' [while t]he language of participation is *we* language: 'Can we?' or 'Is that good for us?'" is untenable ("Liberal Democracy and the Costs of Consent," in Nancy L. Rosenblum, ed., *Liberalism and the Moral Life* [Cambridge: Harvard University Press, 1989], 65).

76. Samuel P. Huntington, *American Politics: The Promise of Disharmony* (Cambridge: Harvard University Press, 1981), 2–3.

77. Miller, *"Democracy is in the Streets,"* 250–51.

78. Tom Hayden, "The Politics of the Movement," in Teodori, *New Left*, 207.

79. Richard Flacks, "Is the Great Society Just a Barbecue?" in Teodori, *New Left*, 195. Miller, *"Democracy is in the Streets,"* 176.

80. Miller, *"Democracy is in the Streets,"* 153, 237, 225.

81. Huntington, *American Politics*, 187.

82. Michael Harrington, *Fragments of the Century* (New York: Saturday Review Press, 1973), 147.

83. Miller, *"Democracy is in the Streets,"* 142.

84. Miller, *"Democracy is in the Streets,"* 152.

85. The classic discussion of these two conceptions of liberty is, of course, Isaiah Berlin's "Two Concepts of Liberty," in Berlin, *Four Essays on Liberty* (New York: Oxford University Press, 1969), 118–72.

86. Miller, *"Democracy is in the Streets,"* 215.

87. Peter Bachrach, quoted in Mark E. Kann, *The American Left: Failures and Fortunes* (New York: Praeger, 1982), 42; emphasis added. The same definition of freedom as participation is used by Benjamin Barber, who argues that representation is incompatible with "full freedom." "Women and men who are not directly responsible through common deliberation, common decision, and common action for the policies that determine their common lives," Barber explains, "are not really free at all—however much they enjoy rights of privacy, property and individuality" ("Undemocratic Party System," 35, 37).

88. Miller, *"Democracy is in the Streets,"* 284.

89. Miller, *"Democracy is in the Streets,"* 270. Bruce Payne, "SNCC: An Overview Two Years Later," in Mitchell Cohen and Dennis Hale, eds., *The New Student Left: An Anthology* (Boston: Beacon Press, 1966), 95.

90. Miller, *"Democracy is in the Streets,"* 225. Also see Kirkpatrick Sale, *SDS* (New York: Random House, 1973), 159.

91. Marcuse, *Essay on Liberation,* quoted in Wini Breines, *Community and Organization in the New Left, 1962–1968: The Great Refusal* (New York: Praeger, 1982), 58.

92. Clayborne Carson, *In Struggle: SNCC and the Black Awakening of the 1960s* (Cambridge: Harvard University Press, 1981), 154. Miller, *"Democracy is in the Streets,"* 270. David J. Garrow, *Bearing the Cross: Martin Luther King, Jr., and the Southern Christian Leadership Conference* (New York: Morrow, 1988), 423. Payne, "SNCC: An Overview," 97.

93. Miller, *"Democracy is in the Streets,"* 282. Paul Jacobs and Saul Landau, *The New Radicals: A Report with Documents* (New York: Vintage, 1966), 31.

94. Huntington, *American Politics,* 172.

Chapter 5

1. Samuel P. Huntington, *American Politics: The Promise of Disharmony* (Cambridge: Harvard University Press, 1981), 4, 33, 41.

2. Luther Martin, quoted in Joshua Miller, "Democracy and the Politics of Experience in Antifederalist Thought," paper presented for delivery at the Annual Meeting of the Western Political Science Association, Sacramento, California, April 12–14, 1984, 15.

3. Huntington actually draws the same distinction. He writes that "the essence of liberalism is freedom from governmental control," and "the essence of egalitarianism is rejection of the idea that one person has the right to exercise power over another" (*American Politics,* 33). But he never pursues the different consequences for authority that stem from these competing cultural visions.

4. Thomas Paine, "Reflections on Titles," quoted in Eric Foner, *Tom Paine and Revolutionary America* (New York: Oxford University Press, 1976), 72.

5. Paine, "Age of Reason," quoted in Foner, *Paine and Revolutionary America* 247–49.

6. Thomas Paine, "Letter to George Washington" (1795), in *The Thomas Paine Reader,* ed. Michael Foot and Isaac Kramnick (Harmondsworth: Penguin, 1987), 502.

7. Paine, "Letter to George Washington," 491.

8. Gordon S. Wood, *The Creation of the American Republic, 1776–1787* (Chapel Hill: University of North Carolina Press, 1969), 482.

9. The following two paragraphs draw from Richard Ellis and Aaron Wildavsky, *Dilemmas of Presidential Leadership: From Washington Through Lincoln* (New Brunswick: Transaction Publishers, 1989), 21–22.

10. See Thomas Sowell, *A Conflict of Visions: Ideological Origins of Political Struggles* (New York: Quill, 1987), 33.

11. Thomas Paine, *Common Sense,* ed. Isaac Kramnick (Harmondsworth: Penguin, 1986), 68.

12. Foner, *Paine and Revolutionary America,* 91.

13. See Sowell, *Conflict of Visions,* 87. Sowell's distinction between the constrained and unconstrained visions closely parallels the distinction between individualism and egalitarianism. A deficiency of Sowell's dichotomy, however, is that the constrained vision conflates hierarchy and individualism, making the hierarchical vision of a Hamilton or Burke indistinguishable from the individualistic vision of a Madison or Adam Smith.

14. A more in-depth analysis may be found in Michael Thompson, Richard Ellis, and Aaron Wildavsky, *Cultural Theory* (Boulder: Westview Press, 1990), 33–36.

15. Huntington, *American Politics*, 37.

16. Foner, *Paine and Revolutionary America*, 122.

17. Foner, *Paine and Revolutionary America*, xvi.

18. See the discussion in Thompson, Ellis, and Wildavsky, *Cultural Theory*, 257.

19. Of the forty-nine convention delegates who supported the Constitution, twenty-three would later align themselves with the Federalist party, and fifteen would become Jeffersonian Republicans. The fifteen Federalists of 1787 who became Republicans were Abraham Baldwin, William Blount, Pierce Butler, John Dickinson, William Few, Nicholas Gilman, William Houstoun, John Langdon, James Madison, Alexander Martin, Charles Pinckney, Edmund Randolph, John Rutledge, Rich Spaight, and George Wythe. Those who supported the Constitution and later aligned with the Federalist party were Richard Bassett, John Blair, Jacob Broom, Gunning Bedford, George Clymer, William Davie, Jonathan Dayton, Oliver Ellsworth, Thomas Fitzsimmons, Nathaniel Gorham, Alexander Hamilton, Jared Ingersoll, William Samuel Johnson, Rufus King, James McHenry, Thomas Mifflin, Gouverneur Morris, William Paterson, Charles Cotesworth Pinckney, George Read, Caleb Strong, George Washington, and James Wilson. This list is drawn from David Hackett Fischer, *The Revolution of American Conservatism: The Federalist Party in the Era of Jeffersonian Democracy* (New York: Harper & Row, 1965), 222n53. The only difference is that Fischer considers Gorham unclassifiable. Although Gorham died in 1796, his Federalist credentials are strong enough—stronger probably than those of Blair, Mifflin, Wilson, and Broom—for him to be included as a future Federalist.

20. Madison to Jefferson, August 10, 1788, quoted in Alpheus Thomas Mason, "The Federalist: A Split Personality," *American Historical Review* 57 (April 1952), 630. Jefferson agreed that "in some parts it is discoverable that the author means only to say what may be best said in defense of opinions in which he did not concur" (Jefferson to Madison, November 18, 1788, in ibid.).

21. *Federalist* No. 68, p. 461. All page references are to the Jacob E. Cooke edition published by Wesleyan University Press in 1961.

22. Gerald Stourzh, *Alexander Hamilton and the Idea of Republican Government* (Stanford: Stanford University Press, 1970), 82.

23. David H. Fischer, "Myth of the Essex Junto," *William and Mary Quarterly* 21 (April 1964), 211.

24. Foner, *Paine and Revolutionary America*, 123.

25. Stourzh, *Alexander Hamilton*, 82. The quotation from Hamilton was made at the New York ratifying convention.

26. Isaac Kramnick, "The 'Great National Discussion': The Discourse of Politics in 1787," *William and Mary Quarterly* 45 (January 1988), 24.

27. See, e.g., *Federalist* No. 15, p. 97.

28. *Federalist* No. 17, pp. 108–9.

29. *Federalist* No. 26, p. 164.

30. Kramnick, "The 'Great National Discussion,'" 26.

31. *Federalist* No. 85, p. 594.

32. *Federalist* No. 17, p. 106. *Federalist* No. 16, p. 102.

33. *Federalist* No. 15, pp. 92, 91.

34. *Federalist* No. 62, p. 420.

35. Lance Banning, "The Practicable Sphere of a Republic: James Madison, the Constitutional Convention, and the Emergence of a Revolutionary Federalism," in Richard Beeman, Stephen Botein, and Edward C. Carter II, eds., *Beyond Confederation: Origins of the Constitution and American National Identity* (Chapel Hill: University of North Carolina Press, 1987), 174, 169; emphasis in original.

36. Kramnick, "The 'Great National Discussion,'" 30–31.

37. Kramnick, "The 'Great National Discussion,'" 31. *Federalist* No. 62, p. 421.

38. Banning, "Practicable Sphere," 171n19. Lance Banning, "The Hamiltonian Madison: A Reconsideration," *Virginia Magazine of History and Biography* 92 (January 1984), 19.

39. *Federalist* No. 45, p. 313. Banning, "Practicable Sphere," 172.

40. Mason, "A Split Personality," 641.

41. Richard M. Pious, *The American Presidency* (New York: Basic Books, 1979), 38. The words are Pious's, not Madison's.

42. Max Farrand, ed., *The Records of the Federal Convention of 1787,* 4 vols. (New Haven: Yale University Press, 1937), 1: 285. A few days later Hamilton elaborated on this point, warning that "we have been taught to reprobate the dangers of influence in the British government, without duly reflecting how far it was necessary to support a good government" (ibid., 1: 381)

43. *Federalist* No. 76, pp. 510–11.

44. Mason, "Split Personality," 638.

45. *Federalist* No. 51, p. 349.

46. *Federalist* No. 6, p. 35. Also see Stourzh, *Alexander Hamilton,* 255n88.

47. *Federalist* No. 76, pp. 513–14.

48. Forrest McDonald tries to reduce this ideological difference between Madison and Hamilton to a psychological one. "At bottom," McDonald argues, "Hamilton and other court-party nationalists trusted themselves and therefore trusted power if it was in their own hands. Madison and other men of his temperament did not trust themselves and therefore did not trust power in anyone's hands" (McDonald, *Novus Ordo Seclorum: The Intellectual Origins of the Constitution* [Lawrence: University Press of Kansas, 1985], 205). We are thus asked to believe that what separated Federalists like Hamilton, Washington, and Gouverneur Morris from their Republican opponents like Madison, Jefferson, and Paine was self-esteem. If only Madison and his cohorts had had a better opinion of themselves, they would have come around to the Federalist view of vesting greater power in a national elite. Are we also to believe that Leninists have higher self-esteem than libertarians? Applied to individuals such psychological explanations should be met with due skepticism; applied to entire groups of people they are absurd.

49. Mason, "A Split Personality," 632–33.

50. Kramnick, "The 'Great National Discussion,'" 31.

51. Irving H. Bartlett, *Wendell Phillips: Brahmin Radical* (Boston: Beacon Press, 1961), 196.

52. Aaron Wildavsky, "A World of Difference—The Public Philosophies and Political Behaviors of Rival American Cultures," in Anthony King, ed., *The New American Political System* (Washington, D.C.: AEI Press, 1990).

53. Richard Hofstadter, "Wendell Phillips: The Patrician as Agitator," in Hofstadter, *The American Political Tradition* (New York: Vintage, 1973), 210.

54. James Brewer Stewart, *Wendell Phillips: Liberty's Hero* (Baton Rouge: Louisiana State University Press, 1986), 183.

55. Irving H. Bartlett, "The Persistence of Wendell Phillips," in Martin Duberman, ed., *The Antislavery Vanguard: New Essays on the Abolitionists* (Princeton: Princeton University Press, 1965), 110. Stewart, *Wendell Phillips,* 122.

56. The New York *Tribune,* quoted in Bartlett, *Wendell Phillips,* 332.

57. Bartlett, "Persistence of Wendell Phillips," 110, 117, 118. Bartlett, *Wendell Phillips,* 161, 190, 207, 303, 310, 314. Stewart, *Wendell Phillips,* 139, 146–50, 240. Also see Irving H. Bartlett, "Wendell Phillips and the Eloquence of Abuse," *American Quarterly* 11 (Winter 1959), 509–20.

58. Bartlett, *Wendell Phillips*, 197.

59. Stewart, *Wendell Phillips*, 161; emphasis added.

60. Bartlett, *Wendell Phillips*, 391. Also Robert D. Marcus, "Wendell Phillips and American Institutions," *Journal of American History* 56 (June 1969), 46.

61. Bartlett, *Wendell Phillips*, 127. Also Stewart, *Wendell Phillips*, 119.

62. Bartlett, *Wendell Phillips*, 116. Also Stewart, *Wendell Phillips*, 117. In Stewart's version, the word is "indignity" rather than "iniquity."

63. Lawrence J. Friedman, *Gregarious Saints: Self and Community in American Abolitionism, 1830–1870* (Cambridge: Cambridge University Press, 1982), 44; also see 236. The strategic side of the abolitionist posture is accented in Aileen S. Kraditor, *Means and Ends in American Abolitionism: Garrison and His Critics on Strategy and Tactics, 1834–1850* (New York: Pantheon, 1969), and in Hofstadter, "Wendell Phillips: The Patrician as Agitator."

64. Bartlett, *Wendell Phillips*, 56.

65. Stewart, *Wendell Phillips*, 69. Phillips did not extend this concern for purity quite as far as did Stephen Foster, Wright, or Garrison, all of whom called for abolitionists to avoid all slave-produced products in the name of purity. "If we do," Phillips jibed, "what shall we do next—dig our fields with our fingers, for iron shovels come from mines where oppression reigns?" (Stewart, *Wendell Phillips*, 118).

66. Stewart, *Wendell Phillips*, 121. Also Bartlett, *Wendell Phillips*, 118.

67. Stewart, *Wendell Phillips*, 121, 150.

68. Bartlett, *Wendell Phillips*, 198.

69. Bartlett, *Wendell Phillips*, 311.

70. Stewart, *Wendell Phillips*, 99, 127.

71. Bartlett, *Wendell Phillips*, 161.

72. Bartlett, *Wendell Phillips*, 275.

73. Stewart, *Wendell Phillips*, 145, 118.

74. Hofstadter, "Wendell Phillips: Patrician as Agitator," 178.

75. Stewart, *Wendell Phillips*, 117–18.

76. Bartlett, "The Persistence of Wendell Phillips," 110.

77. Stewart, *Wendell Phillips*, 108–9, 82–96.

78. Stewart, *Wendell Phillips*, 130, 129.

79. Bartlett, *Wendell Phillips*, 195.

80. Bartlett, *Wendell Phillips*, 114.

81. See Jean H. Baker, *Affairs of Party: The Political Culture of Northern Democrats in the Mid-Nineteenth Century* (Ithaca: Cornell University Press, 1983), 184.

82. Stewart, *Wendell Phillips*, 67–68, 74, 99, 126–27. Also see Richard Ellis and Aaron Wildavsky, "A Cultural Analysis of the Role of Abolitionists in the Coming of the Civil War," *Comparative Studies in Society and History* 32 (January 1990), esp. 94–95; and Friedman, *Gregarious Saints*.

83. The question of race is an obvious and egregious exception. Frontier farmers, and Douglas, drew firm boundaries between black and white. Phillips, in contrast, desired to wipe out racial distinctions and grant blacks equal status within a redeemed, noncoercive community.

84. Robert W. Johannsen, *Stephen A. Douglas* (New York: Oxford University Press, 1973), 712.

85. Stewart, *Wendell Phillips*, 119.

86. Johannsen, *Stephen Douglas*, 279.

87. David M. Potter, *The Impending Crisis, 1848–1861* (New York: Harper & Row, 1976), 338. Johannsen, *Stephen Douglas*, 427, 440. Damon Wells, *Stephen Douglas: The*

Last Years (Austin: University of Texas Press, 1971), 67, 69–70. In *Antislavery Origins of the Civil War in the United States* (Ann Arbor: University of Michigan Press, 1939), Dwight Lowell Dumond describes Douglas as a "master of ambiguity" (101).

88. Irving H. Bartlett, "New Light on Wendell Phillips: The Community of Reform, 1840–1880," *Perspectives in American History* 12 (1979), 41–43.

89. Wells, *Stephen Douglas*, 56.

90. Robert W. Johannsen, "Stephen A. Douglas and the American Mission," in Johannsen, *The Frontier, the Union, and Stephen A. Douglas* (Urbana: University of Illinois Press, 1989), 96.

91. To the Editor of the Concord (New Hampshire) *State Capitol Reporter*, February 16, 1854, in Robert W. Johannsen, ed., *The Letters of Stephen A. Douglas* (Urbana: University of Illinois Press, 1961), 289.

92. Robert W. Johannsen, *Frontier Politics and the Sectional Conflict: The Pacific Northwest on the Eve of the Civil War* (Seattle: University of Washington Press, 1955), 133.

93. Friedrich A. Hayek, "The Use of Knowledge in Society," in Chiaki Nishiyama and Kurt R. Leube, eds., *The Essence of Hayek* (Stanford: Hoover Institution Press, 1984), 214, 217.

94. Bartlett, "Persistence of Wendell Phillips," 112. Stewart, *Wendell Phillips*, 118, 68, 138. Douglas, in contrast, believed that men like Phillips, who fixed their eyes unswervingly on unalterable truths, were a menace to democracy. "The man is only consistent," Douglas argued, "who follows out his principles and adapts his measures to them in view of the conditions of things he finds in existence in the period of time when it is necessary to make the application" (Johannsen, *Stephen Douglas*, 440).

95. Baker, *Affairs of Party*, 193.

96. See especially Harry Jaffa, *Crisis of the House Divided: An Interpretation of the Issues in the Lincoln–Douglas Debates* (Garden City, N.Y.: Doubleday, 1959); and *Equality and Liberty: Theory and Practice in American Politics* (New York: Oxford University Press, 1965).

97. Johannsen, *Stephen Douglas*, 446; also see 149, 457, 462.

98. Baker, *Affairs of Party*, 146. Bruce Collins, "The Ideology of the Antebellum Northern Democrats," *Journal of American Studies* 11 (April 1977), 106. J. David Greenstone, "Political Culture and American Political Development: Liberty, Union, and the Liberal Bipolarity," *Studies in American Political Development* 1 (1986), 29–32.

99. Douglas at Chicago, July 9, 1858, in Paul M. Angle, ed., *Created Equal? The Complete Lincoln–Douglas Debates of 1858* (Chicago: University of Chicago Press, 1958), 17.

100. Baker, *Affairs of Party*, 188.

101. Lawrence Frederick Kohl, *The Politics of Individualism: Parties and the American Character in the Jacksonian Era* (New York: Oxford University Press, 1989), 109.

102. Douglas at Springfield, July 17, 1858, in Angle, *Created Equal?*, 61.

103. Potter, *Impending Crisis*, 342. Douglas's position is not so very different from that of Milton Friedman, who argues that "on the whole, the formal restrictions on governmental activity should be most severe at the federal level, less so at the state level and least of all at the local level" (quoted in A. Craig Waggaman, "Milton Friedman and the Recovery of Liberalism," in Mark J. Rozell and James F. Pontuso, eds. *American Conservative Opinion Leaders* [Boulder: Westview Press, 1990], 165).

104. Baker, *Affairs of Party*, 189. Greenstone, "Political Culture and American Political Development," 30.

105. Wells, *Stephen Douglas*, 59.

106. Johannsen, *Stephen Douglas*, 24.

107. Stewart, *Wendell Phillips*, 210.

108. Statement, April 14, 1861; and To Virgil Hickox, May 10, 1861, in Johannsen, *Letters of Douglas*, 509, 512. Also see Johannsen, *Stephen Douglas*, 841–70, and To Ninety-Six New Orleans Citizens, November 13, 1860, in *Letters of Douglas*, 499–500.

109. Before the war, Phillips was fond of telling audiences that "the genuine Yankee . . . does not need a government" (Bartlett, *Wendell Phillips*, 204).

110. On the abolitionists' embrace of "righteous violence," see Lewis Perry, *Radical Abolitionism: Anarchy and the Government of God in Antislavery Thought* (Ithaca: Cornell University Press, 1973), 231–67; and Friedman, *Gregarious Saints*, 196–222.

111. Bartlett, *Wendell Phillips*, 238–39. Also see George M. Fredrickson, *The Inner Civil War: Northern Intellectuals and the Crisis of the Union* (New York: Harper & Row, 1965), 69.

112. Stewart, *Wendell Phillips*, 148. Bartlett, *Wendell Phillips*, 247. Bartlett, "Persistence of Wendell Phillips," 113.

113. Bartlett, *Wendell Phillips*, 252–53, 290. The treatment Lincoln received at the hands of Phillips and other abolitionists is covered in greater depth in Ellis and Wildavsky, *Dilemmas of Presidential Leadership*, 188–93.

114. See Aaron Wildavsky, "A Cultural Theory of Leadership," in Bryan D. Jones, ed., *Leadership and Politics: New Perspectives in Political Science* (Lawrence: University Press of Kansas, 1989), 87–113; *The Nursing Father: Moses as a Political Leader* (Birmingham: University of Alabama Press, 1984); and Ellis and Wildavsky, *Dilemmas of Presidential Leadership*.

115. Albert O. Hirschman, *Exit, Voice, and Loyalty: Responses to Decline in Firms, Organizations, and States* (Cambridge: Harvard University Press, 1970). As a number of critics have pointed out (and Hirschman himself has acknowledged), this categorization leaves out such responses as passivity, acquiescence, inaction, and resignation. All of these responses belong to the fatalist. The importance of such fatalistic responses are documented in Freek Bruinsma, "The (Non)Assertion of Welfare Rights: Hirschman's Theory Applied," *Acta Politica* 15 (July 1980), 357–83; and Lena Kolarska and Howard Aldrich, "Exit, Voice, and Silence: Consumers' and Managers' Responses to Organizational Decline," *Organization Studies* 1 (1980), 41–58.

116. Albert O. Hirschman, *Rival Views of Market Society* (New York: Viking, 1986), 78, 89. Hirschman, *Exit, Voice, and Loyalty*, 4, 108.

117. Hirschman, *Exit, Voice, and Loyalty*, 30, 4.

118. Hirschman, *Exit, Voice, and Loyalty*, 78.

119. Hirschman, *Exit, Voice, and Loyalty*, 106, 112.

120. Hirschman, *Exit, Voice, and Loyalty*, 107. The quotation by Frederick Jackson Turner is from "The Significance of the Frontier in American History," in Turner, *The Frontier in American History* (New York: Holt, 1947), 38.

121. Ray Allen Billington, *America's Frontier Heritage* (New York: Holt, Rinehart and Winston, 1966), 190. The individualist ethos is superbly captured by Fredrika Bremer, a European visitor to America during the 1840s. The typical citizen, she observed, was "a young man (no matter if he be old) who makes his own way in the world in full reliance on his own power, stops at nothing, shrinks from nothing, finds nothing impossible, tries everything, has faith in everything, hopes everything, goes through everything, and comes out of everything—ever the same. If he fails, he immediately gets up again and says, 'No matter!' If he is unsuccessful, he says, 'Try again!' 'Go ahead!' and he begins over again, undertaking something else, and never stopping until he succeeds" (Rush Welter, *The Mind of America, 1820–1860* [New York: Columbia University Press, 1975], 131).

122. Johannsen, *Stephen Douglas*, 11, 15, 8, 36–37.

123. Mary Douglas and Baron Isherwood, *The World of Goods* (New York: Basic Books, 1979), 43.

124. Billington, *America's Frontier Heritage*, 190. Perry, *Radical Abolitionism*, 93. Staughton Lynd, "The New Radicals and 'Participatory Democracy,'" in Massimo Teodori, ed., *The New Left: A Documentary History* (Indianapolis: Bobbs-Merrill, 1969), 229.

125. On free schools, see Ann Swidler, *Organization Without Authority: Dilemmas of Social Control in Free Schools* (Cambridge: Harvard University Press, 1979), and Allen Graubard, *Free the Children: Radical Reform and the Free School Movement* (New York: Pantheon, 1973). On freedom of choice, see John E. Chubb and Terry M. Moe, *Politics, Markets, and America's Schools* (Washington, D.C.: Brookings Institution, 1990).

126. See Victor A. Thompson, *Modern Organization* (New York: Knopf, 1961).

127. See Joyce Rothschild-Whitt, "The Collectivist Organization: An Alternative to Rational-Bureaucratic Models," *American Sociological Review* 44 (August 1979), 509–27; and Leonard Davidson, "Countercultural Organizations and Bureaucracy: Limits on the Revolution," in Jo Freeman, ed., *Social Movements of the Sixties and Seventies* (New York: Longman, 1983), 166–67.

128. Hirschman makes a similar point in commenting on what he calls the "cop-out" movements of the 1960s. "By making their exit so spectacular, by oddly combining deviance with defiance, they are actually closer to voice than was the case for their pilgrim, immigrant, and pioneer forebears" (*Exit, Voice, and Loyalty*, 108).

129. Bartlett, *Wendell Phillips*, 96, 106–7.

130. Bartlett, *Wendell Phillips*, 121–22; emphasis in orginal.

131. Stewart, *Wendell Phillips*, 196.

132. Bartlett, *Wendell Phillips*, 315, 129. Stewart, *Wendell Phillips*, 68, 74.

133. Mary Douglas, "The Problem of Evil," in Douglas, *Natural Symbols: Explorations in Cosmology* (London: Barrie & Rockliff, 1970), 107–24.

134. Hirschman explains that most nineteenth-century Americans "who departed from their communities had no thought of improving them thereby or of fighting against them from the outside; they were immigrants rather than émigrés, and soon after the move 'couldn't care less' about the fate of the communities whence they came" (*Exit, Voice, and Loyalty*, 108).

135. The breadth of the "voice" category has prompted a number of attempts to refine the concept. Perhaps the best known of these is Guillermo O'Donnell's distinction between vertical and horizontal voice, i.e., communicating opinions from citizens to authorities as opposed to communication among citizens ("On the Fruitful Convergences of Hirschman's *Exit, Voice, and Loyalty* and *Shifting Involvements*: Reflections from the Recent Argentine Experience," in Alejandro Foxley, Michael S. McPherson, and Guillermo O'Donnell, eds., *Development, Democracy, and the Art of Trespassing: Essays in Honor of Albert O. Hirschman* (Notre Dame: University of Notre Dame Press, 1986).

136. Hirschman, *Rival Views of Market Society*, 80, 91. Hirschman, *Exit, Voice, and Loyalty*, 107. The quote by Hartz is from *The Liberal Tradition in America*, 65. If the existence of the western frontier advantaged individual exit over collective voice, individualism over egalitarianism, then the "closing of the frontier" may be thought to have gone some distance in correcting that imbalance. That at least has been the thesis of a number of historians. The Populists' aggressive challenge to authority, for instance, has often been explained in this manner. With the frontier closed, John D. Hicks explained in his classic *The Populist Revolt*, "the restless and discontented voiced their sentiments more and fled from them less" (*The Populist Revolt: A History of the Farmers' Alliance and the People's Party* [Minneapolis: University of Minnesota Press, 1931], 95). And Turner, in

his 1910 presidential address to the American Historical Association, argued that the growth of socialism as well as such progressive reforms as primary elections, popular election of senators, and the initiative, referendum, and recall were all "the sequence to the extinction of the frontier" (Frederick Jackson Turner, "Social Forces in American History," in Turner, *Frontier in American History*, 321).

137. Drew McCoy, *The Elusive Republic: Political Economy in Jeffersonian America* (Chapel Hill: University of North Carolina Press, 1980), 199–200. Timothy Dwight quoted in Welter, *Mind of America*, 308.

138. Stourzh, *Alexander Hamilton*, 193. Also see the comments by Gouverneur Morris at the Constitutional Convention, in Farrand, *Records of the Federal Convention*, 2:52.

139. Greenstone, "Political Culture and American Political Development," 33. John H. Schroeder, *Mr. Polk's War: American Opposition and Dissent, 1846–1848* (Madison: University of Wisconsin Press, 1973), 28.

140. Daniel Feller, *The Public Lands in Jacksonian Politics* (Madison: University of Wisconsin Press, 1984), 190. Also see Welter, *Mind of America*, 306–12, and Daniel Walker Howe, *The Political Culture of the American Whigs* (Chicago: University of Chicago Press, 1979).

141. Frederick Merk, *Manifest Destiny and Mission in American History* (New York: Vintage, 1963), 40. Also see Major L. Wilson, "The Concept of Time and the Political Dialogue in the United States, 1828–48," *American Quarterly* 19 (Winter 1967), 619–44; Thomas R. Hietala, *Manifest Design: Anxious Aggrandizement in Late Jacksonian America* (Ithaca: Cornell University Press, 1985), esp. 6–7, 112–13, 258, 297; and McCoy, *Elusive Republic*, esp. 200–204.

142. Stourzh, *Alexander Hamilton*, 193. Hietala, *Manifest Design*, 121–22.

Chapter 6

1. Important recent studies accenting the persistence and political importance of hierarchical social relations and values in the United States include Karen Orren, *Belated Feudalism: Labor, the Law, and Liberal Development in the United States* (Cambridge: Cambridge University Press, 1991); Mark E. Kann, *On the Man Question: Gender and Civic Virtue in America* (Philadelphia: Temple University Press, 1991); Rogers M. Smith, "The Tocquevillian Thesis Reconsidered," *American Political Science Review* (forthcoming); and Denis J. Brion, "The Meaning of the City: Urban Redevelopment and the Loss of Community," *Indiana Law Review* 25 (1992), 685–740. Kann carries revisionism too far, however, when he suggests that "individualism is no more than a whisper in the lives of most Americans" (*On the Man Question*, 32).

2. Actually this has been evident for some time. As early as 1961, Lee Benson criticized Hartz for having "followed Tocqueville too closely and exaggerated the extent to which feudalism and the canon law have been absent, the liberal idea present, men born equal, and democracy has prevailed in America." Benson continued, "by the 1840s, however, . . . the nation had essentially completed the transformation from a liberal aristocratic republic to a populistic democracy" (*The Concept of a Jacksonian Democracy* [Princeton: Princeton University Press, 1961], 274).

3. Joyce Appleby, "The Radical *Double-Entendre* in the Right to Self-Government," in Margaret Jacob and James Jacob, eds., *The Origins of Anglo-American Radicalism* (London: George Allen & Unwin, 1984), 279. Elsewhere, Appleby cautions that "probably no historian ever completely subscribed to the old belief that America was born free, rich, and modern, but approximations of this view abound in texts" ("Value and Society," in

Jack P. Greene and J. R. Pole, eds., *Colonial British America: Essays in the New History of the Early Modern Era* [Baltimore: Johns Hopkins University Press, 1984], 305).

4. James A. Henretta, "The Slow Triumph of Liberal Individualism: Law and Politics in New York, 1780–1860," in Richard O. Curry and Lawrence B. Goodehart, eds., *American Chameleon: Individualism in Trans-National Context* (Kent: Kent State University Press, 1991), 89. Also see Melvin Yazawa, *From Colonies to Commonwealth: Familial Ideology and the Beginnings of the American Republic* (Baltimore: Johns Hopkins University Press, 1985); and Gordon S. Wood, *The Radicalism of the American Revolution* (New York: Knopf, 1991), pt. 1.

5. Rowland Berthoff and John M. Murrin, "Feudalism, Communalism, and the Yeoman Freeholder: The American Revolution Considered as a Social Accident," in Stephen G. Kurtz and James H. Hutson, eds. *Essays on the American Revolution* (Chapel Hill: University of North Carolina Press, 1973), 264. Berthoff and Murrin concede that "in terms of genuine feudalism the revival was . . . grossly imperfect, more a matter of economic profit for the proprietor than of mutual obligations between lord and man or landlord and community that might have harmonized the relationship" (271).

6. Jack P. Greene and J. R. Pole, "Reconstructing British-American Colonial History: An Introduction," in Greene and Pole, *Colonial British America*, 15. Also see Wood, *Radicalism*, 16–18.

7. Berthoff and Murrin, "The American Revolution Considered as a Social Accident."

8. Appleby, "Value and Society," in Greene and Pole, *Colonial British America*, 305.

9. Richard Ashcraft, *Revolutionary Politics and Locke's Two Treatises of Government* (Princeton: Princeton University Press, 1986), 166.

10. Alan Macfarlane, *The Origins of English Individualism: The Family, Property and Social Transition* (Oxford: Basil Blackwell, 1978), 163. Also see Alan Macfarlane, *The Culture of Capitalism* (Oxford: Basil Blackwell, 1987); Appleby, "Value and Society," 309–10; and Wood, *Radicalism*, 12–15. Brian Tierney makes a parallel point with respect to medieval political thought. "All through the Middle Ages," writes Tierney, "there were two attitudes—not just one—to the problem of equality and inequality. One could emphasize that hierarchical ranking was necessary in an ordered society; or one could emphasize that, because all men shared a common humanity, they were all by nature equal, and also by nature free (for no one had a natural right to dominate his equals). . . . The two attitudes coexisted side by side in the twelfth and thirteenth centuries and they still coexisted in the seventeenth and eighteenth centuries" ("Hierarchy, Consent, and the 'Western Tradition,'" *Political Theory* 15 [November 1987], 650).

11. Michael Walzer, *The Revolution of the Saints* (Cambridge: Harvard University Press, 1965), 169.

12. I take the term "English ways" from David Grayson Allen, *In English Ways: The Movement of Societies and the Transferal of English Local Law and Custom to Massachusetts Bay in the Seventeenth Century* (Chapel Hill: University of North Carolina Press, 1981).

13. Louis Hartz, *The Liberal Tradition in America* (New York: Harcourt, 1955), 3.

14. The analysis of Bernard Bailyn suggests that even this view needs amending. See *The Peopling of British North America: An Introduction* (New York: Knopf, 1986), chap. 1.

15. It is true, however, that many of Virginia's Royalist immigrants emigrated from England during the 1650s to escape the persecution of the Puritan oligarchy established after Charles I was beheaded.

16. David Hackett Fischer, *Albion's Seed: Four British Folkways in America* (New York: Oxford University Press, 1989), 212–25. Bernard Bailyn, "Politics and Social Structure in Virginia," in James Morton Smith, ed., *Seventeenth-Century America: Essays in Colonial History* (Chapel Hill: University of North Carolina Press, 1959), 98. The Virgin-

ians' attempt to "fulfill English virtues" rather than "escape from English vices" is stressed in Daniel J. Boorstin, *The Americans: The Colonial Experience* (New York: Random House, 1958), 99–104. Also see Dickson D. Bruce, Jr., *The Rhetoric of Conservatism: The Virginia Convention of 1829–1830 and the Conservative Tradition in the South* (San Marino, Calif.: Huntington Library, 1982), xiii.

17. Hartz, *Liberal Tradition*, 147.

18. Henry Hughes, *Treatise on Sociology, Theoretical and Practical* (New York: Negro University Press, 1968; originally published 1854), 83.

19. Fischer, *Albion's Seed*, 28, 227, 210–12, 232–36, 377.

20. Fischer, *Albion's Seed*, 243–46.

21. Fischer, *Albion's Seed*, 233; also see 235–36.

22. Rhys Isaac, *The Transformation of Virginia, 1740–1790* (Chapel Hill: University of North Carolina Press, 1982), 61.

23. Isaac, *Transformation of Virginia*, 64.

24. Fischer, *Albion's Seed*, 360–64.

25. Fischer, *Albion's Seed*, 399–401, 404–5. The notion of a hierarchical scale could also work in a more "progressive" direction when punishment pertained to the neglect of public responsibilities. Those at the top of the social scale were expected to set an example, and failure to do so brought a larger fine. So, for instance, the higher one's station the heavier the punishment for failing to comply with mustering requirements (Isaac, *Transformation of Virginia*, 105).

26. Isaac, *Transformation of Virginia*, 38–39, 352. Also see 66–67, 354; and Fischer, *Albion's Seed*, 264–74.

27. Darrett B. Rutman and Anita H. Rutman, *A Place in Time: Middlesex County, Virginia, 1650–1750* (New York: Norton, 1984), 129.

28. James Oakes, *The Ruling Race: A History of American Slaveholders* (New York: Knopf, 1982), 5. Fischer, *Albion's Seed*, 398.

29. Isaac, *Transformation of Virginia*, 113–14.

30. Fischer, *Albion's Seed*, 358.

31. Isaac, *Transformation of Virginia*, 43. Fischer, *Albion's Seed*, 355. Also see Peter Laslett, *The World We Have Lost: England Before the Industrial Age*, 2nd ed. (New York: Scribner's, 1971), 30, 33.

32. Isaac, *Transformation of Virginia*, 43. Rutman and Rutman, *A Place in Time*, 129.

33. Isaac, *Transformation of Virginia*, 56. Fischer, *Albion's Seed*, 387. Also see Isaac, *Transformation of Virginia*, 161–62.

34. Isaac, *Transformation of Virginia*, 90.

35. Fischer, *Albion's Seed*, 387. Isaac, *Transformation of Virginia*, 65.

36. Isaac, *Transformation of Virginia*, 131–32. Jack P. Greene, *The Quest for Power: The Lower Houses of Assembly in the Southern Royal Colonies, 1689–1776* (Chapel Hill: University of North Carolina Press, 1963), 25.

37. Fischer, *Albion's Seed*, 314–16.

38. Isaac, *Transformation of Virginia*, 42. Oakes, *Ruling Race*, 4.

39. Fischer, *Albion's Seed*, 220–21. Also see 274, 279, 321.

40. Isaac, *Transformation of Virginia*, 39–40.

41. Fischer, *Albion's Seed*, 279.

42. Fischer, *Albion's Seed*, 280. Also see Isaac, *Transformation of Virginia*, 20–21.

43. Fischer, *Albion's Seed*, 365–67. Also see Isaac, *Transformation of Virginia*, 39.

44. Fischer, *Albion's Seed*, 225.

45. Laslett, *World We Have Lost*, 38.

46. Isaac, *Transformation of Virginia*, chap. 9.

47. Fischer, *Albion's Seed*, 355–59. Isaac, *Transformation of Virginia*, 44.

48. Those historians wishing to accent the difference between the hierarchical Old World and the fluid, individualistic New World have too often underestimated the mobility, geographic as well as social, of the Old World. The past several decades of research in European social history, as Bernard Bailyn points out, has shown that "the traditional society of early modern Europe was a mobile society—a world in motion" (*Peopling of North America*, 20). The upward and downward social mobility of preindustrial England is stressed in Laslett, *World We Have Lost*, esp. 34–35.

49. Bailyn, *Peopling of North America*, 128.

50. Bailyn, *Peopling of North America*, 100–101. Oakes, *Ruling Race*, 110–11.

51. Isaac, *Transformation of Virginia*, 116.

52. Oakes, *Ruling Race*, 7.

53. Oakes, *Ruling Race*, 205.

54. Fischer, *Albion's Seed*, 384, 222. Also see Bailyn, *Peopling of North America*, 102.

55. Rhys Isaac, "Evangelical Revolt: The Nature of the Baptists' Challenge to the Traditional Order in Virginia, 1765–1775," *William and Mary Quarterly* 31 (July 1974), 353–55. Isaac, *Transformation of Virginia*, 173, 291, 315.

56. Isaac, "Evangelical Revolt," 356. Isaac, *Transformation of Virginia*, 356, 292–93.

57. Isaac, *Transformation of Virginia*, 265, 321; also see 255, 259.

58. Isaac, *Transformation of Virginia*, 251. Also see Richard Ellis and Aaron Wildavsky, *Dilemmas of Presidential Leadership: From Washington Through Lincoln* (New Brunswick: Transaction Publishers, 1989), 19–20.

59. Isaac, *Transformation of Virginia*, 320.

60. Isaac, *Transformation of Virginia*, 322.

61. Hartz, *Liberal Tradition*, 151, 153.

62. Hartz, *Liberal Tradition*, 153.

63. Isaac, *Transformation of Virginia*, 309.

64. Larry E. Tise, *Proslavery: A History of the Defense of Slavery in America, 1701–1840* (Athens: University of Georgia Press, 1987), esp. 15–16, 23–24.

65. Tise, *Proslavery*, 25, 30, 38.

66. Oakes, *Ruling Race*, 194, 208; emphasis added. Also see Tise, *Proslavery*.

67. The most forceful presentation of the view that slaveholders adhered to a "market culture" is Oakes, *Ruling Race*. Eugene D. Genovese is the leading spokesman for the position that slaveholders lived in and defended a paternalistic way of life. See especially his *The World the Slaveholders Made: Two Essays in Interpretation* (New York: Pantheon, 1969); and *Roll, Jordan, Roll: The World the Slaves Made* (New York: Pantheon, 1974), bk. 1.

68. J. Mills Thornton III, *Politics and Power in a Slave Society: Alabama, 1800–1860* (Baton Rouge: Louisiana State University Press, 1978).

69. A preliminary attempt to differentiate between regions can be found in Richard J. Ellis, "Legitimating Slavery in the Old South: The Effect of Political Institutions on Ideology," *Studies in American Political Development* 5 (Fall 1991), 340–51.

70. Samuel P. Huntington, *The Soldier and the State: The Theory and Practice of Civil–Military Relations* (Cambridge: Harvard University Press, 1957), 211.

71. Huntington, *Soldier and the State*, 217–19. Also see John Hope Franklin, *The Militant South, 1800–1861* (Cambridge: Harvard University Press, 1956), chap. 8.

72. Huntington, *Soldier and the State*, 213.

73. Huntington, *Soldier and the State*, 213.

74. Hartz, *Liberal Tradition*, 172, 176.

75. Huntington, *Soldier and the State*, 218, 214.

76. Daniel J. Elazar, *American Federalism: A View from the States* 2nd ed. (New York: Crowell, 1972). I analyze Elazar's categories in greater depth in chapter 9.

77. Tise, *Proslavery*, 16–18. Oakes, *Ruling Race*, 3.

78. Tise, *Proslavery*, 30–32.

79. Tise, *Proslavery*. Also see Oakes, *Ruling Race*, 196, 198, 280n7.

80. Oakes, *Ruling Race*, 200–201.

81. James M. Banner, Jr., *To the Hartford Convention: The Federalists and the Origins of Party Politics in Massachusetts, 1789–1815* (New York: Knopf, 1970). Also see David Hackett Fischer, *The Revolution of American Conservatism: The Federalist Party in the Era of Jeffersonian Democracy* (New York: Harper & Row, 1965); and Linda K. Kerber, *Federalists in Dissent: Imagery and Ideology in Jeffersonian America* (Ithaca: Cornell University Press, 1970).

82. Banner, *To the Hartford Convention*, 53. Also see David Hackett Fischer, "The Myth of the Essex Junto," *William and Mary Quarterly* 21 (April 1964), 191–235, esp. 201.

83. Banner, *To the Hartford Convention*, 53.

84. Banner, *To the Hartford Convention*, 65, 57.

85. Banner, *To the Hartford Convention*, 54–55.

86. Jonathan Jackson, Stephen Higginson, and Alden Bradford, cited in Banner, *To the Hartford Convention*, 54–55. Also see Fischer, *Revolution of American Conservatism*, 49.

87. Banner, *To the Hartford Convention*, 60–61.

88. Banner, *To the Hartford Convention*, 58.

89. Fischer, "Essex Junto," 203.

90. See Daniel Walker Howe, *The Unitarian Conscience: Harvard Moral Philosophy, 1805–1861* (Cambridge: Harvard University Press, 1970); and Daniel Walker Howe, *The Political Culture of the American Whigs* (Chicago: University of Chicago Press, 1979), esp. chap. 9.

91. Max Weber, "Politics as a Vocation," in *From Max Weber: Essays in Sociology* ed. H. H. Gerth and C. Wright Mills (New York: Oxford University Press, 1946), 109.

92. *Plunkitt of Tammany Hall*, recorded by William L. Riordan (New York: Dutton, 1963; originally published 1905), 3.

93. Arthur Mann, "Introduction," in *Plunkitt of Tammany Hall*, xviii. Also see Ari Hoogenboom, "Civil Service Reform and Public Morality," in H. Wayne Morgan, ed., *The Gilded Age*, rev. ed. (Syracuse: Syracuse University Press, 1970), 78.

94. Ray Ginger, *Age of Excess: The United States from 1877 to 1914* (New York: Macmillan, 1965). Thomas C. Cochran and William Miller, *The Age of Enterprise: A Social History of Industrial America* (New York: Harper & Row, 1942). The most common designation, "The Gilded Age," derives from the novel of that name by Mark Twain and Charles Dudley Warner.

95. Ari Hoogenboom, *Outlawing the Spoils: A History of the Civil Service Reform Movement, 1865–1883* (Urbana: University of Illinois Press, 1961), 21. John Tomsich, *A Genteel Endeavor: American Culture and Politics in the Gilded Age* (Stanford: Stanford University Press, 1971), 84.

96. Hoogenboom, *Outlawing the Spoils*, 21. Gerald W. McFarland, *Mugwumps, Morals and Politics, 1884–1920* (Amherst: University of Massachusetts Press, 1975), 30, 37.

97. McFarland, *Mugwumps, Morals and Politics*, 26. James McLachlan, "American Colleges and the Transmission of Culture: The Case of the Mugwumps," in Stanley Elkins and Eric McKitrick, eds., *The Hofstadter Aegis: A Memorial* (New York: Knopf, 1974),

192–93. Geoffrey Blodgett, *The Gentle Reformers: Massachusetts Democrats in the Cleveland Era* (Cambridge: Harvard University Press, 1966), 21. Hoogenboom, *Outlawing the Spoils*, 192, 21.

98. Hoogenboom, *Outlawing the Spoils*, 192, 16. Tomsich, *Genteel Endeavor*, 76. Geoffrey Blodgett, "Reform Thought and the Genteel Tradition," in Morgan, *The Gilded Age*, 70.

99. Hoogenboom, *Outlawing the Spoils*, 196, 21, ix.

100. Blodgett, *Gentle Reformers*, 20–21, 34. McFarland, *Mugwumps, Morals and Politics*, 36. McLachlan, "American Colleges and the Transmission of Culture," 186. Also see Dorothy Ross, "Socialism and American Liberalism: Academic Social Thought in the 1880s," *Perspectives in American History* 11 (1977–1978), 17, 73–74n114; and Howe, *The Unitarian Conscience*.

101. Blodgett, "Reform Thought and the Genteel Tradition," 69. Also see Stow Persons, *The Decline of American Gentility* (New York: Columbia University Press, 1973), 157.

102. Gordon S. Wood, "The Massachusetts Mugwumps," *New England Quarterly* 33 (December 1960), 435–51, quotation on 436.

103. Ari Hoogenboom, "Civil Service Reform and Public Morality," in Morgan, *The Gilded Age*, 77.

104. According to Hoogenboom, the "leadership of the [civil service] reform movement was . . . left to a group of professional men almost as hostile to the grasping capitalist as to the dominant spoilsman" (*Outlawing the Spoils*, 196).

105. Richard Hofstadter, *The Age of Reform: From Bryan to F.D.R.* (New York: Vintage: 1955), 141. Also see Blodgett, "Reform Thought and the Genteel Tradition," 60–61; and Blodgett, *Gentle Reformers*, 38–39.

106. Charles Francis Adams, *Charles Francis Adams, 1835–1915: An Autobiography* (Boston: Houghton Mifflin, 1916), 190.

107. Vincent DeSantis, *The Shaping of Modern America: 1877–1920,* 2nd ed. (Arlington Heights, Ill.: Forum Press, 1989), 81.

108. Blodgett, "Reform Thought and the Genteel Tradition," 61.

109. Hoogenboom, "Civil Service Reform and Public Morality," 80–81. Also see William R. Brock, *Investigation and Responsibility: Public Responsibility in the United States, 1865–1900* (Cambridge: Cambridge University Press, 1984), 17.

110. The *Century*, quoted in Tomsich, *Genteel Endeavor*, 107.

111. Tomsich, *Genteel Endeavor*, 97, 77, 89, 101.

112. Tomsich, *Genteel Endeavor*, 98–99. Also see Hoogenboom, "Civil Service Reform and Public Morality," 80–81; and Blodgett, *Gentle Reformers*, 21.

113. These words are from the charter of the American Social Science Association, founded in 1865. Quoted in Blodgett, "Reform Thought and the Genteel Tradition," 73.

114. Blodgett, *Gentle Reformers*, 22–23, 50.

115. See, e.g., David P. Thelen, *The New Citizenship: Origins of Progressivism in Wisconsin, 1885–1900* (Columbia: University of Missouri Press, 1972), 138–39.

116. Blodgett, "Reform Thought and the Genteel Tradition," 70–71, 66–67.

117. Tomsich, *Genteel Endeavor*, 94.

118. Blodgett, "Reform Thought and the Genteel Tradition," 68. Also see George M. Fredrickson, *The Inner Civil War: Northern Intellectuals and the Crisis of the Union* (New York: Harper, 1965), esp. 205–8.

119. Thomas K. McCraw, *Prophets of Regulation: Charles Francis Adams, Louis D. Brandeis, James M. Landis, Alfred E. Kahn* (Cambridge: Harvard University Press, 1984).

120. Tomsich, *Genteel Endeavor*, 111, 107, 97.

121. Blodgett writes that "some [Mugwumps] were openly nostalgic for a pre-

Jacksonian society of recognized ranks and gradations" ("Reform Thought," 60). Hoogenboom agrees that civil service "reformers wished to return to the attitudes of the good old days before Jacksonian democracy and the industrial revolution—days when men with their background, status, and education were the unquestioned leaders of society" (*Outlawing the Spoils*, 197). "Their thought," echoes Tomsich, "naturally gravitated back to the days of John Quincy Adams" (*Genteel Endeavor*, 77; also see 87).

122. Frederic J. Stimson, *My United States* (1931), quoted in Blodgett, *Gentle Reformers*, 32; emphasis added.

123. Hoogenboom, *Outlawing the Spoils*, 41.

124. Robert Bellah and his colleagues explain that the reformist vision of the "Establishment" Progressives (of whom "Theodore Roosevelt was perhaps [the] classic embodiment in political life") "sought to spread a cosmopolitan ethic of *noblesse oblige* and public service," and "emphasized paternal relationships of reciprocal, though unequal, duties" (Robert N. Bellah, Richard Madsen, William M. Sullivan, Ann Swidler, and Steven M. Tipton, *Habits of the Heart: Individualism and Commitment in American Life* [Berkeley: University of California Press, 1985], 258–60). Also see Carolyn Webber and Aaron Wildavsky, *A History of Taxation and Expenditure in the Western World* (New York: Simon & Schuster, 1986), which documents "the reemergence of hierarchy" that was envisioned by the progressive reformers in the areas of public administration and budgeting (400–411).

125. McFarland, *Mugwumps, Morals and Politics*, 35, 38–39; also see 50. Also see William E. Nelson, *The Roots of Bureaucracy, 1830–1900* (Cambridge: Harvard University Press, 1982), chap. 4.

126. See Morris Janowitz, *The Professional Soldier: A Social and Political Portrait* (New York: Free Press, 1960); and Lewis H. Lapham, "Military Theology," *Harper's Magazine* 243 (July 1971), 73–85.

127. Samuel A. Stouffer, Edward A. Suchman, Leland C. DeVinney, Shirley A. Star, and Robin M. Williams, Jr., *The American Soldier*, 4 vols. (Princeton: Princeton University Press, 1949), esp. vol. 1, chaps. 2–5. Throughout American history, supporters of hierarchical values have looked to the military for help. Horace Bushnell, a leading mid–nineteenth-century theologian, for instance, praised the army as "a vast and mighty schooling of authority" where "nothing goes by consent, or trust, or individual sovereignty" (Fredrickson, *Inner Civil War*, 140). Also see Mark E. Kann, "Individualism, Civic Virtue, and Gender in America," *Studies in American Political Development* 4 (1990), 67–68.

128. Aaron Wildavsky, "Choosing Preferences by Constructing Institutions: A Cultural Theory of Preference Formation," *American Political Science Review* 81 (March 1987), 11. Also see Aaron Wildavsky, *The Rise of Radical Egalitarianism* (Washington, D.C.: American University Press, 1991), 227–32.

129. Frances FitzGerald, *Cities on a Hill: A Journey Through Contemporary American Cultures* (New York: Simon & Schuster, 1986), 140–41. Also see Robert Lerner, Stanley Rothman, and Robert S. Lichter, "Christian Religious Elites," *Public Opinion* 11 (March/April 1989), 54–58.

130. Robert E. Lane, for instance, was struck by "the relatively low salience of the father in the American [family] scheme." Lane asked the men of Eastport, "'Who made the important decisions in your parents' household?' One replied that they were jointly made, two that their fathers made the important decisions, and twelve testified that mother was boss" (*Political Ideology: Why the American Common Man Believes What He Does* [New York: Free Press, 1962], 271).

131. Relevant is Jennifer L. Hochschild's finding that "except for a few people most of the time, and most of the people a few times," her respondents shared "a commitment

to a principle of equality in the socializing domain" (*What's Fair? American Beliefs About Distributive Justice* [Cambridge: Harvard University Press, 1981], 106).

132. Robert S. Lynd and Helen Merrell Lynd, *Middletown: A Study in Modern American Culture* (New York: Harcourt, Brace, 1929). Also see Theodore Caplow, Howard M. Bahr, Bruce A. Chadwick, Reuben Hill, and Margaret Holmes Williamson, *Middletown Families: Fifty Years of Change and Continuity* (Minneapolis: University of Minnesota Press, 1982), a follow-up study done fifty years later, which found some modifications in family patterns but stressed that overall "today's Middletown is still the same place the Lynds studied in the 1920s."

133. Lynd and Lynd, *Middletown*, 143–44. On changes over time, see Duane F. Alvin, "From Obedience to Autonomy: Changes in Traits Desired in Children, 1924–1978," *Public Opinion Quarterly* 52 (Spring 1988), 33–52, and the sources cited therein.

134. John W. Gardner, *Excellence: Can We Be Equal and Excellent Too?*, rev. ed. (New York: Norton, 1984), 19–21. Hochschild, *What's Fair?*, 71.

135. Louis Dumont, *Essays on Individualism: Modern Ideology in Anthropological Perspective* (Chicago: University of Chicago Press, 1986), 17. Karl Polanyi, *The Great Transformation: The Political and Economic Origins of Our Time* (Boston: Beacon, 1957). Robert H. Wiebe, *The Search for Order, 1877–1920* (New York: Hill & Wang, 1967). Oliver E. Williamson, *Markets and Hierarchies, Analysis and Antitrust Implications: A Study in the Economics of Internal Organization* (New York: Free Press, 1975). Alfred D. Chandler, Jr., *The Visible Hand: The Managerial Revolution in American Business* (Cambridge: Harvard University Press, 1977).

136. Victor A. Thompson, *Modern Organization* (New York: Knopf, 1961).

137. Louis Dumont, *Homo Hierarchicus* (Chicago: University of Chicago Press, 1970).

138. Herbert McClosky and John Zaller, *The American Ethos: Public Attitudes Toward Capitalism and Democracy* (Cambridge: Harvard University Press, 1984), 66.

139. Stouffer, *American Soldier*, 1:374. Sixty-seven percent of officers assented to the same statement.

140. McClosky and Zaller, *American Ethos*, 116. Also see Lane, *Political Ideology*.

141. Hochschild, *What's Fair?*, 198, 202, 198, 200. Compare this with the comments of the Catholic housekeeper from Somerville, Massachusetts, interviewed by Robert Coles. The woman explains, "I tell my kids to obey the teacher and listen to the priest; and their father gives them a whack if they cross him." She is appalled by what she sees in the "fancy" Cambridge house she cleans. "The house is full of talk even early in the morning. He's read something that's bothered him, and she's read something that's bothered her. They're both ready to phone their friends. The kids hear all that and they start complaining about what's bothering *them*—about school, usually. They're all so *critical*. . . . The kids speak back to their parents, act as fresh and snotty as can be. I want to scream sometimes when I hear those brats talking as if they know everything" (Christopher Lasch, *The True and Only Heaven: Progress and Its Critics* [New York: Norton, 1991], 494).

142. Hochschild, *What's Fair?*, 198–99. Compare with the respondent interviewed in *The Authoritarian Personality*, who explained that "there have to be the ditch diggers. . . . Certain people were cut out for certain things" (Theodor W. Adorno, Else Frenkel-Brunswick, Daniel J. Levinson, and R. Nevitt Sanford, *The Authoritarian Personality*, 2 vols. [New York: Harper, 1950], 1: 414).

143. Hochschild, *What's Fair?*, 264–65, 196, 119–20. "I don't think Nelson Rockefeller should sit out in his front yard in his lawnchair with a beer," explains Pamela McLean. "He wouldn't look good" (120).

144. Studs Terkel, *The Great Divide: Second Thoughts on the American Dream* (New York: Pantheon, 1988), 242, 45–46, 310.

145. Kristin Luker, *Abortion and the Politics of Motherhood* (Berkeley: University of California Press, 1984), 159. Similarly, Pamela Johnston Conover and Virginia Gray conclude that "basic conflict over values and life-styles" underlies competing positions on abortion and the ERA (*Feminism and the New Right: Conflict over the American Family* [New York: Praeger, 1983], 126). Also see Marilyn Falik, *Ideology and Abortion Policy Politics* (New York: Praeger, 1983).

146. Luker, *Abortion and the Politics of Motherhood*, 196. Similarly, Robert J. Spitzer found that 84 percent of his sample of Right to Life activists were Catholics (*The Right to Life Movement and Third Party Politics* [New York: Greenwood Press, 1987], 85).

147. Luker, *Abortion and the Politics of Motherhood*, 160–62, 173.

148. Byron E. Shafer, "The New Cultural Politics," *P.S.* 18 (Spring 1985), 224.

149. Rebecca E. Klatch, *Women of the New Right* (Philadelphia: Temple University Press, 1987), 4, 9. Also see Rebecca Klatch, "Coalition and Conflict Among Women of the New Right," *Signs: Journal of Women in Culture and Society* 13 (Summer 1988), 671–94.

150. Fred Kerlinger, *Liberalism and Conservatism: The Nature and Structure of Social Attitudes* (Hillsdale, N.J.: Lawrence Erlbaum Associates, 1984), esp. 150–52, 164. Also see Pamela Johnston Conover and Stanley Feldman, "The Origins and Meaning of Liberal/Conservative Self-Identifications," *American Journal of Political Science* 25 (November 1981), 617–45.

151. McClosky and Zaller, *The American Ethos*, 221–22. Still, many social conservatives in America clearly remain uncomfortable with free market capitalism. New right spokesman Paul Weyrich, for instance, envisions "the essence of the new right" to be a "morally based conservatism," not free market economics. "Big corporations," he explains, "are as bad as big government." "Laissez faire is not enough"; people need "some higher value" than the pursuit of wealth (Lasch, *The True and Only Heaven*, 505–6).

152. On the Whig dilemma of "hierarchy without a hierarch," see Ellis and Wildavsky, *Dilemmas of Presidential Leadership*, 116–20.

153. Paul M. Sniderman with Michael Gray Hagen, *Race and Inequality: A Study in American Values* (Chatham, N.J.: Chatham House, 1985), 73–76.

154. George F. Will, at least until recently, has been one of the few contemporary American conservatives who matches the two in a way that sounds recognizably European. So, for instance, Will condemns those conservatives who harbor "a frivolous hostility toward the state" and who lack "the traditional conservative appreciation of the dignity of the political vocation and the grandeur of its responsibilities" (*The Pursuit of Happiness, and Other Sobering Thoughts* [New York: Harper & Row, 1978], xvi). Also see Mark J. Rozell, "George F. Will's 'Tory' Conservatism," in Mark J. Rozell and James F. Pontuso, eds., *American Conservative Opinion Leaders* (Boulder: Westview Press, 1990), 13–28.

Chapter 7

1. Underreported but not ignored completely. See, for example, Edward C. Banfield, *The Moral Basis of a Backward Society* (New York: Free Press, 1958); and Kai T. Erikson, *Everything in Its Path: Destruction of Community in the Buffalo Creek Flood* (New York: Simon & Schuster, 1976). As both of these examples suggest, the study of fatalism is greatly dependent on the methods of participant observation and interviewing. This, however, is of no help to the historian concerned with past historical eras in which no participants survive.

2. Kenneth M. Stampp, *The Peculiar Institution: Slavery in the Ante-Bellum South* (New York: Knopf, 1956), 430.

3. Gilbert Osofsky, ed., *Puttin' On Ole Massa* (New York: Harper & Row, 1969), 12.

4. Bertram Wyatt-Brown, *Yankee Saints and Southern Sinners* (Baton Rouge: Louisiana State University Press, 1985), 166, 167. Also see Eugene D. Genovese, *Roll, Jordan, Roll: The World the Slaves Made* (New York: Pantheon, 1974), 70–75; and Richard Ellis and Aaron Wildavsky, "A Cultural Analysis of the Role of Abolitionists in the Coming of the Civil War," *Comparative Studies in Society and History* 32 (January 1990), 105–6.

5. See David Brion Davis, "Slavery and the Post–World War II Historians," *Daedalus*, Spring 1974, 1–16, esp. 2; Stanley M. Elkins, *Slavery: A Problem in American Institutional and Intellectual Life,* 3rd ed. (Chicago: University of Chicago Press, 1976), 9ff; and Eugene D. Genovese, "Ulrich Bonnell Phillips & His Critics," in Genovese, *In Red and Black: Marxian Explorations in Southern and Afro-American History* (New York: Vintage, 1972), esp. 261.

6. Quoted in Eugene D. Genovese, "Race and Class in Southern History: An Appraisal of the Work of Ulrich Bonnell Phillips," in Genovese, *In Red and Black*, 291–92.

7. Ulrich Bonnell Phillips, *American Negro Slavery* (Baton Rouge: Louisiana State University Press, 1966), 287, 309.

8. Phillips, *American Negro Slavery*, 291, 324.

9. Phillips, *American Negro Slavery*, 327–28.

10. Phillips, *American Negro Slavery*, 329.

11. Phillips, *American Negro Slavery*, 308.

12. Richard Hofstadter, "U. B. Phillips and the Plantation Legend," *Journal of Negro History* 29 (April 1944), 124.

13. Frederick Douglass, *My Bondage and My Freedom* (New York: Dover, 1969; first published 1855), 150.

14. Stampp, *Peculiar Institution*, 141.

15. Stampp, *Peculiar Institution*, 341.

16. Elkins, *Slavery*, 37. The influence of the consensus theory of Richard Hofstadter (Elkins's doctoral adviser) and Louis Hartz is very much in evidence here. Also important is Frank Tannenbaum, *Slave and Citizen: The Negro in the Americas* (New York: Vintage, 1946).

17. Elkins, *Slavery*, 93.

18. Elkins, *Slavery*, 101.

19. George M. Fredrickson and Christopher Lasch, "Resistance to Slavery," *Civil War History* 13 (December 1967), 315–29.

20. Elkins, *Slavery*, 23.

21. Elkins, *Slavery*, 86.

22. Robert William Fogel and Stanley L. Engerman, *Time on the Cross: The Economics of American Negro Slavery* (Boston: Little, Brown, 1974), 231, 232.

23. Fogel and Engerman, *Time on the Cross*, 231, 232, 41, 149.

24. Fogel and Engerman, *Time on the Cross*, 258.

25. Paul A. David and Peter Temin, "Capitalist Masters, Bourgeois Slaves," in Paul A. David, Herbert G. Gutman, Richard Sutch, Peter Temin, and Gavin Wright, *Reckoning with Slavery: Critical Essays in the Quantitative History of American Negro Slavery* (New York: Oxford University Press, 1976), 38; also see 48, 56, 74, 97, 340–43. Also see Elkins, *Slavery*, 300.

26. Fogel and Engerman, *Time on the Cross*, 145. Herbert G. Gutman and Richard Sutch, "Sambo Makes Good, or Were Slaves Imbued with the Protestant Work Ethic," in David et al., *Reckoning with Slavery*, 58–59.

27. Gutman and Sutch, "Sambo Makes Good," 59.

28. Douglass, *My Bondage*, 81–82.

29. Fogel and Engerman, *Time on the Cross*, 232.

30. Stampp, *Peculiar Institution*, 177.

31. Douglass, *My Bondage*, 219; also see 205ff.

32. Stampp, *Peculiar Institution*, 164, 165.

33. George P. Rawick, *From Sundown to Sunup: The Making of the Black Community* (Westport, Conn.: Greenwood Publishing, 1972), xix.

34. John W. Blassingame, *The Slave Community: Plantation Life in the Antebellum South*, rev. ed. (New York: Oxford University Press, 1979), 105–6.

35. Thomas L. Webber, *Deep Like the Rivers: Education in the Slave Quarter Community, 1831–1865* (New York: Norton, 1978), xii–xiii, 205–6, 215, 222, 223, 149. Angela Davis agrees that "domestic life in the slave quarters" was characterized by "sexual equality." Slaves, Davis continues, "transformed that negative equality which emanated from the equal oppression they suffered as slaves into a positive equality: the egalitarianism characterizing their social relations" (*Women, Race, and Class* [New York: Random House, 1981], 18).

36. Genovese, *Roll, Jordan, Roll*, 472, 659. Other quotations by Genovese are taken from David Herbert Donald's review of *Roll, Jordan, Roll* ("Writing About Slavery," *Commentary* 59 [January 1975], 87). The quotation by Stampp is from Kenneth M. Stampp, "A Humanistic Perspective," in David et al., *Reckoning with Slavery*, 15. Also see George M. Fredrickson, *The Black Image in the White Mind* (New York: Harper & Row, 1971), 101–2.

37. Rawick and Levine, quoted in Elkins, *Slavery*, 280–81.

38. See Elkins, *Slavery*, 285–86; and Robert William Fogel, *Without Consent or Contract: The Rise and Fall of American Slavery* (New York: Norton, 1989), 187.

39. Isaiah Berlin, "Russian Populism," reprinted in Berlin, *Russian Thinkers* (Harmondsworth: Penguin, 1978), 211.

40. Paul A. David and Peter Temin, "Slavery: The Progressive Institution?," in David et al., *Reckoning with Slavery*, 170.

41. The importance of the size of the plantation is brought out in Fogel, *Without Consent or Contract*, esp. 182–85.

42. Peter Kolchin, *Unfree Labor: American Slavery and Russian Serfdom* (Cambridge: Harvard University Press, 1987), 196, 200, 205–7.

43. Kolchin, *Unfree Labor*, 233, 156; also see 357.

44. See Kenneth M. Stampp, "Rebels and Sambos: The Search for the Negro's Personality in Slavery," *Journal of Southern History* 37 (August 1971), 367–92; Stampp, "A Humanistic Perspective," 29; and Blassingame, *Slave Community*.

45. Eugene D. Genovese, "Rebelliousness and Docility in the Negro Slave: A Critique of the Elkins Thesis," reprinted in *In Red and Black*, 95. Elkins was not unaware of this. See Elkins, *Slavery*, 86–87, 103–4, 137–38.

46. Blassingame, *Slave Community*, 320. Also see Elkins, *Slavery*, 279.

47. Stampp, *Peculiar Institution*, 149–50, 209.

48. Stampp, *Peculiar Institution*, 149–51.

49. Stampp, *Peculiar Institution*, 73. Kolchin, *Unfree Labor*, 348–49.

50. Kolchin, *Unfree Labor*, 201.

51. Kolchin, *Unfree Labor*, 331.

52. Erving Goffman, *Asylums: Essays on the Social Situation of Mental Patients and Other Inmates* (Garden City, N.Y.: Anchor, 1961), 60. In *The Prison Community* (New York: Holt, Rinehart and Winston, 1958), Donald Clemer found that "The prisoner's world is an atomized world. . . . Trickery and dishonesty overshadow sympathy and cooperation. . . . It is world of 'I,' 'me,' and 'mine,' rather than 'ours,' 'theirs,' and 'his'" (297–98).

Also see Michael Flusche, "Joel Chandler Harris and the Folklore of Slavery," *Journal of American Studies* 9 (December 1975), 362; and Fredrickson and Lasch, "Resistance to Slavery."

53. Kolchin, *Unfree Labor*, 278. Douglass, *My Bondage*, 269. Osofsky, *Puttin' On Ole Massa*, 19.

54. Kolchin, *Unfree Labor*, 289, 286. Also see "Narrative of the Life and Adventures of Henry Bibb, an American Slave," in Osofsky, *Puttin' On Ole Massa*, 72.

55. Kolchin, *Unfree Labor*, 289. Gerald W. Mullin, *Flight and Rebellion: Slave Resistance in Eighteenth-Century Virginia* (New York: Oxford University Press, 1972), 34.

56. Kolchin, *Unfree Labor*, 291.

57. Blassingame, *Slave Community*, 127–28.

58. Flusche, "Folklore of Slavery," 358–60.

59. Flusche, "Folklore of Slavery," 359.

60. Flusche, "Folklore of Slavery," 360.

61. Flusche, "Folklore of Slavery," 358.

62. Elkins, *Slavery*, 283.

63. Elkins, *Slavery*, 283.

64. Flusche, "Folklore of Slavery," 360–61.

65. The term "amoral individualism" is from Banfield, *Moral Basis of a Backward Society*, 83.

66. "Narrative of William Wells Brown, a Fugitive Slave," in Osofsky, *Puttin' On Ole Massa*, 198–200.

67. Flusche, "Folklore of Slavery," 361.

68. Kolchin, *Unfree Labor*, 230–32.

69. Flusche, "Folklore of Slavery," 361.

70. Genovese, *Roll, Jordan, Roll*, 638.

71. Douglass, *My Bondage*, 34.

72. Douglass, *My Bondage*, 175.

73. Douglass, *My Bondage*, 184, 183.

74. Because fatalism is a learned response to a hostile and arbitrary social environment, we should not be surprised to find, as Stampp does, that "the majority [of slaves], as they grew older, lost hope and spirit" (*Peculiar Institution*, 91).

75. This thesis is developed in the work of Paul Veyne and discussed in Jon Elster, *Making Sense of Marx* (Cambridge: Cambridge University Press, 1985), 20–21, 505. Also see Michael Thompson, Richard Ellis, and Aaron Wildavsky, *Cultural Theory* (Boulder: Westview Press, 1990), 153.

76. "And so," Stampp goes on to explain, a slave "tended to be a fatalist and futilitarian, for nothing else could reconcile him to his life" (*Peculiar Institution*, 361).

77. Douglass, *My Bondage*, 36.

78. Kolchin, *Unfree Labor*, 351.

79. "Narrative of William Wells Brown."

80. Genovese, *Roll, Jordan, Roll*, 638.

81. Stampp, *Peculiar Institution*, 149.

82. Stampp, "Rebels and Sambos," 386.

83. Many slaveholders who understood the reciprocal relation between slavery and fatalistic submission vehemently opposed hiring out. "No higher evidence can be furnished of [hiring out's] baneful effects," wrote a South Carolinian, "than the unwillingness it produces in the slave, to return to the regular life and domestic control of the master" (Stampp, *Peculiar Institution*, 147).

84. Elkins, *Slavery*, 137. Kolchin, *Unfree Labor*, 348. Stampp, *Peculiar Institution*, 147.

85. Douglass, *My Bondage*, 142, 145, 147, 157–59, 138–39, 272.

86. Douglass, *My Bondage*, 279.

87. Elkins, *Slavery*, 138.

88. Elkins, *Slavery*, 138. Fredrickson and Lasch, "Resistance to Slavery," 225. Kolchin, *Unfree Labor*, 318, 255.

89. Much confusion has surrounded the relationship between suicide and fatalism. Genovese has quite correctly rejected the claim that suicide represented a form of resistance to slavery. There is no evidence that slaves understood their suicides in this way. But if this interpretation of suicide as resistance is unfounded, Genovese errs in drawing attention to the low levels of suicide among slaves as a warning against "hasty generalizations" about fatalism (*Roll, Jordan, Roll*, 639). For this neglects, as Durkheim showed, that there are many social routes to suicide; fatalistic suicide is not only not the only route to suicide but may not even be the most common one. A cultural analysis of Durkheim's discussion of suicide may be found in Thompson, Ellis, and Wildavsky, *Cultural Theory*, 137–40.

90. Robert Smalls, in John W. Blassingame, ed., *Slave Testimony: Two Centuries of Letters, Speeches, Interviews, and Autobiographies* (Baton Rouge: Louisiana State University Press, 1977), 377. Also see Fogel, *Without Consent or Contract*, 197.

91. Peter Kolchin writes that "although most antebellum slaves longed for freedom, they had little reason to believe that it was attainable" (*Unfree Labor*, 320), and Kenneth Stampp explains that for "the masses of slaves . . . freedom could have been little more than an idle dream" (*Peculiar Institution*, 97).

92. "Narrative of the Life and Adventures of Henry Bibb," 69.

93. Banfield, *Moral Basis of a Backward Society*, 64, 58–59, 88, 109. Also see the analysis of Banfield in Thompson, Ellis, and Wildavsky, *Cultural Theory*, 223–27.

Chapter 8

1. John P. Diggins, "Thoreau, Marx, and the 'Riddle' of Alienation," *Social Research* 39 (Winter 1972), 571.

2. Stanley Edgar Hyman, "Henry Thoreau in Our Time," reprinted in Wendell Glick, ed., *The Recognition of Henry David Thoreau: Selected Criticism Since 1848* (Ann Arbor: University of Michigan Press, 1969), 338.

3. Sacvan Bercovitch, *The American Jeremiad* (Madison: University of Wisconsin Press, 1978), 187.

4. Henry David Thoreau, *Walden*, ed. J. Lyndon Shanley (Princeton: Princeton University Press, 1971), 19, 328. Subsequent citations refer to this edition.

5. Henry Seidel Canby, *Thoreau* (Boston: Houghton Mifflin, 1939), 447.

6. Staughton Lynd, *Intellectual Origins of American Radicalism* (New York: Pantheon, 1968), 94.

7. Lynd, *Intellectual Origins*, 95.

8. Karl Marx, "On the Jewish Question," in *Early Writings*, tr. G. Benton (Penguin: Harmondsworth, 1975), 229.

9. Leonard N. Neufeldt, "Henry David Thoreau's Political Economy," *New England Quarterly* 57 (September 1984), 361.

10. Richard Drinnon, "Thoreau's Politics of the Upright Man," reprinted in Glick, *Recognition of Thoreau*, 374.

11. Louis Hartz, *The Liberal Tradition in America* (New York: Harcourt, 1955). Seymour Martin Lipset, *The First New Nation: The United States in Historical and Comparative Perspective* (New York: Basic Books, 1963). Samuel P. Huntington, *American Politics: The Promise of Disharmony* (Cambridge: Harvard University Press, 1981).

12. Mary Douglas, "Cultural Bias," in Douglas, *In the Active Voice* (London: Routledge & Kegan Paul, 1982), 232–33. Indispensable for an understanding of the hermit or autonomous category is Michael Thompson, "The Problem of the Centre: An Autonomous Cosmology," in Mary Douglas, ed., *Essays in the Sociology of Perception* (London: Routledge & Kegan Paul, 1982). Also see Michael Thompson, Richard Ellis, and Aaron Wildavsky, *Cultural Theory* (Boulder: Westview Press, 1990). Douglas discusses Thoreau as an example of the hermit category in "Cultural Bias," 234–38.

13. Joseph Wood Krutch, *Henry David Thoreau* (New York: William Sloane, 1948), 77. Brook Farm was an egalitarian community established in the 1840s just outside Boston. All residents received the same wages, worked the same number of hours, and paid the same room and board.

14. Thoreau, *Walden*, 3.

15. Ralph Henry Gabriel, *The Course of American Democratic Thought: An Intellectual History Since 1815* (New York: Ronald Press, 1940), 47. Also see Robert A. Gross, "Culture and Cultivation: Agriculture and Society in Thoreau's Concord," *Journal of American History* 69 (June 1982), 42–55.

16. Thoreau, *Walden*, 7.

17. Thoreau, *Walden*, 71, 72, 171.

18. Lawrence J. Friedman, *Gregarious Saints: Self and Community in American Abolitionism, 1830–1870* (Cambridge: Cambridge University Press, 1982), 44. Thoreau, *Walden*, 130.

19. Thoreau, *Walden*, 135.

20. Friedman, *Gregarious Saints*, 66. Thoreau, *Walden*, 144.

21. Douglas, "Cultural Bias," 234.

22. Thoreau, *Walden*, 323, 267–68, 111, 310. A perceptive discussion of Thoreau's conception of friendship can be found in Diggins, "Thoreau, Marx, and Alienation," 580–81.

23. Thompson, Ellis, and Wildavsky, *Cultural Theory*, chap. 2.

24. Thoreau, *Walden*, 9, 21. Ralph Waldo Emerson, "Thoreau," in Walter Harding, ed., *Thoreau: A Century of Criticism* (Dallas: Southern Methodist University Press, 1954), 25.

25. Thoreau, *Walden*, 91, 28.

26. Po Chu-i, "A Mad Poem Addressed to My Nephews and Nieces," *Chinese Poems*, tr. Arthur Waley (London: George Allen & Unwin, 1976), 185. Reprinted by permission of Harper Collins.

27. Thoreau, *Walden*, 5, 68, 16.

28. Thoreau, *Walden*, 36, 69.

29. Thoreau, *Walden*, 15, 36.

30. Thoreau, *Walden*, 329, 14.

31. Thoreau, *Walden*, 14, 329.

32. Thoreau, *Walden*, 70, 19.

33. Thoreau, *Walden*, 115, 64, 63, 329–30.

34. Thoreau, *Walden*, 208, 196, 199, 173.

35. Thoreau, *Walden*, 6–7, 46.

36. Thoreau, *Walden*, 129, 131–32, 86, 131, 85, 42, 157.

37. Thoreau, *Walden*, 138, 210, 228–32, 228.

38. Thoreau, *Walden*, 294.

39. On the social construction of memory, see Mary Douglas's corpus, especially *How Institutions Think* (Syracuse: Syracuse University Press, 1986), her introduction to Maurice Halbwachs, *The Collective Memory* (New York: Harper & Row, 1980), and "Three Types of Public Memory," typescript. On the hermit's conception of time, see Douglas, "Cultural Bias," 234; and Thompson, "The Problem of the Centre," 318, 320.

40. Thoreau, *Walden*, 97, 314, 88. Also see R. W. B. Lewis, *The American Adam* (Chicago: University of Chicago Press, 1955), 20–27.

41. Thoreau, *Walden*, 9, 10–11, 57.

42. Thoreau, *Walden*, 54, 5, 326, 97.

43. Thoreau, *Walden*, 326–27.

44. Thoreau, *Walden*, 111–12.

45. Thoreau, *Walden*, 95, 96, 92, 96, 22, 21, 52.

46. Thoreau, *Walden*, 96, 97–98.

47. Quoted in Michael Meyer, "Introduction" to Henry David Thoreau, *Walden and Civil Disobedience*, ed. Michael Meyer (Harmondsworth: Penguin, 1983), 14.

48. Thoreau, *Walden*, 321.

49. Emerson, "Thoreau," 25, 30, 37.

50. Thoreau, *Walden*, 96.

51. Emerson wrote of Thoreau: "He understood the matter in hand at a glance, and saw the limitations and poverty of those he talked with, so that nothing seemed concealed from such terrible eyes" ("Thoreau," 30).

52. Thoreau, *Walden*, 330, 6, 331, 332, 5.

53. Thoreau, *Walden*, 84, 74, 71.

54. Meyer, "Introduction," 34.

55. The famed Tibetan hermit Milarepa, who lived in a Himalayan cave on naught but boiled nettles, bumped up against the physical limitations on autonomy. In an attempt to rid himself of his last ties with the outside world, he deliberately broke the pot in which he cooked his nettles. The result, notes Michael Thompson, was to destroy "his already tenuous life-support system" ("A Three-Dimensional Model," in Douglas, ed., *Essays in the Sociology of Perception*, 37).

56. This quotation appears on the dust jacket of *Walden and Other Writings of Henry David Thoreau*, ed. Brooks Atkinson (New York: Modern Library, 1965).

Chapter 9

1. A similar understanding is reached by Thomas Sowell in *A Conflict of Visions: Ideological Origins of Political Struggles* (New York: Quill, 1987), esp. 224–25.

2. As Isaiah Berlin stresses, "where ends are agreed, the only questions left are those of means, and these are not political but technical, that is to say, capable of being settled by experts or machines like arguments between engineers or doctors" ("Two Concepts of Liberty," in Berlin, *Four Essays on Liberty* [New York: Oxford University Press, 1969], 118; also see Berlin, *The Crooked Timber of Humanity: Chapters in the History of Ideas* [New York: Knopf, 1991], esp. 1–19; and Bernard Crick, *In Defense of Politics*, 2nd ed. [Harmondsworth: Penguin, 1982]).

3. For a modest attempt to do something like this, see Michael Thompson, Richard Ellis, and Aaron Wildavsky, *Cultural Theory* (Boulder: Westview Press, 1990), chap. 15, entitled "Hard Questions, Soft Answers." Also see Richard J. Ellis, "The Case For Cultural Theory: A Reply to Friedman," *Critical Review* (forthcoming).

4. Here I have heeded Bernard Barber's warning that "one tendency in new schools is to pay more attention to the development of ideas than to communicating them as effectively as possible to others [and defining] . . . for others how they are a continuation of older developments" ("Structural-Functional Analysis: Some Problems and Misunderstandings," *American Sociological Review* 21 [April 1956], 129).

5. The term was coined by Robert E. Shalhope in "Toward a Republican Synthesis:

The Emergence of an Understanding of Republicanism in American Historiography," *William and Mary Quarterly* 29 (January 1972), 49–80.

6. John M. Murrin, "Gordon S. Wood and the Search for Liberal America," *William and Mary Quarterly* 44 (July 1987), 600. Gordon S. Wood, *The Creation of the American Republic, 1776–1787* (Chapel Hill: University of North Carolina Press, 1969), 606. Michael Lienesch, *New Order of the Ages: Time, the Constitution, and the Making of Modern American Political Thought* (Princeton: Princeton University Press, 1988), 3. Stephen Watts, *The Republic Reborn: War and the Making of Liberal America, 1790–1820* (Baltimore: Johns Hopkins University Press, 1987), xvii. Daniel Joseph Singal, "Beyond Consensus: Richard Hofstadter and American Historiography," *American Historical Review* 89 (October 1984), 1000. Also see Gordon S. Wood, "Ideology and the Origins of Liberal America," *William and Mary Quarterly* 44 (July 1987), 634–40; and Jean V. Matthews, *Toward a New Society: American Thought and Culture, 1800–1830* (Boston: Twayne, 1991). More ambiguous is John L. Brooke's recent study of Massachusetts political culture from the beginning of the eighteenth century up to the Civil War. Brooke "finds ample room for both republican and liberal visions in the narrow confines of Worcester County," but he admits that "in broad overview, [his] study describes a grand reversal of ideological fortunes, hinging on the triumph of Lockean constitutionalism in revolutionary Massachusetts" (*The Heart of the Commonwealth: Society and Political Culture in Worcester County, Massachusetts, 1713–1861* [Cambridge: Cambridge University Press, 1989], xiii).

7. Isaac Kramnick, "Republican Revisionism Revisited," *American Historical Review* 87 (June 1982), 664. An exception is John Patrick Diggins, who denies that classical republicanism ever played an important role in American political thought (*The Lost Soul of American Politics: Virtue, Self-Interest, and the Foundations of Liberalism* [Chicago: University of Chicago Press, 1984]).

8. John M. Murrin writes, "all of us have insisted, and still do insist, that there was a transition, a before and after. We do not believe that America was born modern. . . . We probably do agree that the shift to modernity was virtually complete by the 1820s" ("Self-Interest Conquers Patriotism: Republicans, Liberals, and Indians Reshape the Nation," in Jack P. Greene, *The American Revolution: Its Character and Its Limits* [New York: New York University Press, 1987], 226). A notable exception to this generalization is labor historians, many of whom have argued that republicanism, rooted in the social experience of the artisan and small farmer, continued to provide an alternative cultural tradition to liberalism throughout the nineteenth century. See the sources cited in Daniel T. Rodgers, "Republicanism: The Career of a Concept," *Journal of American History* 79 (June 1992), esp. 26–31. Rodgers shows how classical republicanism's conceptual life is extended forward in time by stretching the original concept of classical republicanism beyond all recognition. For an early effort to trace classical republican ideas well into the latter half of the nineteenth century, see Dorothy Ross, "The Liberal Tradition Revisited and the Republican Tradition Addressed," in John Higham and Paul K. Conkin, eds., *New Directions in American Intellectual History* (Baltimore: Johns Hopkins University Press, 1979), 116–31.

9. See Thompson, Ellis, and Wildavsky, *Cultural Theory*, esp. pt. 2. A critique of the tradition–modernity dichotomy is developed at greater length in *Cultural Theory*, esp. at 104, 163, 166, 184, 186.

10. See Lance Banning, "Jeffersonian Ideology Revisited: Liberal and Classical Ideas in the New American Republic," *William and Mary Quarterly* 43 (January 1986), 3–19, and Joyce Appleby, "Republicanism in Old and New Contexts," ibid., 20–34, as well as Lance Banning, *The Jeffersonian Persuasion: Evolution of a Party Ideology* (Ithaca: Cornell

University Press, 1978), and Joyce Appleby, *Capitalism and a New Social Order: The Republican Vision of the 1790s* (New York: New York University Press, 1984).

11. See Richard Ellis and Aaron Wildavsky, *Dilemmas of Presidential Leadership: From Washington Through Lincoln* (New Brunswick: Transaction, 1989), chap. 4.

12. Banning, "Jeffersonian Ideology Revisited," 17.

13. Jefferson, quoted in Robert M. Johnstone, Jr., *Jefferson and the Presidency: Leadership in the Young Republic* (Ithaca: Cornell University Press, 1978), 45.

14. One of the few books to pay systematic attention to the ideological divisions within the Jeffersonian party is Richard E. Ellis, *The Jeffersonian Crisis: Courts and Politics in the Young Republic* (New York: Oxford University Press, 1971). Also suggestive in this regard is John Ashworth, "The Jeffersonians: Classical Republicans or Liberal Capitalists?" *Journal of American Studies* 18 (December 1984), 425–35.

15. Daniel Walker Howe, "European Sources of Political Ideas in Jeffersonian America," *Reviews in American History* 10 (December 1982), 34. Also see Harry S. Stout's insistence on the need to distinguish between "the classical (deferential) theory of republicanism" and "the more radical egalitarian 'republicanism'" ("Religion, Communications, and the Ideological Origins of the American Revolution," *William and Mary Quarterly* 34 [October 1977], 524n).

16. See Jeffrey Isaac, "Republicanism vs. Liberalism? A Reconsideration," *History of Political Thought* 9 (Summer 1988), 349–77; James T. Kloppenberg, "The Virtues of Liberalism: Christianity, Republicanism, and Ethics in Early American Political Discourse," *Journal of American History* 74 (June 1987), 9–33; and J. David Greenstone, "Political Culture and American Political Development: Liberty, Union, and the Liberal Bipolarity," *Studies in American Political Development* 1 (1986), 6. Given the diversity of cultural biases to be found within what is termed "classical republicanism," it is little wonder that scholars have found evidence of classical republican ideas in some of the most improbable places—from working-class radicalism (see Sean Wilentz, *Chants Democratic: New York City and the Rise of the American Working Class, 1788–1850* [New York: Oxford University Press, 1984]) to conservative Whiggery (Daniel Walker Howe, *The Political Culture of the American Whigs* [Chicago: University of Chicago Press, 1979]) to (perish the thought) Adam Smith (see Donald Winch, *Adam Smith's Politics: An Essay in Historiographic Revision* [Cambridge: Cambridge University Press, 1978]).

17. Isaac Kramnick, "Republican Revisionism Revisited," 662.

18. The term "Lockean monolith" is used by J. G. A. Pocock in "Between Gog and Magog: The Republican Thesis and the *Ideologia Americana*," *Journal of the History of Ideas* 48 (April–June 1987), 341.

19. Bernard Bailyn, "New England and a Wider World: Notes on Some Central Themes of Modern Historiography," in David D. Hall and David Grayson Allen, eds., *Seventeenth-Century New England* (Charlottesville: University Press of Virginia, 1984), 328. Perry Miller and Thomas H. Johnson, *The Puritans* (New York: Harper & Row, 1963), 5. T. H. Breen extends this line of thought, suggesting that the aim of the "ethnohistorian" is to "interpret the cultural 'conversations' of the past in terms that the participants themselves would have comprehended" ("Creative Adaptations: Peoples and Cultures," in Jack P. Greene and J. R. Pole, eds., *Colonial British America: Essays in the New History of the Early Modern Era* [Baltimore: Johns Hopkins University Press, 1984], 197).

20. Geoffrey Blodgett, "A New Look at the Gilded Age: Politics in a Cultural Context," in Daniel Walker Howe, ed., *Victorian America* (Philadelphia: University of Pennsylvania Press, 1976), 101–2. Similarly, in his study of New England Puritanism, David D. Hall explains that he "tried to view the preachers' rhetoric from within, to reconstruct the way they saw themselves and their values in relation to society" (*The Faithful Shep-*

herd: A History of the New England Ministry in the Seventeenth Century [Chapel Hill: University of North Carolina Press, 1972], xi). Hartz's methodological position on the need for "getting outside your subject" (*Liberal Tradition*, 29) runs counter to this Geertzian contextualist position that the analyst should strive to get inside the subject.

21. Skinner and Pocock's historicist approach, while marking a revolutionary departure in the field of political philosophy, was a commonplace to most practicing historians. "The new history of political thought," as David Wooton comments, represented little more than the belated "application of the methods and values of professional history to the history of ideas" ("Preface" to *Divine Rights and Democracy: An Anthology of Political Writings in Stuart England* [Harmondsworth: Penguin, 1986], 11–12). Little wonder that Skinner's contextualist analysis, like Geertz's "thick description," found a receptive audience among historians predisposed already toward the particular context.

22. Quentin Skinner, "A Reply to My Critics," in James H. Tully, ed. *Meaning and Context: Quentin Skinner and His Critics* (Princeton: Princeton University Press, 1989), 234.

23. Quentin Skinner, "Meaning and Understanding in the History of Ideas," *History and Theory* 8 (1969), 53.

24. Skinner, "Meaning and Understanding," 53.

25. J. G. A. Pocock, *Virtue, Commerce, and History: Essays on Political Thought and History, Chiefly in the Eighteenth Century* (Cambridge: Cambridge University Press, 1985), 8.

26. Skinner, "Meaning and Understanding," 52.

27. A similar note of caution was recently sounded by the historian Gordon S. Wood, who notes that "the new contextualist and antiquarian-minded history writing can be carried too far" ("Struggle Over the Puritans," *New York Review of Books*, November 9, 1989, 34).

28. Pocock, *Virtue, Commerce, and History*, 290. Isaac Kramnick, "The 'Great National Discussion': The Discourse of Politics in 1787," *William and Mary Quarterly* 45 (January 1988), 4.

29. Appleby, "Republicanism in Old and New Contexts," esp. 26–31, quotations on 31, 29.

30. Ruth Benedict, *Patterns of Culture* (Boston: Houghton Mifflin, 1934).

31. Margaret Mead, *Soviet Attitudes Toward Authority: An Interdisciplinary Approach to Problems of Soviet Character* (New York: McGraw-Hill, 1951). Ruth Benedict, *The Chrysanthemum and the Sword: Patterns of Japanese Culture* (Boston: Houghton Mifflin, 1946). Geoffrey Gorer, *The American People: A Study in National Character* (New York: Norton, 1948). Margaret Mead, *And Keep Your Powder Dry: An Anthropologist Looks at America* (New York: Morrow, 1942).

32. Ronald Walters, "Signs of the Times: Clifford Geertz and Historians," *Social Research* 47 (Autumn 1980), 556. Also relevant are Carl N. Degler, "Remaking American History," *Journal of American History* 67 (June 1980), esp. 13–17; and Thomas Bender, "Wholes and Parts: The Need for Synthesis in American History," *Journal of American History* 73 (June 1986), 120–36.

33. The material in this section is adapted from Ellis and Wildavsky, *Dilemmas of Presidential Leadership*, 222–25, with the permission of Transaction Publishers, New Brunswick, New Jersey.

34. Samuel P. Huntington, *American Politics: The Promise of Disharmony* (Cambridge: Harvard University Press, 1981).

35. Huntington, *American Politics*, 4.

36. Eric Foner, *Free Soil, Free Labor, Free Men: The Ideology of the Republican Party*

Before the Civil War (New York: Oxford University Press, 1970). Lawrence J. Friedman, *Gregarious Saints: Self and Community in American Abolitionism, 1830–1870* (Cambridge: Cambridge University Press, 1982). Richard Ellis and Aaron Wildavsky, "A Cultural Analysis of the Role of Abolitionists in the Coming of the Civil War," *Comparative Studies in Society and History* 32 (January 1990), 89–116.

37. Jason DeParle, "Eclipsed in the Reagan Decade, Ralph Nader Again Feels the Public Glare," *New York Times*, September 21, 1990, p. A11.

38. Huntington, *American Politics*, 96–97.

39. This distinction is made in Wilson Carey McWilliams, "On Equality as the Moral Foundation for Community," in Robert H. Horowitz, ed., *The Moral Foundations of the American Republic* (Charlottesville: University Press of Virginia, 1977), 183–213, esp. 185–86.

40. Huntington, *American Politics*, 41.

41. Huntington, *American Politics*, 172.

42. Seymour Martin Lipset, *The First New Nation: The United States in Historical and Comparative Perspective* (New York: Norton, 1979; originally published 1963).

43. Lipset, *The First New Nation*, xxxiii.

44. Lipset, *The First New Nation*, 102.

45. Lipset, *The First New Nation*, xxxiii.

46. Lipset, *The First New Nation*, 101.

47. Lipset, *The First New Nation*, 104, 106. The study cited is Lee Coleman, "What Is American? A Study of Alleged American Traits," *Social Forces* 19 (May 1941), 492–99.

48. For a more fully elaborated critique of the left–right distinction, see Ellis and Wildavsky, *Dilemmas of Presidential Leadership*, 3–5; and Aaron Wildavsky, "Choosing Preferences by Constructing Institutions: A Cultural Theory of Preference Formation," *American Political Science Review* 81 (March 1987), 3–21.

49. J. David Greenstone, "The Transient and the Permanent in American Politics: Standards, Interests, and the Concept of 'Public,'" in Greenstone, ed., *Public Values and Private Power in American Politics* (Chicago: University of Chicago Press, 1982), 6. Greenstone, "Political Culture and American Political Development," 2.

50. Greenstone, "Political Culture and American Political Development," 27. Berlin, "Two Concepts of Liberty."

51. Greenstone, "Political Culture and American Political Development," 29, 32.

52. Greenstone explains: "Although the central tenets of American liberalism are widely shared, they are interpreted in different ways. And these differences of interpretation may be of fundamental rather than secondary importance. Admittedly, there is no set of criteria that can indisputably distinguish between vital and trivial political differences. But in this case, we have competing conceptions of freedom and community as well as divergent positions on ethics, epistemology, and the human personality. And these alternatives would seem to form a fundamental, indeed bipolar, opposition, unless we simply stipulate that there can be no fundamental conflicts between two liberal or bourgeois social philosophies" (Greenstone, "Political Culture and American Political Development," 28).

53. Massachusetts governor John Davis, quoted in Ronald P. Formisano, *The Transformation of Political Culture: Massachusetts Parties, 1790s–1840s* (New York: Oxford University Press, 1983), 271.

54. Greenstone, "Political Culture and American Political Development," 33.

55. Hartz, *Liberal Tradition*, 60. Also see Greenstone, "Political Culture and American Political Development," 4.

56. Hartz, *Liberal Tradition*, 264, 260. It is also possible to defend Hartz's interpretation by keeping the focus on the U.S. position as a welfare state "laggard" (see, e.g.,

Anthony King, "Ideas, Institutions, and the Policies of Government: A Comparative Analysis," *British Journal of Political Science* 3 [July 1973], 191–313, and 3 [October 1973], 409–23). The problem with this view is that in the area of governmental regulation, especially regarding public health and safety, America is a leader rather than a laggard. Moreover, in many areas in which America has traditionally lagged behind Europe, the gap has been closing (see Charles Lockhart, *Gaining Ground: Tailoring Social Programs to American Values* [Berkeley: University of California Press, 1989]). Also, Richard Rose has pointed out that relative to Pacific Rim nations the U.S. welfare state is not exceptional ("How Exceptional Is the American Political Economy?" *Political Science Quarterly* 104 [Spring 1989], 91–115).

57. James Morone, *The Democratic Wish: Popular Participation and the Limits of American Government* (New York: Basic Books, 1990), 1, 4, 7.

58. Morone, *Democratic Wish*, 8.

59. Morone, *Democratic Wish*, 8, 29.

60. Morone, *Democratic Wish*, 336, 332. "In the radically different dense environment of contemporary government," Morone explains, "large interests and organizations— private and public, domestic and foreign—have grown interdependent. The actions of one affect numerous others, often those that seem only distantly related. Problems in such a setting are often not amenable to the simple striving of individual interests pressing ahead; there are too many consequences for too many others. Moreover, it is more difficult for government agents to assist one interest without repercussions for others. *Decision making in such settings is far more complicated, marked by multiple and often unforeseen consequences*" (330–31). Morone's premise is that increasing complexity and uncertainty increase the need for long-term planning and thus greater state autonomy. He fails to confront the counterargument that incrementalism becomes more rather than less necessary as the policy environment becomes denser and more unpredictable. It is precisely when we can't predict the consequences of our actions that incrementalism is arguably most rational. On this point, see James D. Thompson and Arthur Tuden, "Strategies, Structures, and Processes of Organizational Decisions," in Thompson, ed., *Comparative Studies in Administration* (Pittsburgh: University of Pittsburgh Press, 1959).

61. See, e.g., Morone, *Democratic Wish*, 9, 137, 141, 272, 275, 331.

62. Morone, *Democratic Wish*, 329.

63. Morone, *Democratic Wish*, 336–37. "The task for the future," Morone writes, "is to replace the longing for imaginary community with a communal imagination that is part of the government" (30).

64. See Thompson, Ellis, and Wildavsky, *Cultural Theory*, 89–90.

65. Morone, *Democratic Wish*, 323, 142, 29.

66. Aaron Wildavsky, "A Cultural Theory of Leadership," in Bryan D. Jones, ed., *Leadership and Politics: New Perspectives in Political Science* (Lawrence: University Press of Kansas, 1989). Also see Richard J. Ellis, "Explaining the Occurrence of Charismatic Leadership in Organizations," *Journal of Theoretical Politics* 3 (July 1991), 305–19.

67. James Q. Wilson, *Political Organizations* (New York: Basic Books, 1973), 324.

68. Aaron Wildavsky, "A World of Difference—The Public Philosophies and Political Behaviors of Rival American Cultures," in *The New American Political System* (Washington, D.C.: AEI Press, 1990), 274; also see Wildavsky, "The Three Cultures: Explaining Anomalies in the American Welfare State," *Public Interest* 69 (Fall 1982), 45–58.

69. See Aaron Wildavsky, "Resolved, that Individualism and Egalitarianism Be Made Compatible in America: Political-Cultural Roots of Exceptionalism," in Wildavsky, *The Rise of Radical Egalitarianism* (Washington, D.C.: American University Press, 1991), 29–48; and Ellis and Wildavsky, *Dilemmas of Presidential Leadership*.

70. Jesse Jackson, for instance, insists that the middle class "really means a relatively small group who traditionally vote. The [Democrats'] focus must be on the unregistered who do not vote, those who did not chose to vote for Dukakis or Bush. That is our strength" (Thomas Edsall, "Clinton Is Managing to Play Both Ends Against the Middle," *Washington Post Weekly*, January 27–February 2, 1992). After the Clarence Thomas hearings, former lieutenant governor of Missouri Harriet Woods, speaking for the National Women's Political Caucus, said that "there's a new phenomenon in this country called political homelessness because people in this country have lost faith in their government" (Bob Dart, "Thomas Picks Up Votes as Hearings End," *The Oregonian*, September 21, 1991, A11).

71. Morone, *Democratic Wish*, 337.

72. The material in this section is adapted from Thompson, Ellis, and Wildavsky, *Cultural Theory*, chap. 13, with permission of Westview Press, Boulder, Colorado.

73. Morone, *Democratic Wish*, 18.

74. Daniel J. Elazar, *American Federalism: A View from the States*, 2nd ed. (New York: Crowell, 1972), 94. Also see Elazar, *Cities on the Prairie: The Metropolitan Frontier and American Politics* (New York: Basic Books, 1970); Elazar and Joseph Zikmund, II, eds., *The Ecology of American Political Culture: Readings* (New York: Crowell, 1975); and Elazar, *Cities of the Prairie Revisited: The Closing of the Metropolitan Frontier* (Lincoln: University of Nebraska Press, 1986).

75. Elazar, *American Federalism*, 99.

76. Elazar, *American Federalism*, 90–91.

77. Elazar himself uses the term "communalism" in giving us an incisive portrait of the difference between egalitarianism and hierarchy. "The distinction they [moralists] make (implicitly at least)," Elazar explains, "is between what they consider legitimate community responsibility and what they believe to be central government encroachment, or between 'communalism' which they value and 'collectivism' which they abhor" (*American Federalism*, 97n10). He also posits that the major difficulty faced by adherents of the moralistic culture "in adjusting bureaucracy to the political order is tied to the potential conflict between communitarian principles and the necessity for large-scale organization" (98).

78. Elazar, *American Federalism*, 97.

79. On the hierarchical nature of Federalist political culture in Massachusetts, see James M. Banner, Jr., *To the Hartford Convention: The Federalists and the Origins of Party Politics in Massachusetts, 1789–1815* (New York: Knopf, 1970); David Hackett Fischer, "The Myth of the Essex Junto," *William and Mary Quarterly* 21 (April 1964), 191–235; and Formisano, *Transformation of Political Culture*.

80. Daniel Walker Howe, *The Political Culture of the American Whigs* (Chicago: University of Chicago Press, 1979), 29–30, and passim.

81. John L. Thomas, ed., *Slavery Attacked: The Abolitionist Crusade* (Englewood Cliffs: Prentice-Hall, 1965), 79; emphasis in original. Blanche Glassman Hersch, *The Slavery of Sex: Feminist Abolitionists in America* (Urbana: University of Illinois Press, 1978), 234, 225, 236. Lewis Perry, *Radical Abolitionism: Anarchy and the Government of God in Antislavery Thought* (Ithaca: Cornell University Press, 1973), 230. Ronald G. Walters, *The Antislavery Apeal: American Abolitionism After 1830* (Baltimore: Johns Hopkins University Press, 1976), 72.

82. Howe, *Political Culture of American Whigs*, 45, 86.

83. John Ashworth, *"Agrarians" and "Aristocrats": Party Political Ideology in the United States, 1837–1846* (Cambridge: Cambridge University Press, 1987), 54–55. Formisano, *Transformation of Political Culture*, 458n1.

84. Friedman, *Gregarious Saints*, 63–64, 27.

85. Ronald G. Walters, *The Antislavery Appeal: American Abolitionism After 1830* (Baltimore: Johns Hopkins University Press, 1976), 43.

86. Walters, *The Antislavery Appeal*, 41–43. Perry, *Radical Abolitionism*, 57, 105.

87. George A. Lipsky, *John Quincy Adams: His Theory and Ideas* (New York: Crowell, 1950), 82. Ashworth, *"Agrarians" and "Aristocrats,"* 60.

88. Thomas, "Antislavery and Utopia," 247. Ronald G. Walters, *American Reformers, 1815–1860* (New York: Hill & Wang, 1978), esp. 69–70.

89. V. O. Key, Jr., *Southern Politics in State and Nation* (New York: Knopf, 1949), 36. The absence of deference in antebellum Alabama is highlighted in J. Mills Thornton III's first-class study, *Power and Politics in a Slave Society: Alabama, 1800–1860* (Baton Rouge: Louisiana State University Press, 1978).

90. Uncertainty about where these categories come from and how they relate to each other is evident in the confusion over how to represent these three categories spatially. Perhaps the most common way, following Ira Sharansky, is to conceive of Elazar's categories as lying on a single continuum, running from moralism to traditionalism, with individualism as the midpoint (Ira Sharansky, "The Utility of Elazar's Political Culture: A Research Note," *Polity* 2 [Fall 1969], 66–83). David Haight agrees that Elazar's categories are best represented as constituting a single continuum, but he argues the continuum runs from moralism to individualism, with traditionalism as the midpoint (David Haight, "Collectivists, Particularists and Individualists: An Emerging Typology of Political Life," paper presented at the Annual Meeting of the Midwest Political Science Association, 1976). More recently, Frederick Wirt suggested a continuum running from traditionalism to individualism, with moralism as the midpoint (Frederick Wirt, "Does Control Follow the Dollar? Value Analysis, School Policy, and State-Local Linkages," *Publius* 10 [Spring 1980], 69–88). Elazar himself has waded in and tried to settle the debate by arguing that a single continuum is inappropriate, and that the relationship is best represented by a triangle (Daniel J. Elazar, "Afterword: Steps in the Study of American Political Culture," *Publius* 10 [Spring 1980], 129; and *Cities of the Prairie Revisited*, 86). No rationale is offered, however, for why political cultures arrange themselves in this triangular fashion.

91. David Harlan, "A People Blinded from Birth: American History According to Sacvan Bercovitch," *Journal of American History* 78 (December 1991), 962. Similarly, Daniel Walker Howe writes: "Bercovitch has, in effect, restated the thesis of Louis Hartz's *Liberal Tradition in America*, substituting Puritanism for the Enlightenment. . . . 'The jeremiad' seems to become a code word for American bourgeois culture in general, much as Hartz used 'Locke' as a kind of shorthand for the same thing" ("Descendants of Perry Miller," *American Quarterly* 34 [Spring 1982], 92). Strangely, Bercovitch pays little attention to Hartz in *The American Jeremiad* (Madison: University of Wisconsin Press, 1978), briefly quoting from him in only one footnote (xiiin). In truth it is unclear whether Bercovitch had even read Hartz, as the source of the one quote is C. Vann Woodward's *Burden of the South* (1968) rather than Hartz's original text.

92. Bercovitch, *American Jeremiad*, 20. By "middle class," Bercovitch explains, he means "the norms we have come to associate with the free-enterprise system" (xiii).

93. Bercovitch, *American Jeremiad*, 28, 19–20, 141.

94. For criticism of Bercovitch's portrayal of the European jeremiad, see Theodore Dwight Bozeman, *To Live Ancient Lives: The Primitivist Dimension in Puritanism* (Chapel Hill: University of North Carolina Press, 1988), 314n6.

95. Bercovitch, *American Jeremiad*, 23, 7.

96. Bercovitch, *American Jeremiad*, 209, 11, 31, 6, 8, xiv, 31.

97. Bercovitch, *American Jeremiad*, 23.

98. Bercovitch, *American Jeremiad*, 22.

99. Paul Ehrlich, Jacques Cousteau, Friends of the Earth, quoted in Mary Douglas and Aaron Wildavsky, *Risk and Culture: An Essay on the Selection of Technical and Environmental Dangers* (Berkeley: University of California Press, 1982), 131, 170.

100. Douglas and Wildavsky, *Risk and Culture*, 123.

101. Stephen Cotgrove, *Catastrophe or Cornucopia: The Environment, Politics and the Future* (Chichester: John Wiley, 1982), 5. Douglas and Wildavsky, *Risk and Culture*, 135–36.

102. Thompson, Ellis, and Wildavsky, *Cultural Theory*, chap. 1. Michael Thompson, "Socially Viable Ideas of Nature: A Cultural Hypothesis," in Erik Baark and Uno Svedin, eds., *Man, Nature, and Technology: Essays on the Role of Ideological Perceptions* (New York: St. Martin's Press, 1988), 57–79.

103. Sacvan Bercovitch, "Emerson, Individualism, and the Ambiguities of Dissent," *South Atlantic Quarterly* 89 (Summer 1990), 636. Sacvan Bercovitch, "The Problem of Ideology in American Literary History," *Critical Inquiry* 12 (Summer 1986), 645. Bercovitch, *American Jeremiad*, 150, 134, 179; also see 159–60, 180, 204–5.

104. Recently Bercovitch seemed to move in this direction, talking of "the multivalence of the symbol of America," the "discrepant value-systems inherent in the symbol of America," and describing "America" as "a symbol that has been made to stand for alternative and sometimes mutually antagonistic outlooks" ("Afterword," in Bercovitch and Myra Jehlen, eds., *Ideology and Classic American Literature* [Cambridge: Cambridge University Press, 1986], 428, 424, 418). But in even more recent writings ("Ambiguities of Dissent"; "Investigations of an Americanist," *Journal of American History* 78 [December 1991], 972–87; and "The A-Politics of Ambiguity in *The Scarlet Letter*," *New Literary History* 19 [Spring 1988], 629–54), Bercovitch shows that his understanding of American ideology and culture is still framed largely in terms of "hegemony" rather than "discrepant value-systems."

105. Bercovitch, *American Jeremiad*, xiii, 194.

106. Bercovitch, *American Jeremiad*, 155.

107. Similarly in Marxist theory, as Jon Elster points out, "all apparently innocent activities, from Sunday picnics to health care for the elderly, are explained through their function for capitalism" ("Marxism, Functionalism, and Game Theory: The Case for Methodological Individualism," *Theory and Society* 11 [July 1982], 456).

108. The problems with structural-functionalism are explored in greater depth in Thompson, Ellis, and Wildavsky, *Cultural Theory*, esp. pt. 2.

109. It should become clear that I disagree with that part of David Harlan's criticism that faults Bercovitch for analyzing Puritan texts for their function in society rather than for their enduring value ("A People Blinded from Birth"). These projects are not mutually exclusive, and I can see no reason why Bercovitch should not pursue the question of how a piece of literature functions within a given historical era, leaving others to pursue the question of what is of enduring value in Puritan thought (although it might be legitimately asked whether students of literature, given that their training is not in history, politics, or social theory, are best suited to address what are fundamentally political, sociological, and historical questions).

110. In addition, Bercovitch (like so many Marxist theorists) is far too attached to sentences in which purposes and intentions are ascribed to disembodied collectives like "the state" or "the dominant culture." Typical is the following passage: "The dominant culture adopts utopia for its own *purposes*. It does not simply endorse the trans-historical ideals of harmony and regeneration; it absorbs and redefines these in ways that support the social system. It molds the 'archetypes' of organic community to fit its distinctive system

of values. It re-creates the Eden of the 'racial unconscious' in its own image. It ritualizes the egalitarian energies of the liminal process in such a way as to control discontent and harness anarchy itself to the social enterprise" (Bercovitch, "Afterword," 433; emphasis added). For a powerful critique of this style of "lazy and frictionless thinking" and the fallacy of positing "a purpose without a purposive actor," see Elster, "Marxism, Functionalism, and Game Theory," 453, 454.

111. Bercovitch, *American Jeremiad*, xiv. Bercovitch approvingly quotes Myra Jehlen, who argues that in America alternative or oppositional ideologies have been "associated with the Old World and thus rejected by the very process of national emergence" (xiiin).

112. Perry Miller seems to me to have had a much surer grasp of this point than Bercovitch, probably because Miller had a much truer sense of the Puritans' ambivalence if not hostility toward a culture of competitive individualism. Although there are critical differences between Miller's and Bercovitch's analyses of Puritanism, they share ironically a deep alienation from the competitive individualism of contemporary American society. Miller was dismayed by "expanding, capitalist, exploiting America," and continually criticized its "ruthless individualism," its "profiteering merchants," and its "manufacturing and huckstering" (Harlan, "A People Blinded from Birth," 959). So, too, Bercovitch regards competitive capitalism in America with a good deal of distaste and sees his own position as one of "extreme marginality." But where Miller challenges capitalism from a more or less hierarchical vantage point, Bercovitch's critique emerges from egalitarianism, from "the Yiddishist–left wing world of my parents, . . . an outpost barricaded from the threat of assimilation by radical politics and belles lettres, an immigrant enclave locked into a romantic-Marxist utopianism" (Bercovitch, "Investigations of an Americanist," 973n1, 974). Both Miller and Bercovitch feel themselves to be outside the dominant competitive individualist tradition, but Miller looks to the Puritans for an alternative set of values with which to critique materialistic capitalism while Bercovitch sees Puritanism as the source of the problem—the very fountainhead of competitive capitalism.

113. Bercovitch, *American Jeremiad*, 179. Harlan, "A People Blinded from Birth," 965.

114. Bercovitch, *American Jeremiad*, 4, 198, 187.

115. As John Patrick Diggins points out, "the Tocqueville–Hartz thesis could readily be endorsed by Marx, Engels, Trotsky, and Gramsci" ("Knowledge and Sorrow: Louis Hartz's Quarrel with American History," *Political Theory* 16 [August 1988], 361; also see Diggins, "Comrades and Citizens: New Mythologies in American Historiography," *American Historical Review* 90 [June 1985], 614–38, and his "Reply," 639–49, to Paul Conkin's comment). The same point was made over two decades ago by Richard Hofstadter: "The idea of consensus is not intrinsically linked to ideological conservatism. In its origins I believe it owed almost as much to Marx as to Tocqueville" (*The Progressive Historians: Turner, Beard, Parrington* [New York: Knopf, 1968], 451). Indeed, one of the earliest postwar formulations of the consensus thesis is to be found in Hofstadter's *The American Political Tradition* (New York: Knopf, 1948), a work thoroughly grounded in the Marxism of the 1930s. Those on the left who have made use of Hartz's liberal consensus thesis include Michael Paul Rogin, *Fathers and Children: Andrew Jackson and the Subjugation of the American Indian* (New York: Knopf, 1975), and N. Gordon Levin, Jr., *Woodrow Wilson and World Politics: America's Response to War and Revolution* (New York: Oxford University Press, 1968). Bercovitch, too, has been misinterpreted by some as celebrating consensus (see, e.g., Nina Baym's review in *Nineteenth Century Fiction* 34 [December 1979], esp. 350; and Stephen J. Stein's review in *American Historical Review* 84 [October 1979], 1142), a misreading foretold with great prescience by Daniel Walker Howe in his review, "Descendents of Perry Miller," 92. Bercovitch attributes the "pattern of misap-

propriations of my work" to "the extreme marginality of my outlook" ("Investigations of an Americanist," 973). This is a somewhat puzzling if not absurd claim coming as it does from a scholar whose citations in the *Social Science Index* among students of early American history are second only to those of Perry Miller (Harlan, "A People Blinded from Birth," 952), and who was recently appointed general editor of the Cambridge History of American Literature.

116. Hartz, *Liberal Tradition*, 32.

117. Hartz, *Liberal Tradition*, 302.

118. Hartz, *Liberal Tradition*, 6. Huntington's criticism that "the consensus theory posits the uniformity of the prairie but not the fury of the tornado that the prairie's very flatness engenders" (*American Politics*, 11) is plainly wrong, at least as it relates to Hartz.

119. Hartz, *Liberal Tradition*, 10, 287, 308.

120. Hartz, *Liberal Tradition*, 11, 15, 308.

121. Hartz, *Liberal Tradition*, 287, 14.

122. Hartz, *Liberal Tradition*, 305.

123. Marvin Meyers is exactly right when he says that Hartz's *Liberal Tradition* is "a study of the unconscious mind of America, . . . the unarticulated premises of the society and culture" ("Louis Hartz, *The Liberal Tradition in America*: An Appraisal," *Comparative Studies in Society and History* 5 [April 1963], 264).

124. Ironically, Hartz's view of culture as an inescapable prison bears a strong resemblance to the conception of culture presented by some of his severest critics, particularly J. G. A. Pocock. Having rescued early America from the grips of an absolute attachment to Lockean liberalism, Pocock proceeds to portray the Founding Fathers and their generation as "enclosed" by a pervasive language of classical republicanism. Early Americans, Pocock suggests, were imprisoned within a linguistic paradigm, unable to "do what they have no means of saying they have done" ("Virtue and Commerce in the Eighteenth Century," *Journal of Interdisciplinary History* 3 [Summer 1972], 122). In *The Machiavellian Movement: Florentine Political Thought and the Atlantic Republican Tradition* (Princeton, N.J.: Princeton University Press, 1975), Pocock speaks of the "singular cultural and intellectual homogeneity of the Founding Fathers and their generation" (507; also see Banning, *Jeffersonian Persuasion*, 41). A classical republican prison replaces Hartz's Lockean liberal prison. In a recent article ("Between Gog and Magog," esp. 342, 344–45) Pocock vehemently denies the charge—leveled by Gordon Wood ("Hellfire Politics," *New York Review of Books*, February 28, 1985), Isaac Kramnick ("The Great National Discussion," 4; "Republican Revisionism Revisited," 631–32), and Joyce Appleby ("Republicanism in Old and New Contexts," 26–31)—that he substitutes a Machiavellian or classical republican monolith for a Lockean liberal monolith. One can certainly point to passages where Pocock seems to envision competing ideologies, paradigms, or languages existing at a single point in time (e.g., "Virtue and Commerce" 122–23; and "The History of Political Thought: A Methodological Enquiry," in Peter Laslett and W. G. Runciman, eds., *Philosophy, Politics and Society*, 2nd ser. [Oxford: Basil Blackwell, 1962], 195), but there are other places where Pocock seems to move beyond the limited task of dethroning the Lockean liberal paradigm in the direction of enthroning classical republicanism as the paradigmatic language or vocabulary of early America. Even Pocock's most sympathetic reviewers have detected in Pocock's work a "tendency to overdetermination" in which "thinkers appear to be making the only conceptual moves available to them" (Ian Hampsher-Monk, "Review Article: Political Languages in Time—The Work of J G. A. Pocock," *British Journal of Political Science* 14 [January 1984], 104).

125. The term is used by Alvin W. Gouldner to describe the Weberian view that bureaucracy is an inevitable and inescapable "iron cage" ("Metaphysical Pathos and the Theory of Bureaucracy," *American Political Science Review* 49 [June 1955], 496–507).

126. The following two paragraphs are based on material from Thompson, Ellis, and Wildavsky, *Cultural Theory*, 69.

127. Michael Thompson and Aaron Wildavsky, "A Poverty of Distinction: From Economic Homogeneity to Cultural Heterogeneity in the Classification of Poor People," *Policy Sciences* 19 (September 1986), 170. Also see Mary Douglas, "Cultural Bias," in Douglas, *In the Active Voice* (London: Routledge & Kegan Paul, 1982), 188–89; and Michael Thompson, Richard Ellis, and Aaron Wildavsky, "Political Cultures," in Mary Hawkesworth and Maurice Kogan, eds., *Encyclopedia of Government and Politics,* 2 vols. (London: Routledge & Kegan Paul, 1992), 1:508.

128. The critique of Parsons is developed at greater length in Thompson, Ellis, and Wildavsky, *Cultural Theory*, esp. 181–83.

Index